民机系统工程与项目管理丛书

总主编　贺东风

商用飞机系统工程
特定领域应用
（第2版）

Systems Engineering for Commercial Aircraft
A Domain-Specific Adaptation

（Second Edition）

【美】斯科特·杰克逊 著

钱仲焱　赵越让 等 译

上海交通大学出版社
SHANGHAI JIAO TONG UNIVERSITY PRESS

内容简介

　　本书立足于民用航空运输系统，自上向下，通过需求驱动和正向设计的理念，将研制过程，如功能分析、需求捕获、设计综合、需求确认和验证等过程集成起来，以更准确、更顺畅地满足客户需要并且实现产品全生命周期成本最优。本书强调系统工程不是在现有的研制过程之上增加一个过程层级，而是更应像"胶水"一样将所有技术过程和技术管理过程衔接粘合。

　　本书面向的读者群体包括航空业主制造商、供应商和规章制定者等，尤其是技术、项目及采购等管理人员；飞机设计和特殊专业工程师(人为因素、可靠性、安全性工程师等)；航空系统工程及技术管理专业的学生；以及诸如美国联邦航空局、欧洲航空安全局、中国民用航空局等政府机构。

(商用飞机系统工程)
© Scott Jackson
All Rights Reserved Authorised translation from the English language edition published by Routledge, a member of the Taylor & Francis Group .Copies of this book sold without a Taylor & Francis sticker on the cover are unanthorized and illegal.
上海市版权合同登记号：09-2016-389.

图书在版编目(CIP)数据

商用飞机系统工程 / (美) 斯科特·杰克逊 (Scott Jackson)著；钱仲焱，赵越让等译.
—上海：上海交通大学出版社，2016
(大飞机出版工程)
ISBN 978-7-313-15046-2

Ⅰ.①商…　Ⅱ.①斯…②钱…③赵…　Ⅲ.①民用飞机—系统工程　Ⅳ.①V271

中国版本图书馆CIP数据核字(2016)第114498号

商用飞机系统工程：特定领域应用(第2版)

著　　者：[美]斯科特·杰克逊		译　　者：钱仲焱　赵越让 等	
出版发行：上海交通大学出版社		地　　址：上海市番禺路951号	
邮政编码：200030		电　　话：021-64071208	
出 版 人：韩建民			
印　　刷：上海万卷印刷有限公司		经　　销：全国新华书店	
开　　本：710mm×1000mm　1/16		印　　张：40.5	
字　　数：811千字			
版　　次：2016年6月第1版		印　　次：2016年6月第1次印刷	
书　　号：ISBN 978-7-313-15046-2/V			
定　　价：168.00元			

版权所有　侵权必究
告 读 者：如发现本书有印装质量问题请与印刷厂质量科联系
联系电话：021-56928277

《商用飞机系统工程：特定领域应用》
（第 2 版）

译校人员

钱仲焱　赵越让　吴卜圣

涂　睿　查振羽　李慧颖　徐　州

郭晋之　魏　博　迟悦臻　唐玉军

吴昊天　贺白羽　禹梦泽　张莘艾

总 序

　　大型民用飞机项目是一项极其复杂的系统工程，是一个国家工业、科技水平和综合实力的集中体现。在当今全球经济环境下，飞机全生命周期活动是分布式的，从单个设计区域分配到各个供应商网络，完成后返回进行最终产品集成。许多政治、经济和技术因素都影响其中的协作过程。全球协作网络中过程、方法和工具的紧密、高效整合，是现代商用飞机成功的关键因素。商用飞机的研制需要将主制造商作为一个复杂系统，从企业的层级上统筹考虑产品系统的设计研发和生产制造，并将供应链管理也纳入系统工程的过程中，用系统工程的视角，组织、整合和利用现有资源，以更加快速、高效地开展企业的生产活动，同时需要在更大的范围内整合资源，将型号研制纳入全球民用空中运输体系的范畴中，以期生产出更优质的、更具竞争力的产品。基于系统工程的项目管理，对研制过程各要素进行整合，以满足客户及适航要求，在有限的时间经费等资源内，打造一款飞行员愿意飞、乘客愿意坐、航空公司愿意买的飞机，是我国民用航空产业面临的主要挑战，同时也是实现项目商业成功和技术成功的必由之路。

　　经过几十年的发展，欧美工业界形成了《ISO/IEC 15288—2008：软件与系统工程——系统生命周期流程》等一系列系统工程工业标准；美国国家航空航天局（NASA）、美国国防部（DoD）、美国联邦航空局（FAA）、美国海军和空军等都制定了本行业的系统工程手册；针对商用航空运输领域，制定了SAE ARP 4754A《民机飞机与系统研制指南》等相关指南。同时，航空工业界也一直在开展系统工程实践，尤其是以波音777项目为代表，首次明确地按照系统工程方法组织人员、定义流程和建立文档规范，并通过组织设计制造团队，实现数字化的产品定义和预装配，从而较大地改进研制过程，提高客户满意度，降低研发成本。其后的波音787项目、空客A350项目更是进一步大量使用最新的系统工程方法、工具，为项目成功带来实实在在的好处。

　　目前，我国在系统工程标准方面，也制定了一些工业标准，但总的来说，还是缺乏一些针对特定领域（比如商用飞机领域）的指南和手册，相较国外先进工业实践还有一定的差距。通过新型涡扇支线飞机和大型干线客机两大型号的积累，我国民机产业在需求管理、安全性分析、变更管理、系统集成与验证和合格审定等方面取得了长足进步，在风险管理、构型管理、供应链管理、项目组织建设等方面也进行了全面的探索，初步形成了以满足客户需求为目的，围绕产品全生命周期，通过产品集成与过程集成实现全局最优的技术和管理体系，并探索出适用商用飞机领域的系统工程是"以满足客户需求为目的，围绕产品全生命周期，通过产品集成与过程集成，实现全局最优的一种跨专业、跨部门、跨企业的技术和管理方法"。

进入到美国的再工业化、德国工业4.0、中国制造2025的时代，各制造强国和制造大国在新一轮工业革命浪潮下，都选择以赛博物理系统为基础，重点推进智能制造，进而实现工业的转型升级。其中一个重要的主题是要解决整个生命周期内工程学的一致性。要让现实世界和虚拟世界在各个层次融合，要在机械制造、电气工程、计算机科学领域就模型达成共识。因此，系统工程方法在这个新的时代变得尤为重要，是使产品、生产流程和生产系统实现融合的基础。对于我国航空工业而言，面对标准的挑战、数据安全的挑战、战略及商业模式挑战、企业组织挑战、过程管理挑战、工具方法（SysLM管理工具）挑战、工业复杂性挑战、系统工程人才培养与教育的挑战，积极推进系统工程，才能为在新一轮的工业革命中实现跨越式发展打好基础。

编著这套丛书的目的，一是介绍国内外民用飞机领域先进的系统工程与项目管理理念、理论和方法，为我国航空领域人员提供一套系统、全面的教材，满足各类人才对系统工程和项目管理知识的迫切需求；二是将民用飞机领域内已有型号的系统工程与项目管理实践的重要成果和宝贵经验，以及专家、学者的知识总结继承下来，形成一套科学化、系统化的理论知识体系；三是提供一套通用技术概念，理清并定义商用飞机领域的所有接口，促进一系列技术标准的制定，推动系统工程和项目管理技术体系的形成，促进整个民机产业工业化和信息化的深度融合。

"民机系统工程和项目管理"丛书编委会由谢友柏院士、汪应洛院士、林忠钦院士、中国商飞公司赵越让副总经理，以及美国南加州大学、清华大学、浙江大学、上海交通大学、中国商飞公司、上海航空器适航审定中心等高校和国内外航空界系统工程与项目管理的专家和学者组建而成，凝结了国内外航空界专业人士的智慧和成果。同时还要感谢参与本丛书编撰工作的所有编著者，以及所有直接或间接参与本丛书审校工作的专家学者的辛勤工作，希望本丛书能为民用飞机产业中各有关单位系统工程和项目管理能力的提升做出应有的贡献。

译 者 序

20 世纪 90 年代以来，世界主流商用飞机制造商为获得竞争优势，纷纷采用复合材料、电传操控、多电系统等先进技术，随之而来的是飞机功能和实现功能的系统之间综合水平和复杂度的不断提高。飞机全生命周期的集成活动面临许多巨大挑战：项目规模庞大，管理协调难度大；众多高端系统，集成复杂度高；客户化程度高，需求管理复杂；人才专业跨度大，沟通协调难；风险巨大，成本高昂，进度难保证。要将系统各类要素有机结合形成一个整体，满足客户需要和适航要求，是各国商用飞机主制造商必须解决的难题。

Scott Jackson 所著《商用飞机系统工程》关注飞机整体，采用自顶向下的设计理念，同时结合（美国）汽车工程师协会《民用飞机及系统研制指南》(SAE ARP 4754A)，多方面、多角度对商用飞机系统工程进行了阐述。涵盖新研飞机、衍生型飞机和改装飞机等不同领域，是系统工程在商用飞机领域应用的最新成果。本书对系统工程需求分析、需求管理、接口管理、研制保证、设计综合、验证及确认等过程进行了全面阐述。针对系统工程在商用飞机领域的实际应用问题，提出如何将系统工程方法和理论融入已有的组织机构的路径和建议，对全面创建基于系统工程方法的技术和管理创新体系工作提供了良好的借鉴。在中国商飞公司领导的支持和鼓励下，我们组织力量对该书进行了翻译，希望对商用飞机这一复杂产品的研制及项目管理起到参考作用。由于书中所涉学科知识分布广泛、工程背景要求繁杂，翻译中存在的不足、甚至错误之处，敬请读者批评指正。

在本书的翻译过程中，得到许多同行和专家的指导和帮助，在此一并表示感谢！首先感谢中国商用飞机有限责任公司领导对本书翻译工作给予的关怀和支持；也对中国商飞公司老专家赵维善老师在本书校审工作中所付出的努力表示衷心的感谢；最后感谢上海交通大学出版社钱方针老师在本书翻译出版等方面给予的悉心指导和帮助。在众位领导、专家和同行的大力帮助和鼓励下，最终完成此版本的翻译工作。

译　者
2016 年 6 月

Preface

There have been many developments in the commercial aircraft domain since the publication of the first edition in 1997. From the technology point of view, there have been many innovations, such as the introduction of composite materials and flight envelope protection, both discussed in Chapter 2, among other developments. With respect to safety, the emergence of the Commercial Aircraft Safety Team (CAST), an international consortium of manufacturers, regulators, employee groups, and airlines has served both to track developments in safety and also to suggest improvements in procedures which have reduced the fatality rate dramatically. From a management point of view, the increased use of outsourcing discussed in Chapter 14, has created a challenge for which greater rigor in supplier management is required. This chapter discusses outsourcing in the context of large-scale system integration (LSSI), an advanced topic in the systems engineering lexicon.

In addition to developments in the commercial aircraft domain since 1997, systems engineering has continued to grow in scope and maturity both as a general concept and also in the commercial aircraft domain. The publication of the *Systems Engineering Body of Knowledge* (SEBoK) edited by Pyster (2012) has expanded the scope of systems engineering into three categories: product systems engineering (PSE), enterprise systems engineering (ESE), and service systems engineering (SSE). The discussion of outsourcing in Chapter 14 falls more into the ESE category. Within the commercial aviation domain two important documents have been published: First is the Federal Aviation Administration (FAA) *Systems Engineering Manual* (2014). Secondly, the Society of Automotive Engineers (SAE) guideline ARP 4754A (2010) lays out in a concise way how systems engineering applies to aircraft development with a focus on safety and certification.

This book is not intended to replace the above standards and guidelines or to be a definitive interpretation of them. Rather it is the intent to be a "pointer" to these

前 言

自本书第一版 1997 年出版以来，商用飞机领域取得了很多新进展。从技术角度来看，出现了很多创新，例如第 2 章讨论的复合材料的使用及飞行包线保护，以及其他的一些进展。关于安全性，出现了商用航空安全小组 (CAST)，一个由制造商、监管机构、员工团体及航空公司组成的国际联合组织，该组织旨在追寻安全性方面的进展，并提出流程改进建议，并已显著降低死亡率。从管理角度来看，第 14 章所述增加使用外包对更严格的供应商管理提出了挑战。第 14 章介绍了大规模系统集成 (LSSI) 背景下的外包，该领域在系统工程中属于前沿课题。

除了自 1997 年以来商用飞机领域取得的进展外，作为一般理念，系统工程在范围和成熟度上同样在商用飞机领域取得了长足的进步。Pyster (2012) 编写的《系统工程知识体系》(*Systems Engineering Body of Knowledge*，SEBoK) 的出版将系统工程的范围扩展为三大类：产品系统工程 (PSE)，企业系统工程 (ESE) 及服务系统工程 (SSE)。第 14 章所述外包的讨论大部分属于企业系统工程的范畴。在商用航空领域，两份重要文献已经出版：第一份是 (美国) 联邦航空局 (FAA)《系统工程手册 (2014)》，第二份是 (美国) 汽车工程师协会 (SAE) 指南 ARP 4754A(2010)，该文件以简要介绍了系统工程如何运用到飞机研制中，并同时关注其安全性和合格审定。

本书无意于取代上述的标准及指南，或者是成为它们的权威解释，而是希望成为这些文件的"指针"，来表明它们如何融入系统工程内容中，以及下文讨

5

documents, to show how they can fit into a systems engineering context, and to adapt to these processes as discussed below. Furthermore, this book is not intended to be a manual or handbook; rather it is intended to be a guide to understanding.

If there is a central theme of this book it is that the commercial aircraft domain requires attention to the adaptation of the systems engineering process to that domain. Chapter 13 is devoted entirely to the challenges of adaptation. You may have noticed that this edition is subtitled *A Domain-Specific Adaptation*. These challenges result from the unique demands of the market and the technologies in that domain. In addition, an important fact is that there is already considerable systems engineering in this domain, and the developer can take advantage of that fact by incorporating only those aspects that do not already exist. That chapter also describes how an *existing* organization can perform systems engineering to maximum advantage.

Another goal of this edition is to persuade the developer that systems engineering is not the burdensome process it is often perceived to be in other domains, but rather a logical approach to system development.

An important issue in modern commercial aircraft development is the existence of risks. Although the first edition devoted a subsection to this subject, this edition expands that discussion to an entire chapter, Chapter 15, to the principles of risk management and typical risks that the developer may encounter. With respect to risks, this book does not mention specific aircraft developers, specific aircraft, or specific incidents except to the extent that they are mentioned in accident reports by, for example, the National Transportation Safety Board (NTSB).

A final topic not discussed extensively in other texts with respect to commercial aircraft is *resilience* in Chapter 16; an exception is Hollnagel et al. (2011). Resilience is different from safety in that while safety is concerned with the prevention of failures, resilience deals with the anticipation, withstanding, and recovery from any kind of adverse disruption.

It is hoped that you will find this edition both useful as well as informative regarding the commercial aircraft domain in the context of systems theory and in particular systems engineering.

Scott Jackson
Irvine, California

论的如何应用这些过程。此外，本书不是手册或参考书，而是希望成为理解系统工程的指南。

如果说本书必须有一个中心主题，那就是需要关注把系统工程过程应用到商用飞机领域。第 13 章整篇讨论系统工程应用的挑战。你可能已经注意到，本版本的副标题"特定领域应用"。这些挑战来自于这个领域独特的市场要求和技术。此外，一个重要的事实是这个领域已经有大量的系统工程，而研制方可以基于这个事实，将现在没有的那些方面纳入进来。第 13 章同时阐述了现有组织如何最大限度地实施系统工程。

本版的另外一个目标是说服研制方，即系统工程不像是其他领域中认为的那样，是一个烦琐的过程，而是系统研制的一种合理方式。

现代商用飞机研制中的一个重要问题就是风险的存在。虽然本书第一版用了一个小节讨论这个问题，但本版将该讨论扩充至一整个章节，第 15 章介绍风险管理的基本原则，以及研制方可能遇到的典型风险。关于风险，本书没有提到具体飞机研制方、具体机型或者具体事故，除非该内容被（美国）国家运输安全委员会 (NTSB) 等机构在事故报告中提及。

关于商用飞机，最后一个话题是其他文献 [Hollnagel 等 (2011) 以外] 没有展开讨论的恢复力（第 16 章）。恢复力与安全性的差异在于安全性关注预防失效，而恢复力处理各种不利扰动的预测、承受及恢复。

希望本版在商用飞机系统理论，以及特定系统工程方面的内容是翔实的，并能对读者有所帮助。

斯科特·杰克逊
尔湾市，加州

Acknowledgments

For the first edition, I continue to be grateful to Jim Kehres who, in about 1963, advised me "to learn more about systems engineering." I am also especially grateful to Gary Burgess who, at the Douglas Aircraft division of McDonnell Douglas, saw the need for systems engineering in commercial aircraft. The following people were especially helpful in the preparation of this book: First, Tom Nagle and Archie Vickers were helpful for their overall knowledge of the systems engineering process. In addition, the following people contributed much: Gary Bartz, certification; Peter Camacho, top-level sizing of aircraft; Darlene Carpenter, electrical; Madrona Geisert, interior systems; Fred Gray, propulsion; Stu Hann, safety and reliability; Don Hanson, aircraft development; Brian Keeley, airframe; Dr Noreen McQuinn, human factors; Christine Ostrowski, maintainability; Mo Piper, avionics and software; Bob Rich, functional analysis; Todd Strong, mechanical systems; Matt Vance, QFD; Steve Wiles, environmental control; and Beth Clark, integration and life-cycle analysis. Finally, my wife Carole provided invaluable advice regarding syntax, diction, and organization.

For the second edition I am indebted to Jim Hines for further insight into functional analysis; John Hart-Smith, composite structures; Ashok Jain for the supplier perspective; Derek Hitchins for systems theory and cluster analysis; Ed Conrow for risk management; Tim Ferris for assistance with resilience analysis; Wellington Oliveira for the regional jet perspective; and Karin Mayer for help with grammar, spelling, and diction.

致 谢

对于第一版，我要继续感谢 Jim Kehres，正是他在大约 1963 年建议我"更多地学习系统工程。"还要特别感谢 Gary Burgess，当时他在麦克唐纳道格拉斯飞机队任职，并洞察到系统工程在商用飞机应用的需要。以下各位在这本书的准备工作中给予了很大的帮助：首先，Tom Nagle 和 Archie Vickers 对系统工程过程的总体认识非常有帮助。此外，下列各位人士也贡献巨大：Gary Bartz，合格审定方面；Peter Camacho，飞机顶层参数设计方面；Darlene Carpenter，电气方面；Madrona Geisert，内饰系统方面；Fred Gray，推进（系统）方面；Stu Hann，安全性及可靠性方面；Don Hanson，飞机研制方面；Brian Keeley，机身方面；Dr Noreen McQuinn，人为因素方面；Christine Ostrowski，可维修性；Mo Piper，航电和软件方面；Bob Rich，功能分析方面；Todd Strong，机械系统方面；Matt Vance，质量功能展开 (QFD)；Steve Wiles，环控方面；以及 Beth Clark，集成及生命周期分析方面。最后，我的妻子 Carole 提供了关于语法，用词及编排方面的宝贵意见。

对于第二版，我要感谢 Jim Hines 对于功能分析的深刻的洞察力；John Hart-Smith，复合材料结构；Ashok Jain，供应商合作视角；Derek Hitchins，系统理论及群分析；Ed Conrow，风险管理；Tim Ferris，援助与恢复力分析；Wellington Oliveira，支线喷气飞机方面；还要感谢 Karin Mayer 在语法、拼写及修辞方面给予的帮助。

Contents

目　录

List of Figures

图表目录

List of Tables

表格目录

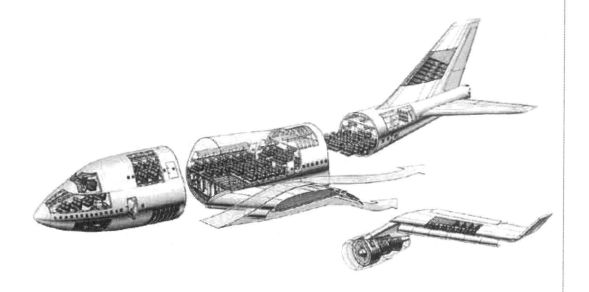

图片来源：Systems Engineering for Commercial Aircraft

1

Introduction

The primary purpose of the book is to provide the reader with the information to apply the systems engineering process to the design of new aircraft, derivative aircraft, and change-based designs. A second purpose is to provide guidance that will allow the reader to adapt this process to the commercial aircraft domain through judicious selection of those aspects that would provide the highest leverage of benefits and the lowest risk of adversities. It is assumed that the reader either already has a basic understanding of the process or can obtain that information from further reading of other sources, such as the ones discussed later in Section 1.3. Although there are many interpretations of systems engineering, the principles discussed are generally universal. This book attempts to stress those which are most relevant to aircraft design.

For brevity, the initials SE will be used for systems engineering throughout this book.

1.1   Definition of a System

A system is anything with many parts, like an airplane, a wrist watch, the human body, or the US government. The parts of a system are hierarchical: that is, the airplane parts can be subdivided into subsystems, sub-subsystems, and so forth. However, the principles described here apply equally well to the design of a subsystem as to a system. The official definition of a system adopted by the International Council on Systems Engineering (INCOSE) is as follows: A system is:

> ... an integrated set of elements, subsystems, or assemblies that accomplish a defined objective. These elements include products (hardware, software, and firmware), processes, people, information, techniques, facilities, services, and other support elements. (2010, p. 5)

第1章 引 言

本书的主要目的在于帮助读者了解设计一款新研飞机、衍生型飞机和改装飞机的系统工程过程。其次，指导读者通过谨慎选择，将系统工程过程应用于商用飞机领域，以实现利益最大化和风险最小化。本书假设读者已经掌握了基本的过程知识，或可通过进一步阅读其他资料来了解其内容，这些资料稍后在 1.3 节中介绍。尽管对于系统工程有着许多不同的理解，但这些资料所讨论的原理是普遍适用的。本书重点关注与飞机设计相关度最高的内容。

为简便起见，本书以大写的 SE 代表系统工程[1]。

1.1 系统的定义

系统是包含许多组成部分的任何事物。例如，一架飞机、一块手表、人体或是美国政府。系统的组成是分层级的。比如，一架飞机可以划分为多个子系统，子系统下还能划分多个子子系统，并以此类推。然而，不论是系统还是子系统的设计，系统工程原理同样适用。国际系统工程协会 (INCOSE) 对系统的正式定义为：

系统是：

以实现既定目标的一组元素、子系统或者组件的集成，这些元素可以是产品（硬件、软件及固件）、过程、人员、信息、技术、设施、服务及其他支持元素 (2010，第 5 页)。

In the commercial aircraft industry the term *system* is normally used for electrical systems, hydraulic systems, and so forth. The term can also refer to the global aviation system. However, in this book *system* will be used in the SE context: that is, for the entire aircraft and its supporting elements. Subordinate elements will be referred to as subsystems, such as the electrical subsystem.

This definition, though, includes non-technical aspects, such as people. It may seem contrary to the classical definition of engineering to include these aspects. However, this definition is consistent with the modern definition of SE which deals with the effort to define such systems.

1.2 Definition of Systems Engineering

It is difficult for two systems engineers to agree on a definition of SE. There are many definitions and many theories on the implementation of SE. We will look at only a few definitions here.

As the discipline began to take form, the search for a definition also began. The need for such a discipline has resulted from a worldwide trend of devoting an increasingly larger portion of the engineering effort towards pre-design requirements definition. SE is a key methodology in that trend.

The Systems Engineering Body of Knowledge (SEBoK) defines three types of SE: product SE (PSE), enterprise SE (ESE), and service SE (SSE). The SE of an aircraft is product SE. Product in this context can broadly be interpreted to include the operators and maintainers. Product SE is the primary, but not exclusive, focus of this book. Enterprise SE includes the developer, the suppliers, and the carriers. This book does incorporate aspects of enterprise SE, for example, in Chapter 14 which discusses large-scale system integration (LSSI) with an emphasis on supplier management, an important element in enterprise SE.

SE is a discipline which has the goal of arranging the parts of a system in such a way that the entire system does something optimally, such as to get from A to B in a minimum time, or at a minimum cost. That is, SE optimizes the system's performance. However, SE goes even further. Major goals of SE are: first, to define the system's requirements so well that the product will never have to be redesigned; secondly, to make the product as reliable as possible; and finally, to make the customer happy. In practice, these goals may seem impractical. However, SE provides some methods which may bring the design closer to the goals.

The official definition of SE adopted by INCOSE is as follows:

在商用飞机行业，系统通常用在电气系统、液压系统等系统上。系统也可以指全球航空系统。但是本书中所述系统属用于系统工程范畴：即整架飞机及其支持元素。次级元素称为子系统，如电气子系统。

上述定义包含了非技术方面，如人员。它似乎有悖于传统的工程定义，然而却与现代系统工程定义一致。

1.2 系统工程定义

任何两个系统工程师很难就系统工程的定义达成一致。系统工程应用中有着许多不同的定义和理论，这里我们只讨论其中的一部分。

随着系统工程这一学科的形成，人们也开始了对系统工程定义的研究。对系统工程学科的需求来源于世界范围内已经形成了的趋势：越来越多的工程工作用在了设计前的需求定义上。系统工程是这种趋势下的关键方法。

系统工程知识体系 (SEBoK) 定义了三种类型的系统工程：产品系统工程 (PSE)、企业系统工程 (ESE) 及服务系统工程 (SSE)。飞机系统工程属于产品系统工程范畴。这里说的产品可以宽泛地解释为包含运营方和维修方。产品系统工程是本书的主要关注点，但不是唯一关注点。企业系统工程包含研制方、供应商及承运人。本书中也包含了企业系统工程的部分内容，例如在第 14 章中，我们在讨论大规模系统集成 (LSSI) 时，着重强调了供应商管理，这是企业系统工程中一个非常重要的元素。

系统工程是一门学科，目标是通过系统部件的合理安排实现整个系统的最优。例如，以最短时间或最低成本实现从 A 到 B 的过程。换言之，系统工程可以优化系统性能，更甚者系统工程有更深层次的作用。系统工程的主要目标是：首先，准确定义系统需求以保证产品无须重新设计；其次，保障产品尽可能可靠；最后，让客户满意。尽管这看起来有些理想化难以成为现实，然而，系统工程提供了一些能使设计尽可能接近上述目标的方法。

> Systems Engineering (SE) is an interdisciplinary approach and means to enable the realization of successful systems. It focuses on defining customer needs and required functionality early in the development cycle, documenting requirements, and then proceeding with design synthesis and system validation while considering the complete problem: operations, cost and schedule, performance, training and support, test, manufacturing, and disposal. SE considers both the business and the technical needs of all customers with the goal of providing a quality product that meets the user needs. (INCOSE 2010)

This definition raises some important points: First, it points out that SE addresses the entire life-cycle of the system, not just the operational phase. This book addresses the life-cycle functions of an aircraft system in Section 3.1. Secondly, it states that SE assigns requirements to people and processes, not just the aircraft.

Section 5.5 discusses the ability and limitations of assigning requirements to people. Section 5.14 discusses how SE can assign requirements to the fabrication and assembly processes.

An earlier definition was given by Simon Ramo (1973) as follows:

> Systems engineering is a branch of engineering that concentrates on the design and application of the whole as distinct from the parts ... looking at a problem in its entirety, taking into account all the facets and all the variables and relating the social to the technical aspects.

Note Remo's inclusion of social factors in the design of a system. This definition indicates the possible breadth of SE.

The term *systems engineering* has also been used to pertain to computer systems, for example, or to subsystems, such as electrical or hydraulic systems. It will be used in this book in the broadest possible sense, that is, to pertain to any system.

The term *system engineering* has also been used. However, SE has become the industry standard. Although treatises have been written breaking down SE into many steps, let's just consider five, for simplicity, as shown in Figure 1.1. Notice the iterative nature of the first three steps. We will discuss each of these throughout the course of this book. The *FAA Systems Engineering Manual* (2014, p. 5) contains a more expanded view of the SE process.

INCOSE 采用的系统工程正式定义如下：

系统工程是一种跨学科的解决方案和方法，有助于实现一个成功的系统。它关注在研制周期早期准确定义客户需要及所需功能，进行需求的文档管理，以及设计综合和系统确认，期间综合考虑下述问题：如运行、成本、进度、性能、培训、支持、试验、制造及处置等。系统工程既关注客户的业务需要也关系客户的技术需要，目标是提供满足用户需要的高质量产品 (INCOSE 2010)。

这个定义指明了几个关键点：首先，它指出系统工程关注的是系统整个生命周期，而不仅仅是运营阶段。本书3.1节阐述了飞机系统[1]的整个生命周期功能。其次，它阐明了系统工程对人和过程指定了需求，而不仅只对飞机指定需求。

本书 5.5 节讨论了将需求分配给人员的能力和限制条件。5.14 节讨论了系统工程如何将需求分配给制造及装配过程。

早期的系统工程定义 (Simon Ramo(1973)) 如下：

系统工程是工程的一个分支，重点关注整体的设计及应用，而非个体……从整体的角度看问题，考虑所有的影响因素，所有变量，同时将社会因素与技术层面相关联。

需要注意的是 Ramo 在系统设计中包含了社会因素，上述定义表明了系统可能达到的广度范围。

系统工程 (Systems Engineering) 这个词也用于计算机系统，或用于子系统，如电气系统、液压系统。本书中系统工程的使用将非常宽泛，适用于任何系统。

也有"单系统工程"(System Engineering) 的说法，而系统工程的说法已经成为工业标准。在一些著作中系统工程被分成若干步骤，简化起见，本书仅考虑分成 5 个步骤，如图 1.1 所示。需要注意的是前 3 个阶段是一个不停迭代的过程。我们将在本书中详尽地阐述。《FAA 系统工程手册》(2014，第 5 页) 对系统工程过程进行了更为详尽的描述。

---

1 为保持与英文字面一致，本书将 Aircraft System 译为"飞机系统"，指包含飞机及运行支持等使能元素在内的更高一级系统。亦可意译为"民用飞机产品系统"，以避免与构成飞机的系统产生歧义。——译者注。

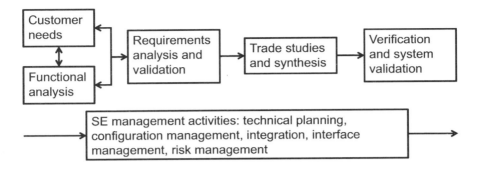

Figure 1.1   Steps in the SE process

Another graphic often used in the description of SE is the famous Vee model. This model is more appropriate when applied to the discussion of requirements; hence, Chapter 4 discusses this model and its implications.

1.3   Historical Background

Early authors to recognize the value of SE and describe the process include A.D. Hall in his book *A Methodology for Systems Engineering* (1962). Other early descriptions of the process include the Army Technical Manual TM 38-760-1, *A Guide to System Engineering* (1973) and the Army Field Manual 770-78, *System Engineering* (1979). The EIA published SYSB-1, *System Engineering* (1989). Other standards followed, culminating in MIL-STD-499B, *Systems Engineering*, which was never formally adopted.

Eventually various professional societies joined together to publish ANSI/EIA 632 (1999), *Processes for the Engineering of a System.* The purpose of ANSI/EIA 632 was to create a standard which would be useful for both military and civilian applications. In addition, the Institute of Electrical and Electronic Engineering (IEEE) has published IEEE Standard 1220-2005, *Application and Management of the Systems Engineering Process* (2005). The standards, in general, pertain to product systems engineering (PSE) as defined in the SEBoK, described above.

The term SE is sometimes used in different contexts to mean different things, even within the commercial aircraft domain. For example, it can be used to mean the development of subsystems, as in the term *avionics systems engineering*. In addition, it is sometimes used to mean software engineering. This book uses the term in the

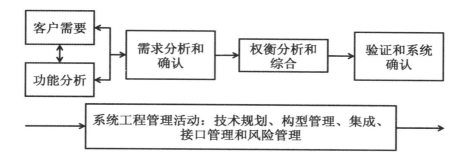

图 1.1　系统工程过程中的步骤

另一个常用的系统工程描述图形是著名的 V 模型。这个模型更适合在讨论需求时应用，本书将在第 4 章中讨论 V 模型及其含义。

1.3　历史背景

较早认识到系统工程价值的作者包括 A. D. Hall，他在《系统工程方法论》(1962) 一书中介绍了系统工程过程。其他较早的介绍系统工程过程的著作还包括《陆军技术手册 TM38-760-1》《系统工程指南》(1973)《陆军战场手册 770-78》《系统工程》(1979)。EIA 发布了 SYSB-1,《系统工程》(1989)。随后，其他标准相继涌现，例如美军标 MIL-STD-499B 系统工程，然而该标准从未正式采用。

最终，各专业领域团体联合发表了 ANSI/EIA 632(1999)《系统工程过程》，ANSI/EIA 632 旨在创立一套能同时适用于军用及民用的标准。此外电气电子工程学会 (IEEE) 发布了 IEEE 标准 1220—2005《系统工程过程的应用及管理》(2005)。

这个标准总体而言是关于产品的系统工程，和前文所述系统工程知识体系的定义一致。

"系统工程"这个词在不同语境下的含义各不相同，即使在商用飞机领域也是如此。例如，系统工程可以指子系统的开发，比如词汇航电系统工程。再比如

broader sense which is consistent with the INCOSE definition discussed above and with the international standard ISO/IEC (2008).

SE is reaching a stage of maturity where its application in specific industries, such as commercial aircraft, can be defined and documented, as described by Petersen and Sutcliffe (1992). The first major guideline is the Society of Automotive Engineers (SAE) publication, ARP 4754 (1996), which applies the principles of SE to the development and certification of commercial aircraft. The later document ARP 4754A (2010) superseded the earlier guideline and contains many aspects of SE compared to the FAA manual with an emphasis on safety and certification.

1.4   Overview of this Book

The overriding SE principle stressed in this book is that the aircraft should be viewed as a whole and not as a collection of parts. Each chapter looks at a different aspect of SE and shows how that aspect would be reflected in the systems engineering of commercial aircraft. Chapter 2 looks at the commercial aircraft industry, describes the levels of aircraft development (new, derivative, and change-based) to which SE would be applied, and shows how the aircraft component architecture fits into the SE hierarchical model. This chapter also describes new technologies which would be applied to aircraft of the future. Chapter 3 introduces the SE concept of functions and shows how to apply functional analysis to the entire life-cycle of the aircraft, to the aircraft as a whole, and to the aircraft's subsystems. The SE concepts of performance requirements and constraints are the subject of Chapter 4. This chapter also shows how these requirements can be allocated to the aircraft's subsystems. Chapter 5 addresses constraints and specialty requirements and focuses on some key aircraft specialty areas, such as weight and reliability. The importance of the human factors aspect of cockpit design and the concept of organizational safety and its importance to the aircraft industry are also discussed in this chapter. Chapter 6 describes the SE concepts of functional and physical interfaces and shows how these concepts would apply to both external and internal aircraft interfaces. Chapter 7 shows how the SE concept of synthesis results in an actual aircraft design during the various stages of functional analysis, architecture development, and trade-offs. This chapter emphasizes the usefulness of quality function deployment (QFD) in this process. It also describes the process for introducing new technology into aircraft development. Chapter 8 shows how an aircraft is synthesized at the top level and how cost constraints are a vital factor in this process. Chapter 9 shows how the subsystems are synthesized and how

有些时候系统工程可以用于软件工程。在本书中，系统工程有着较为宽泛的定义，和前文所述 INCOSE 定义和国际标准 ISO/IEC(2008) 一致。

系统工程在某些领域如商用飞机领域已相当成熟，如彼得森和萨克利夫描述的 (1992) 那样，已完成定义并形成文件。首部重要指南是汽车工程师学会发布的 ARP 4754(1996)，该指南将系统工程原理运用于商用飞机研制及合格审定过程。后来，ARP 4754A(2010) 取代了原有版本，并增加了很多系统工程方面的内容，与 FAA 的手册相比，ARP 4754A 更强调安全性与合格审定。

1.4 全书总览

本书重点强调的一个最重要的系统工程原理就是应将飞机视为一个整体，而不是各个零部件的拼凑。全书各章关注系统工程的不同方面，以及这个方面如何在商用飞机系统工程中反映出来。第 2 章着眼于商用飞机行业，阐述了不同类别的飞机研制 (新研飞机、衍生型飞机和改装飞机) 所对应的系统工程应用，以及飞机的部件架构是如何与系统工程分层模型相匹配的。此外，在这一章我们还讨论了未来飞机可能采用的新技术。第 3 章介绍了系统工程的功能概念，阐明了功能分析如何运用于飞机的全寿命周期、运用于整架飞机及飞机各个子系统。第 4 章的讨论主题是性能需求和约束系统工程的概念，介绍了如何将需求分配到飞机子系统。第 5 章介绍了约束及专业需求，重点关注飞机某些专业领域，如重量和可靠性等，该章强调了驾驶舱设计充分考虑人为因素的必要性，并介绍了组织安全理念及其在飞机行业中的重要性。第 6 章介绍了系统工程中功能及物理接口概念，讨论了这些概念在飞机内外接口中的应用。第 7 章着眼于系统工程中综合的概念，介绍了如何通过不同阶段的功能分析、架构开发和权衡分析实现真正的飞机设计。本章强调了质量功能展开 (QFD) 在整个过程中的作用；并介绍了飞机研制中的新技术引入过程。第 9 章主要描述了如何实现子系统的综合，以及这

this synthesis reflects the functions which have been allocated at the subsystem level. Chapter 10 describes how certification guidelines incorporate the SE philosophy and how safety analysis and software development fit into this philosophy. Chapter 11 stresses the importance of the verification of all requirements in aircraft development and shows how the SE verification concepts of test, demonstration, analysis, and examination fit into the verification of aircraft requirements. Finally, Chapter 12 discusses the key SE management and control activities, such as design reviews and configuration management, and emphasizes the importance of these activities to the aircraft development process. Chapter 13, first, provides a set of rules that will aid the commercial aircraft enterprise adapt the SE processes to that domain. In addition, Chapter 13 explains the roles that different organizations within the enterprise may play in the SE process. These include both technical and managerial organizations. Chapter 14 explains the commercial aircraft enterprise in a LSSI context with an emphasis on the supply chain. Chapter 15 explains the concept of risk management and discusses many risks that may be encountered in this domain, ways to anticipate them, and ways to mitigate them. Finally, Chapter 16 discusses the newly developing discipline of resilience and how it applies to the commercial aircraft domain.

1.5   Roadmap for Applying Systems Engineering to Commercial Aircraft

With these principles in mind, we can now show how the SE steps flow together for commercial aircraft development, as shown in Figure 1.2. This figure also provides the chapter and section numbers for a description of each step. As in Figure 1.1, this figure begins and ends with customer requirements and verification. But it then expands the core steps of functional and requirements analysis and synthesis to illustrate the following principles: ① the flow down from top- to subsystem- level analysis, and ② the treatment of new, existing, and certification-based SE.

By necessity, this figure is somewhat oversimplified. As discussed in Section 1.2, capturing customer requirements and developing system functions are two interrelated steps, both of which occur at the beginning of the SE process. Secondly, system synthesis occurs throughout the SE process. It is any step which leads to an aircraft design. Hence, both functional analysis and architectural development are part of this process.

种综合如何体现已分配给该子系统层级。第 10 章讨论的是合格审定指南如何与系统工程理论相结合，安全性评估和软件开发如何与系统工程理论相匹配。第 11 章强调了飞机研制中需求验证的重要性，描述了系统工程中的验证，包括：试验、演示、分析和检查如何与飞机需求验证工作相匹配。最后，第 12 章讨论了重要的系统工程管理和控制活动。例如设计评审和构型管理，强调了这些活动在飞机研制过程中的重要性。第 13 章首先提出了一系列的规则，帮助商用飞机公司将系统工程过程应用于该领域；其次，在第 13 章中介绍了企业内的不同组织机构，包括技术及管理机构在系统工程过程中可以发挥的作用。第 14 章介绍了处于大规模系统集成 (LSSI) 环境中的商用飞机企业，重点强调了供应链。第 15 章介绍了风险管理的概念，讨论了该领域内可能遇到的诸多风险、风险预测方法及风险缓解方法。最后，第 16 章介绍了恢复力这一新兴学科，及其在商用飞机领域的应用。

1.5 商用飞机系统工程应用路线图

商用飞机研制中系统工程过程之间的关系可以简单表示如下，如图 1.2 所示。图中每一步都标注有章节号，在相应章节中可以找到具体的描述内容。与图 1.1 类似，图 1.2 同样开始于客户需求，终止于验证。但图 1.2 对功能和需求分析以及综合的核心步骤进行了展开说明，主要说明如下几个原则：①工程处理方法自上向下，到子系统的分析过程；②新研、已有和基于合格审定的系统工程处理方法。

出于某些考虑，本文对图 1.2 进行了过度简化。如 1.2 节所述，客户需求的捕获过程和系统功能开发是两个反复迭代的过程，都开始于系统工程过程初期。其次，系统综合贯穿于整个系统工程过程当中，存在于实现飞机设计的任一步骤。因此，功能分析和架构设计都属于综合的一部分。

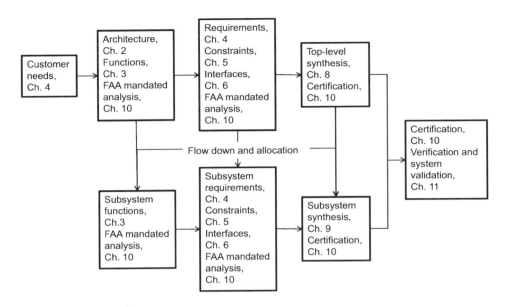

Figure 1.2   Roadmap to systems engineering for commercial aircraft

1.6   Summary of Themes

This is not a systems engineering manual. As pointed out above, it is more of a *how* and a *why* guide rather than a *what* guide. If the reader wants a what guide, then authoritative sources, such as the *FAA Systems Engineering Manual* (2014), might be more appropriate. The guidance provided in this book is based more on an understanding of what should be done and the practical aspects of actually getting it done. With these thoughts in mind, let's look at some of the themes to be elaborated in later chapters. These themes are not in the order of importance; they are all important.

Theme 1—Adaptation

To execute all the processes in a manual such as the *FAA Systems Engineering Manual* might be perceived as an onerous and burdensome task. This book provides guidance on how these processes can be performed more efficiently without sacrificing quality. For example, Chapter 4 shows how requirements can be screened to determine which requirements are important and which ones can be ignored. The essence of requirements screening is risk, that is, the risk of ignoring requirements.

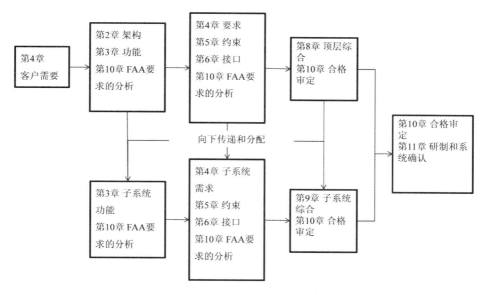

图1.2 商用飞机系统工程路线图

1.6 主题小结

本书不是一本系统工程手册。正如前文提到的那样，本书更多关注系统工程"是怎样的"、"为什么是那样"、而不是关注"是什么"。如果读者对"是什么"感兴趣，可以选择一些权威著作如《FAA系统工程手册》(2014)。本书假定读者对系统工程需要做什么，以及实际过程中怎么做已经有所了解。有了这些概念，下面介绍后面几章将详述的一些主题。注意：这些主题不分先后次序，因为它们都很重要。

主题 1 - 应用

在执行诸如《FAA系统工程手册》等手册的所有过程时，大家会发现工作任务非常烦琐。本书帮助大家了解如何在保证质量的前提下高效执行所有过程。例如，第4章介绍了如何进行需求筛选，决定哪些需求很重要，哪些需求可以忽略掉。需求筛选的本质是有风险的，即忽略需求的风险。第12章介绍了设计评

Chapter 12 suggests ways to reduce the costly and time consuming process of design reviews. In short, all SE processes can be streamlined if appropriate attention is given to the risk of streamlining them.

Theme 2—SE as an integrated technical–managerial process

All systems engineering standards and textbooks involve the execution of tasks which can be considered both technical and managerial. This concept is counter to many conventional organizations in which SE is regarded as solely technical. Chapter 13 provides a tour of a typical organization containing both technical and managerial departments. This chapter explains the role of each department, both technical and managerial. This chapter also provides options regarding ways to bring the technical and managerial functions together in an integrated enterprise.

Theme 3—The importance of risk handling and management

No process permeates all aspects of an engineering organization than risk management. Any process that is performed imperfectly invites the specter of risk. No one has greater responsibility in this process than the program manager as explained in both Chapters 13 and 15. This is because the pressures to ignore or minimize the concentration on risks are enormous, and the consequences of doing so can be substantial. Also, as explained above in the discussion of Theme 1— Adaptation, risk plays an important role in that function.

Theme 4—The systems view

Essential to the discussion of systems engineering in any domain is an understanding of what a system is and the many manifestations of it. From the basic definition of a system in Section 1.1, one can see a system from many degrees of breadth, from the avionics system, to an entire aircraft system, to the entire air transport system. Chapter 14 also discusses the entire supply chain as a system of systems. Only with the systems view can one begin to answer the question of how each one of the systems functions as a whole and performs a useful function.

审环节如何降低成本及缩短时间。简而言之，所有的系统工程过程都可以进行剪裁使用，前提是需要充分考虑剪裁所带来的风险。

主题 2 - 系统工程是一个综合了技术和管理的过程

所用的系统工程标准及教材都涉及相关任务的执行，这些任务既涉及技术也兼顾管理。这个概念和许多传统组织的想法是不同的，它们只考虑了系统工程的技术性。在第 13 章中，我们阐述了一个典型的包含了技术及管理部门的组织是如何运作的。这一章对每个技术和管理部门的作用进行了介绍，同时在这一章读者还能了解到在一个综合化的企业中如何将技术和管理职能整合在一起。

主题 3 - 风险应对和管理的重要性

在工程组织中风险管理无处不在。稍有差池就可能带来各种风险。在风险管理过程中，项目经理承担了最重要的责任，第 13 章和 15 章分别介绍了相关内容。对风险的任何疏忽都可能会造成严重的后果。对于主题 1- 应用来说，风险管理尤为重要。

主题 4 - 系统观点

讨论任意领域的系统工程的本质在于要理解它是一个什么系统及系统的表现。从 1.1 节中对系统的基本定义来看，每个人对系统都有着不同的理解，从航电系统到整个飞机系统或到整个航空运输系统。第 14 章也讲到，整个供应链也可以视为许多系统中的一个系统。只有采用系统的观点，我们才能更准确地回答如下问题：各个系统如何分工合作，完成一个有用的功能。

Theme 5—Added rigor

SE is not, as one might expect, an added layer of processes on top of the existing design processes. Some have described it as the glue that holds all the other processes together. In the Final Comments to the first edition of this book, it is described just as *common sense*. But in the end it must be concluded that SE is the process that adds rigor to the design, integration, and operation of a commercial aircraft.

主题 5 - 增加严苛度

系统工程并不是在已有的设计过程外新增的一个过程。有人说系统工程好比胶水，将其他所有的过程黏接在一起。在本书的第一版的最后评论中，系统工程被称为常识。最终，系统工程可归纳为：增加商用飞机设计、集成及运行增加严苛度的过程。

2

Commercial Aircraft

Although the term *commercial aircraft* generally refers to jet-powered aircraft carrying large numbers of passengers for long distances, the SE principles outlined in this book also apply to freight-carrying aircraft and smaller propeller-driven, or *commuter*, aircraft as well. In addition, these principles also apply to *general aviation*, that is, small privately owned aircraft. Also included in commercial aircraft are regional jet aircraft which usually have fewer than 100 passengers and make trips internal to a country rather than internationally.

There are, indeed, many similarities to some classes of military aircraft which have missions of carrying passengers and cargo over specified distances. We show in Section 2.3, for example, that the ATA Specification 100 (1989) aircraft component hierarchy used in the commercial aircraft industry is also used in military practice (MIL-STD-1808B, 2007). The main difference is that the military hierarchy adds specific military categories, such as provisions to carry weapons. Other requirements unique to military aircraft include, for example, the need to provide protection from enemy weapons.

The main differences between commercial and military aircraft, however, lie in the types of requirements and the types of customers which generate the requirements. We will see in Section 3.1 that there are two types of commercial customers, the aircraft market and specific airline customers. While the military customer is normally a single customer with specific mission requirements, commercial aircraft requirements are strongly driven by economic requirements. In Chapter 8 we will discuss economic requirements as part of the top- level synthesis.

20

2.1 The Commercial Aircraft Industry

The commercial aircraft industry throughout the world is evolving to a small group of large companies. Some of the risks in this sector include the financial cost of new product development, and the financial health of airlines and other customers. Large

第2章 商用飞机

尽管商用飞机通常是指搭载大量乘客进行远程飞行的喷气式飞机，但本书所述的系统工程原理也适用于货运飞机、小型螺旋桨驱动飞机或通勤飞机。另外，这些原理也适用于通用航空，即小型私人飞机，以及商用飞机领域的支线喷气飞机，支线喷气飞机通常执行国内航线，而非国际航线，其乘客通常不到100人。

事实上，商用飞机与某些执行将乘客和货物运送到规定距离任务的军用飞机有许多相似之处。例如，根据本书2.3节介绍，用于民用飞机工业的飞机部件架构《ATA规范100》(1989)也用于军用领域(MIL-STD-1808B，2007)。其主要区别在于，军机中增加了特定的军用类别，如携带武器的装置，军用飞机其他特有的需求还包括：防止敌方武器攻击等。

商用飞机与军用飞机最主要的区别是需求的类型及提出这些需求的客户类型。本书3.1节将会介绍，商用客户分为两类，即飞机市场和具体的航空公司客户。商用飞机的需求很大程度上来源于经济性需求。军用飞机通常只有一个客户，有着特定的任务要求。本书将在第8章进一步讨论经济性需求，该需求是顶层综合的一部分。

2.1 商用飞机行业

全球商用飞机行业不断发展，形成了少数几家大公司占主导的格局。该行业的风险包括新产品研制的财务成本风险，以及航空公司和其他客户的财务健康

established firms have been led to merge to accept the high costs and risks of doing business. This environment and the complex nature of aircraft development make commercial aircraft a prime subject for the application of SE.

2.2   Levels of SE Application

Aircraft development is conducted at three broad levels. Each level demands different aspects of SE. The levels are as follows:

Level 1—New aircraft

The development of new aircraft allows SE to be applied in a *blank slate* fashion: that is, to start from the inception of the requirements for an aircraft and the development of initial concepts. Discussions with *launch* customers are held, and analyses of range, number of passengers, noise, emissions, and other top-level requirements are developed. Economic, technical, and regulatory criteria are analyzed against all potential concepts. Requirements are allocated to the aircraft subsystems. Major components will remain the same for all customers except for those changed below as discussed in Section 2.2.

Level 2—Derivative aircraft

A derivative aircraft utilizes major components of existing aircraft as the basis for the development of an aircraft which meets some new requirements. The derivative aircraft may have increased performance or carry more or fewer passengers than the baseline aircraft. The challenge of SE is to develop requirements and to synthesize and verify solutions to those requirements within the specified constraints of the baseline aircraft. These are aircraft for which major components, primarily airframe, are used from previous models. These components will remain the same for all customers. The development of a derivative aircraft, as opposed to a new aircraft, can result in considerable savings in development and tooling costs, and lower prices for the customer.

Level 3—Change-based aircraft

A change-based aircraft is an aircraft for a specific customer which may have a

风险。为了更好地应对业务经营时遇到的高额成本及风险问题，大型公司纷纷兼并。大环境和飞机研制的复杂性使得商用飞机成为系统工程应用的一个主要案例。

2.2 系统工程应用水平

飞机研制主要包括 3 大类，每一类对系统工程的要求各不相同。具体描述如下：

类别 1 - 新研飞机

在新研飞机项目中，系统工程开始于一张白纸：系统工程开始于飞机顶层需求捕获及初步概念方案论证。与启动客户开展关于航程、乘客数量、噪声、排放及其他顶层需求的讨论。基于经济性、技术和监管要求，对所有潜在的概念方案进行分析，并将需求分配给飞机各子系统。对于所有的客户来说，大多数部件将保持不变，2.2 节中所讨论的改装飞机除外。

类别 2 - 衍生型飞机

衍生型飞机是指以现有飞机主要部件为基础研制的、满足新需求的飞机。与基准飞机相比，衍生型飞机可能在性能方面有所提高，或运载更多（或更少）的乘客。系统工程的挑战来自于需求开发，并在基准飞机的规定约束范围内综合和验证这些需求。衍生型飞机的主要部件（主要是机体）来源于之前的型号。与新研飞机相比，衍生型飞机在研制成本和工装成本上更为经济，对客户而言，价格更加低廉。

类别 3 - 改装飞机

改装飞机是指针对某一特定客户的飞机，它可能要求进行大量的更改。虽

large number of requested changes. Although each change may be small in itself, the SE methodology should be applied to each change to assure that the performance requirements of the affected subsystems are met and that the accumulated changes for the whole aircraft allow the aircraft to meet its own performance requirements. Individual customers provide requirements for specific changes, or the aircraft manufacturer initiates internally generated requirements for design improvements for economic or other reasons.

Another idea frequently encountered in the industry is that derivative and change-based designs must meet less rigorous requirements than new aircraft. On the contrary, all aircraft are subject to *the top-level requirements*.

2.3　Aircraft Architecture

A key SE principle is that commercial aircraft should be considered *as a whole* and not as a collection of parts which can be independently developed and integrated. Requirements flow-down is dependent on viewing the aircraft architecture as a hierarchy in which lower-level elements, such as the subsystems, are subordinate to the aircraft. The aircraft itself is subordinate to a higher-level system called the aircraft system, which includes the aircraft and all its supporting systems.

The main reason for using the graphical depiction of an aircraft architecture is to illustrate its hierarchical nature, which corresponds to the hierarchy of the functions discussed in Chapter 3 and requirements discussed in Chapter 4. However, the hierarchy is not a hardware description; it is an abstract depiction of the aircraft. It is merely a set of *buckets* into which requirements can be placed. When the aircraft hierarchy is defined, one of the first steps, along with aircraft system-level functions discussed in Section 3.2, towards aircraft system synthesis will have begun. The hardware selection is the final step in system synthesis discussed in Chapter 7. Secondly, studying the aircraft architecture helps us understand what is really included in the aircraft *system*, which we will see below. This hierarchical architecture of aircraft fits within the SE model.

As we will see in Section 7.1, another principle is that the aircraft hierarchy is not permanent. As the design evolves, so will the hierarchy. Trade-offs may result in a new and improved hierarchy for a specific customer.

Concrete systems vs. abstract systems

Before we can discuss the aircraft hierarchy, it is necessary to understand the

然每一项更改本身可能很小，但仍应采用系统工程方法，以保证满足受影响子系统的性能需求，并保证整加飞机的累积更改仍能使这架飞机满足其自身的性能需求。个别客户可能要求特定的更改，飞机制造商也可能出于经济性或其他因素考虑提出设计改进的需求。

飞机行业中经常听到的另一种说法是，衍生型飞机和改装飞机设计需求的严格程度比新研飞机低。事实相反，所有飞机都应满足顶层需求。

2.3　飞机架构

将商用飞机看做一个整体，而不是可以独立研制和集成的部件的集合，这是系统工程的重要概念。需求自上而下的传递以飞机架构的不同层级为基础，下级的元件，例如子系统元件隶属于飞机。而飞机本身隶属于称为飞机系统的上一级系统，飞机系统包括飞机及其所有的支持系统。

采用图形化的方式进行飞机架构描述主要是为了表明其层次化特性，与第3章中讨论的功能性层级和第4章中的需求保持一致。层级是对飞机的抽象，而不是硬件描述。这些层级只是一系列可以存放需求的"篮子"。完成飞机层级定义后，从飞机系统级功能开始（见第3.2节），一直到飞机系统综合的工作将逐步开展。硬件选择是系统综合的最后一步，该内容在第7章中介绍。另外，研究飞机架构有助于工程师真正了解飞机系统及其内部构成。飞机的层级架构符合系统工程模型。

本书7.1节介绍的另一个系统工程原理是，飞机的层级不是永恒不变的。随着设计的进展，层级也在不断演变。由于特殊客户的要求，经过权衡研究后，可能会产生新的和改进的飞机层级。

实体系统 vs 抽象系统

在讨论飞机层级前，必须了解实体系统与抽象系统之间的区别。简单来说，

difference between a concrete system and an abstract system. In simple terms a concrete system is the system you can touch: You can touch the wheel; you can touch the wing, and so forth. An abstract system is a mental model, that is, it is a depiction of the system that is the aircraft, for the purpose of analysis. That is, an abstract system in anthropocentric view of a system. SE has adopted this hierarchical mental model as the primary method of depicting systems for analysis. This method is not totally random. It results from the fact that it is possible to view subsystems as clusters of components that exist at various levels of the abstract hierarchy. It is convenient that the traditional SE practice of viewing systems as hierarchies has also been adopted by the Air Transport Association of America (ATA) whose hierarchical depiction of an aircraft is shown below.

From an aircraft point of view, it is important to point out that there are other ways of depicting an aircraft system. Avionics specialists, for example, sometimes depict the avionics system as web rather than a hierarchy. This is because that system can be described at a single level rather than multiple levels. Neither method is correct nor incorrect. It is simply a matter of convenience.

The aircraft hierarchy and the Air Transport Association (ATA) index

The fact that the ATA has adopted this hierarchical view of the aircraft shows that the SE concept of an abstract hierarchy is already an accepted concept in the aircraft industry. Figure 2.1 shows a typical hierarchy of the aircraft system and its subordinate elements. This type of hierarchy is called a specification tree (or *spec tree* for short) since a specification should be written for each element as discussed in Section 12.6. The hierarchical breakdown of the aircraft system and the aircraft allows the flow-down of requirements to all subsystems and components in the classical SE manner.

In the ATA index the major aircraft divisions subsystems because they are not really subsystems; that is, they do not all perform a key function. Rather, they are collections of subsystems which group together technologically.

The arrangement of the ATA Specification 100 (1989) chapter numbers in Figure 2.1 is not a feature of the ATA index. Its purpose is to create a hierarchy useful for SE analysis and yet retain the ATA identification numbers. No components or ATA numbers are lost in this process. There is no requirement to break down the aircraft elements as Figure 2.1 does. There is also no requirement to make it correlate to the ATA index. However, doing so is a convenient way to follow the general SE practice of creating a hierarchical breakdown. Tying the breakdown to the ATA index creates a structure familiar to most aircraft engineers.

实体系统是那些可以触摸到的系统，例如：你可以触摸到起落架轮轴和机翼等。抽象系统是一种思维模型，它是一个对飞机的系统抽象描述，主要用于分析，即抽象系统是以人为中心的视角产生系统。系统工程采用这种分层次的思维模型作为系统描述和分析的主要方法。这种方法并非随意而定的，而是因为子系统可以视为位于飞机不同抽象层级的部件群。非常方便的是，美国航空运输协会(ATA)也采用了传统系统工程将系统视为层次结构的做法，其关于飞机的层次化描述方式如下所述。

需要指出的是，从飞机角度出发，还有其他一些方法描述飞机系统。例如，航电专家有时将航电系统描述为网络结构，而不是层次结构。因为该系统在一个层级内就可以描述清楚，不需要分层描述。方法没有对错之分，关键在于选择一种最便捷的方式。

飞机层级和ATA索引

ATA采用分层方式描述飞机，表明系统工程的抽象层级概念已被飞机行业广泛认可。图2.1是飞机系统及其下级元素的典型架构，这种类型的架构称为规范树，因为每一个元素对应一个规范(具体内容参见12.6节)。飞机系统及飞机的层次化分解使得可以将需求以经典的系统工程方式、从上至下传递到所有子系统及部件中。

由于ATA索引按飞机主要部段划分子系统，而这些子系统并非真正的子系统，因此它们并不全部执行某一关键功能。它们只不过是因技术原因而组合在一起的一些子系统的集合。

图2.1所示的《ATA规范100》(1989)章节号的编排不是ATA索引的特征。其目的是创建一个对系统工程分析有用，同时保留ATA标识号的层级结构。整个过程中任何部分或ATA索引号都不会丢失。这不是要求必须将飞机元件按图2.1所示方式分解，也不是要求必须将其与ATA索引对应，然而这样做有助于采用系统工程方法创建一个分层级的飞机架构。按照ATA索引进行分解会产生一个

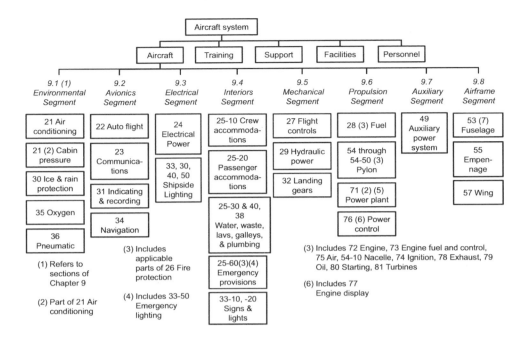

Figure 2.1   Generic aircraft system architecture and ATA chapter correlation

A goal of integrated product development (IPD) is to make the organizational structure and the aircraft hierarchy the same. The IPD process discussed in Section 12.3 divides the aircraft into *product centers*. Each product center is managed by an integrated product team (IPT). For example, a product center might be the wing. The wing IPT would be responsible for the entire development of the wing and all of the internal components. Hence, the program organization would be organized in accordance with the product breakdown. Thus, Figure 2.1 would represent both the product hierarchy and the program organization chart in which each box would represent an IPT at each organizational level. Alternatively, in an IPD environment the hierarchy shown in Figure 2.1 might be very different depending on the organizational structure chosen.

The aircraft system

The aircraft system consists of more than the aircraft (the flight vehicle) itself. The aircraft system consists of the following four elements, of which the aircraft is only one. Each of these subsystems can also be broken down into its subordinate components.

大多数飞机工程师都熟悉的结构。

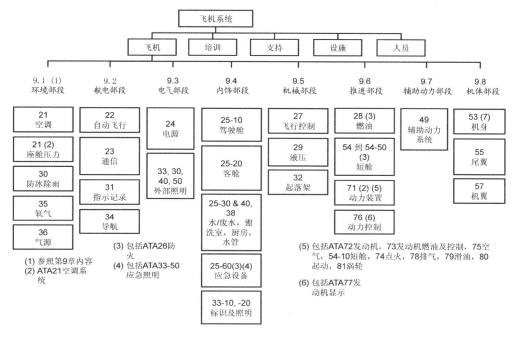

图 2.1　飞机系统架构及对应 ATA 章节

集成产品研制 (IPD) 的目标是让组织架构和飞机层级相匹配。本书 12.3 节将讨论 IPD 过程，该过程将飞机划分为不同的产品中心。每个产品中心由一个集成产品管理团队 (IPT) 来管理。例如，产品中心可以是机翼，机翼 IPT 将负责机翼整体及其内部部件的研制。因此，项目组织架构应根据产品的分解结构来规划和安排。由此来看，图 2.1 既是产品层级架构，也是项目的组织架构，每个方块代表各自组织层级的 IPT 团队。另外，在 IPD 环境中，图 2.1 所示的层次结构也可能因为选择了不同的组织架构而不同。

飞机系统

飞机系统远不止飞机 (飞行器) 本身，飞机系统包含如下四个元素，飞机只

The aircraft The aircraft is the flight vehicle. The aircraft segments are described in Section 2.3.

Training equipment There are two types of training equipment normally included in training equipment. Flight crew training equipment includes simulators and any other equipment needed for flight training. Maintenance training equipment includes such equipment as mock-ups.

Support equipment There are two principal types of support equipment, on-aircraft and off-aircraft. Examples of off-aircraft support equipment are the ground electrical and hydraulic supplies. An example of an on-aircraft support equipment item is built-in test equipment (BITE).

Facilities The facilities of importance can either be buildings where aircraft are stored and maintained, or they can be specialized structures, such as the common airport skyways.

Personnel As discussed in Section 5.5, people can be considered to be a part of the aircraft system. It would be impractical, and perhaps unrealistic, to view people themselves as hierarchical systems. In a rudimentary fashion people might be considered to consist of a physical subelement and a cognitive subelement.

Aircraft segments

There are many ways to break down the aircraft into its subordinate elements. The following breakdown is typical:

1. *Environmental segment*—This segment includes air conditioning, ice and rain protection, cabin pressure, pneumatic supply, and oxygen supply equipment.
2. *Avionics segment*—This segment includes the communications, navigation, indicating and recording, and auto flight equipment. This segment might be more appropriately called the aircraft management segment since it includes the communications subsystem and the aircraft monitoring functions found in the indicating and recording subsystem.
3. *Electrical segment*—This segment includes electrical power and shipside lighting.
4. *Interiors segment*—This segment includes crew accommodations; passenger accommodations; water, waste, lavatories, galleys, and plumbing; emergency

是其中一个。每个子系统又可以向下分解到各从属部件。

飞机：飞机是一种飞行器，2.3 节对飞机各部段进行了介绍。

培训设备：培训设备通常包括两大类。飞行机组培训设备包括模拟器及飞行培训所需的任何其他设备。维修培训设备包括模拟样机等设备。

支持设备：支持类设备主要有两大类，机上支持设备和机下支持设备。例如：地面电源和液压源属于机下支持设备，机内检测装置 (BITE) 属于机上支持设备。

设施：重要的设施可以是停放和维修飞机的场所，或者专门的结构，如一般的机场天桥。

人员：如 5.5 节讨论的，人可以看作是飞机系统的一部分。将人本身看作是一个分层次的系统可能不切实际，或不太现实。在较前卫的思想中，人可以看作是由物理子元素和认知子元素组成的系统。

飞机部段

飞机分段有很多种方法，典型的分段方法如下：

1. 环境部段：包括空调、防冰除雨、座舱压力、气源及供氧设备。

2. 航电部段：包括通信、导航、指示记录、自动飞行设备。航电部段称为飞机管理部段可能更为合适，因为它包括通信子系统，以及具有飞机监控功能的指示记录子系统。

3. 电气部段：包括电源及外部照明。

4. 内饰部段：包括驾驶舱设备、客舱设备、水、废水、盥洗室、厨房及水管、应急设备、内部标识及内部照明。

provisions; and interior signs and lights.

5. *Mechanical segment*—This segment includes landing gears, flight controls, hydraulic power, and cargo loading equipment.

6. *Propulsion segment*—This segment includes the engine pod and its components, fuel components, engine pylons, and thrust management equipment.

7. *Auxiliary segment*—This segment includes any auxiliary power supply, for example, for generating electrical or pneumatic power.

8. *Airframe segment*—This segment includes the wing, fuselage, and empennage.

Chapter 9 discusses the synthesis process as applied to each one of these segments.

Alternative hierarchies

As noted above, the hierarchy shown in Figure 2.1 is only one of many possible hierarchies. The development of an aircraft lends itself to many possible hierarchies. Following are two of the more obvious possible subsystems which might appear in an alternative hierarchy; each of these possible subsystems would contain portions or all of other subsystems shown in this figure:

1. *Cabin subsystem*—Although a cabin subsystem would be dominated by features of the passenger accommodations subsystem described in Section 9.4, it would contain portions or all of other subsystems. For example, it could also contain elements of the following subsystems: air conditioning described in Section 9.1, cabin pressure described in Section 9.1, oxygen described in Section 9.1, internal communications described in Section 9.2, electrical power described in Section 9.3, water, waste, lavatories, galleys, and plumbing described in Section 9.4, emergency provisions described in Section 9.4, and signs and lights described in Section 9.4.

2. *Cockpit subsystem*—In addition to all of the avionics segment subsystems described in Section 9.2, a cockpit subsystem would contain elements of air conditioning described in Section 9.1, cabin pressure described in Section 9.1, oxygen described in Section 9.1, electrical power described in Section 9.3, crew accommodations described in Section 9.4, water, waste, lavatories, galleys, and plumbing described in Section 9.4, emergency provisions described in Section 9.4, signs and lights described in Section 9.4, flight controls described in Section 9.5, and power control described in Section 9.6.

5. 机械部段：包括起落架、飞行控制装置、液压源及货物装载设备。

6. 推进部段：包括发动机短舱及其部件、燃油部件、发动机吊挂及推力管理设备。

7. 辅助动力部段：包括任何辅助动力源。例如，提供电源和或气源的辅助动力源。

8. 机体部段：包括机翼、机身及尾翼。

第9章将介绍应用于上述各部段的综合过程。

其他层级架构

飞机的研制也许会出现很多种可能的架构，图2.1所示的层级架构只是诸多层级架构中的一种。下面介绍的两个子系统不同于图2.1所示的架构，每个子系统都可能包含图2.1中其他的部分或全部的子系统。

1. 客舱子系统：虽然座舱主要包含第9.4节中描述的乘客起居设施子系统，但它还包含部分的或全部的其他子系统。例如：座舱子系统内也可以包含如下子系统：9.1节的空调、座舱压力、氧气；9.2节的内部通信；9.3节的电源；9.4节的水/废水、盥洗室、厨房及水管、应急设备、内部标识及内部照明设备。

2. 驾驶舱子系统：除了9.2节所述的航电部段的子系统外，驾驶舱子系统还可能包含如下子系统的部分：9.1节的空调、座舱压力、氧气；9.3节的电源、9.4节的机组内饰、水/废水、盥洗室、厨房和水管、应急设备、内部标识及内部照明设备；9.5节的飞行控制装置及9.6节的动力控制。

2.4 Advanced Technologies on Aircraft

To meet design requirements for reduced weight, noise, and emissions, robust systems, and safe and economic operation, many advanced technologies are routinely incorporated into commercial aircraft, for example, heads-up displays (HUD), voice recognition, global positioning system (GPS) receivers, point-to-point inertial navigators, reconfigurable instrument displays based entirely on digital video displays, Doppler radar, fly-by-wire (FBW) or fly-by-light (FBL), and real-time computer fault detection and isolation. Composite material technology is key to weight reduction. Some of these technologies are discussed below, as described in part by Kehlet (1995). We will see in Section 7.6 how the SE process evaluates new technologies for incorporation into the aircraft design.

Advanced subsonic transports

For subsonic transports, key advanced technology applications include: center of gravity (c.g.) management systems, for example, with vertical stabilizer tanks; composite primary and secondary structures; supercritical wings with high-load alleviation, hybrid laminar flow control, and high-lift systems; advanced turbofan engines, FBW and power-by-wire (PBW); titanium landing gears; aluminum–lithium or metallic composite fuselage structures; and stability augmentation. Many, if not most, of these technologies are being incorporated into today's designs and will be routine in future aircraft.

Advanced supersonic transports

For supersonic transports, key advanced technology applications include synthetic vision, sidestick control, advanced lightweight materials, mixed flow turbofan engines, negative static margin, mixed compression inlets, arrow wing for supersonic cruise efficiency, FBL and PBW flight controls, and auto control in pitch. Technology requirements for advanced supersonic aircraft are discussed by *Aerospace Engineering* (1994) and Kehlet (1995).

2.4　飞机上的先进技术

为了满足减轻重量、降低噪声、减少排放、系统鲁棒性、运行安全性及经济性等设计需求，商用飞机领域引入了越来越多的先进技术，例如，平视显示器 (HUD)，语音识别，全球定位系统 (GPS) 接收机，点对点惯性导航仪，完全基于数字视频显示的可重构数字仪表显示器、多普勒雷达，电传操纵 (FBW) 或光传操纵 (FBL) 和实时计算机故障检测和隔离。复合材料技术是飞机减重的关键技术。下面讨论其中一些技术，如 Kehlet(1995) 所述。7.6 节将介绍系统工程过程如何评估这些新技术在飞机设计中的应用。

先进的亚声速运输机

对于亚声速运输机，关键先进技术的应用包括：重心 (CG) 管理系统，例如，采用垂直安定面油箱；复合材料的主、次结构；带有高载荷减缓技术的超临界机翼，混合层流控制，及高升力系统；先进涡轮风扇发动机，电传操纵及电传动力 (PBW)；钛合金起落架；铝锂合金或金属复合材料机身结构，以及增稳装置。即使不是大多数，但其中许多新技术正在引入到现今的飞机设计中，也将成为未来飞机的常规技术。

先进的超声速运输机

对于超声速运输机，关键先进技术的应用包括：综合视景，侧杆控制，先进的轻质材料，混流式涡轮风扇发动机，负静余量，混合压缩进气口，改进超声速巡航效率的箭翼，FBL 和 PBW 飞行控制装置及俯仰自动控制。《航空航天工程》(1994) 和 Kehlet(1995 年) 研究讨论了先进的超声速飞机的技术要求。

Airframe technology

Improvements in the technology of aircraft structures will come from advanced materials and integration techniques. Integrated computer codes will allow the aerodynamics and strength aspects to be addressed simultaneously. Thus computer-aided design becomes an advanced tool for the evaluation of physical interfaces within the SE framework. The use of composite materials allows for both a decrease in weight and an increase in performance through higher-aspect ratios. Advanced machining techniques allow for the design to minimize parts.

According to Lin et al. (2013) "one of the main weaknesses of laminate composite structures" is delamination. Lin et al. provide an overview of techniques to alleviate this weakness. Although composite structures do not suffer from some of the weaknesses of metal structures, such as metal fatigue, delamination is a major weakness. Hence, before adopting this type of structure, the developer needs to address these techniques.

Aerodynamic improvements

New aerodynamic techniques include the use of pressure sensitive paint and computational fluid dynamics (CFD). These technologies will allow multipoint wing design that attains the lowest cruise drag characteristics and the highest realistic buffet onset boundary. Another goal is efficient aerodynamic profiles for wings with large high by-pass ratio (HBPR) engines. Another effort is the aerodynamically efficient but low-cost high lift systems. Studies have shown (Martínez-Val, 1994) that greater range and payload capability can be achieved with a third horizontal surface, or *canard*, located on the forward fuselage. Another concept of aircraft configuration is the blended wing-body (BWB), which resembles a large manta ray. This concept is described further below.

Noise control

Active noise control can reduce cabin noise without severe weight penalties. This technique introduces a secondary noise source of comparable amplitude but opposite in phase to the primary noise in order to cancel out the primary noise. It controls noise over a wide range of frequencies to counteract both engine noise and boundary layer noise. It will be especially important to control boundary layer noise on

机体技术

飞机结构技术的改进将来自于先进材料和集成技术。集成计算机代码可同时兼顾空气动力学和强度方面的要求。因此，计算机辅助设计成为在系统工程框架内评估物理接口的先进工具。复合材料的使用使得可以通过更大的展弦比减轻重量，并提高性能。先进的加工技术使得最少化部件的设计成为现实。

据 Lin 等 (2013) 的研究，"层压板复合材料结构的主要弱点之一"是分层。Lin 等针对缓解这一弱点完成一份技术综述。复合材料结构虽然能够避免一些金属结构的缺点，例如金属疲劳，但是复合材料的分层是其一大弱点。因此，采用这种类型的结构之前，研制方需要解决这些技术问题。

空气动力学的改进

新的空气动力学技术包括使用压力敏感涂料和计算流体力学 (CFD)。这些技术将允许多点机翼的设计，达到最低的巡航阻力特性和高仿真的抖振补偿边界。另一个目标是带有大型高涵道比 (HBPR) 发动机的机翼的高效气动翼型。另外一项是空气动力学高效、但低成本的高升力系统。Martínez-Val (1994) 的研究表明，利用位于前机身的第三个水平表面或鸭翼，可以实现更大的航程和商载能力。飞机构型的另一个概念是翼身融合体 (BWB)，类似大的蝠鲼。这个概念将在下面进一步描述。

噪声控制

主动噪声控制可以降低座舱内噪声，而不会导致飞机大幅增重。这种技术引入了振幅与主噪声相当、但相应与主噪声相反的二次噪声源，以抵消主噪声。该技术控制的噪声覆盖很宽的频率范围，以抵消发动机噪声和边界层噪声。控制边界层噪声对于高速民用运输机 (HSCT) 而言尤为重要。

the high-speed civil transport (HSCT).

Fly-by-Light (FBL), Fly-by-Wire (FBW), and Power-by-Wire (PBW) technologies

FBL introduces multiplex photonically-based subsystems into the aircraft. FBL reduces wiring weight, reduces exposure to electromagnetic interference (EMI) hazards, and simplifies certification by eliminating the need for full aircraft subsystem tests. PBW and FBW result in weight savings and eliminate the need for engine bleed air and variable speed drives for secondary power subsystems. FBL, FBW, and PBW result in higher-reliability, lower-maintenance costs, and lighter weight.

FBW is an enabling technology for flight envelope protection as described below.

Synthetic vision capabilities

Synthetic vision enables pilots to use visual imagery and guidance cues to penetrate weather and compensate for low levels of illumination. These subsystems would use satellite-based navigation, imaging sensors, and high-resolution displays to operate with a greater degree of autonomy.

Propulsion controlled aircraft

An aircraft controlled by thrust modulation rather than control surfaces would be more able to survive catastrophic events, including terrorist actions, and perform better in partial failure conditions. Following the 1989 Sioux City DC-10 accident as previously described by Jackson (2010, pp. 78–79), there was much discussion of the possibility of propulsion control. In this incident the pilot was able to maintain partial control by using the propulsion control mechanisms. However, the NTSB (1990) did not recommend implementation of propulsion control but rather focused on other preventive measures.

Autonomous cargo handling

Improved methods for airlift cargo handling (IMACH) are an integrated group

光传操纵 (FBL)、电传操纵 (FBW) 及电传动力 (PBW) 技术

光传操纵 (FBL) 将基于多路复用光学的子系统引进飞机设计，该技术 FBL 可以减轻布线重量，降低暴露于电磁干扰 (EMI) 的危害，并因无须对全机子系统试验而简化合格审定过程。电传动力 (PBW) 和电传操纵 (FBW) 可以减轻飞机重量，且不需要用于次级动力子系统的发动机引气及变速传动装置。光传操纵 (FBL)、电传操纵 (FBW) 及电传动力 (PBW) 技术可以提高可靠性、降低维修成本并减轻飞机重量。

电传操纵 (FBW) 是一项飞行包线保护 (如下所述) 的使能技术。

综合视景能力

综合视景能使驾驶员使用可视化图像和导引手段来洞察天气和补偿低水平照明条件。这些子系统可能使用基于卫星的导航、成像传感器及高分辨率显示器，以使操作有更大程度的自主权。

推进控制飞机

通过推力调节，而不是操纵面控制的飞机在灾难性事件 (包括恐怖事件) 中存活概率更高，并能在局部故障状态下更好地运行。如前面杰克逊 (2010 年，第 78, 79 页) 所述，在 1989 年苏城 DC-10 事故之后，有很多关于推进控制能力的讨论。在这一事件中，驾驶员能通过使用推进控制机制而维持部分控制。然而，国家运输安全委员会 (NTSB)(1990) 不建议实施推进控制，而是专注于其他预防措施。

自动货物装卸

空运货物装卸改进方法 (IMACH) 是改进货物装卸各大技术的集成。这些方法侧重于搬运功能，主要考虑对大型的及更复杂的货物的搬运，以及自动化特点。

of technologies for improving cargo handling. These methods focus on handling functions, on the handling of large and more complex loads, and on automation features. IMACH can completely automate the movement of palletized loads. This improvement can achieve reductions in turnaround time, a prime cost driver for airline customers.

High Speed Civil Transport (HSCT)

The drive towards an HSCT has focused on many new advanced technologies. These technologies include advanced propulsion systems and advanced materials which can manage the temperatures associated with flight on the order of Mach 2.4. Research is being conducted in aerodynamics and technology integration, propulsion, structures and materials, flight deck systems, and key environmental issues, including sonic boom, airport and community noise, and emissions. Research has also begun on a hypersonic transport (HST) (*Aerospace Engineering*, 1996), for which the demands are even greater at Mach 5.0.

Human factors

Human factors have long been critical to aircraft design, especially in flight deck layout. Key demands include new techniques for solving the contradictory hazards of high work load and pilot boredom. Use of advanced visualization tools as part of the SE process in combination with design tools provides cost-effective, rapid prototyping to evaluate and adjust the design. We will see in Section 5.5 how SE integrates human factors into the requirements process.

Advanced design tools

The complexity of aircraft design led the aircraft industry to develop sophisticated design and simulation tools. In many cases such tools provide design information faster and more cheaply than total reliance on wind tunnel testing. These tools, combined with advanced visualization techniques, are sufficiently mature to provide a basis for research as well as aircraft design and engineering. There is a strong synergism between the computer-aided design tools and the training for each aircraft type. The modeling of flight characteristics for new aircraft types is so accurate that type checkout for pilots based only on trainer experience is anticipated. Simulation has

IMACH 可以使托盘货物的移动完全自动化。这项改进可缩短地面转场时间，对于航空公司客户而言，转场时间是一项主要的成本动因。

高速民用运输机 (HSCT)

新兴的先进技术一直是推动高速民用运输机 (HSCT) 发展的关键。这些技术包括先进推进系统及先进的材料，它可以管理与飞行马赫数 2.4 量级相关的温度条件。业界正在进行研究空气动力学和技术集成、推进、结构和材料、驾驶舱系统以及关键的环境问题，包括音爆、机场和社区噪声及排放领域。对高超声速运输机 (HST)(《航空航天工程》，1996) 的研究也已开始，其速度要求甚至高达 5.0 马赫。

人为因素

人的因素早已是飞机设计的关键因素，特别是在驾驶舱布局方面。其关键要求包括用于解除高负荷工作和驾驶员厌倦之间相互矛盾的危害的新技术。与设计工具相结合，采用先进的可视化工具作为系统工程过程一部分，可以提供成本效益高且快速的打样方法，以评估和调整设计。我们将在第 5.5 节介绍系统工程如何将人为因素整合为需求。

先进设计工具

飞机设计的复杂性驱使航空工业开发先进的设计和仿真工具。在许多情况下，这些工具提供的设计信息比完全依赖风洞试验更快、更经济。结合先进的可视化技术，这些工具都足够成熟，可为研究及飞机的设计和工程提供基础。计算机辅助设计工具与针对每个机型的培训之间存在很强的协同作用。对新机型而言，飞行特性的建模十分准确，以至于可以预期仅有模拟机经验的驾驶员便可获得型号等级认可。在商用飞机行业，仿真已被确立为系统工程的主要验证技术之一。

been established as one of the primary SE verification techniques in the commercial aircraft industry.

Flight envelope protection

In recent years a new system has emerged called *flight envelope protection*. According to Airbus,

> Fly-by-wire [the use of digital rather than mechanical interfaces] enhances safety by allowing the programming of the flight envelope protection, which enables pilots to fly the aircraft freely but prevents any abnormal operations, such as stalling, flying too fast, or overstressing. (Airbus, 2013)

Other manufacturers have favored other approaches to safety. To date there is no universal agreement as to which is the most effective approach. However, as seen in Section 10.4, the international safety consortium CAST has recommended that some aspects of flight envelope protection be implemented on all new aircraft.

Blended-Wing-Body (BWB)

Originally conceived by McDonnell Douglas prior to its merger with Boeing in 1997, the blended-wing-body is a concept with much promise for carrying more passengers or cargo more economically. It is similar to a flying wing except that it consists of a wing and fuselage that blend together smoothly. The BWB showed advantages over conventional aircraft in operating costs, fuel efficiency, gross weight, and nitrous oxide emissions.

In spite of its promise, the BWB still has many challenges. In a NASA report Bowers (2000) provides a comprehensive summary of both the advantages and disadvantages of this concept. One disadvantage is that the plane would be too large for current gates. Folding wings were investigated and determined to be unacceptable. Other challenges included structures and materials, aero-structural integration, aerodynamics, controls, aero-structural integration, propulsion- airframe integration, systems integration, and infrastructure.

At the time of his report, Bowers states that NASA was still committed to supporting research on the BWB. For its part, Boeing has not announced any

飞行包线保护

近年来出现了一个新的系统，称为飞行包线保护。据空客公司介绍，

电传操纵【使用数字接口而不是机械接口】通过允许飞行包线保护编程增强安全性，此举使得驾驶员能自如地驾驶飞机，而防止任何非正常运行，如失速、飞行速度过快，或过载（空中客车公司，2013)。

其他制造商曾倾向于采用其他安全性方法，但迄今为止关于最有效的途径还没有普遍一致的意见。然而，如 10.4 节所述，国际安全联盟 CAST 建议，应在所有新飞机上考虑飞行包线保护的一些方面。

翼身融合体 (BWB)

在 1997 年与波音公司合并之前，麦道公司最初设想的翼身融合体是一个预期能更经济地运送更多乘客或货物的概念。它类似于一个飞翼，只是它包含了机翼和机身，而且机翼和机身很平滑地结合在一起。与传统飞机相比，BWB 在运营成本、燃油效率、总重量和氮氧化物的排放量方面更具优势。

尽管这个概念很有前景，但是，BWB 仍面临许多挑战。在美国航空航天局(NASA) 的报告里，Bowers(2000) 对这一概念的优点和缺点进行了全面的总结。该概念的一个缺点是，对于目前的航站楼登机口而言，这种飞机体积过大。经过研究发现，机翼折叠也不可行。这类飞机的其他的挑战包括结构和材料、空气动力学、控制、气动结构一体化、推进机体结构集成、系统集成及保障设施。

在他提出这份报告时，Bowers 指出，美国航空航天局仍承诺支持对 BWB 的研究。就其本身而言，波音公司还没有宣布将这个概念引入商业实践的任何意向。

intention to introduce this concept into commercial practice.

2.5   Aircraft Manufacturing Processes

The aircraft industry has in recent years seen a shift from virtually hand- made to mass produced aircraft. Although many steps in the manufacturing process of commercial aircraft parallel those for motor vehicles, design and manufacture of commercial aircraft are even more complex. Although aircraft manufacturing is a low-volume process, its complexity arises from dependency on highly integrated high-technology subsystems, use of advanced materials, detailed specifications, and extremely rigorous testing. SE is especially suited to addressing these issues.

2.6   Trends in Commercial Aviation

Economic and regulatory pressures

Pressures in the commercial aircraft domain come from all directions, both economic and regulatory. Economic pressures come primarily from the airline customers. Of course, every airline wants to carry as many passengers as far as possible and as cheaply as possible. These pressures have resulted in more efficient engines and radically different designs, such as the blended-wing-body discussed above. Another trend is toward more efficient flight paths. One concept being looked into is the 4D trajectory. According to Skybrary (2014),

> The 4D trajectory of an aircraft consists of the three spatial dimensions plus time as a fourth dimension. This means that any delay is in fact a distortion of the trajectory as much as a level change or a change of the horizontal position. Tactical interventions by air traffic controllers rarely take into account the effect on the trajectory as a whole due to the relatively short look-ahead time (in the order of 20 minutes or so).

Among the many benefits of the 4D trajectory are optimal operations for airlines, reduced cost and time, reduced emissions, and reduced load on controllers.

Other pressures especially on engine makers are the pressures to reduce carbon footprint and engine noise.

Trends in component procurement

2.5 飞机制造工艺

近年来，飞机制造业逐渐由手工制造向大规模生产转变。虽然商用飞机的生产过程许多方面与汽车非常类似，但是商用飞机的设计和制造更加复杂。飞机制造是一个小批量生产过程，它的复杂性由以下几方面引起：依赖于高度集成的高科技子系统、先进材料的使用、详细说明的规范和极其严格的试验。系统工程可在解决这些问题时发挥重要作用。

2.6 商用航空发展趋势

经济性及监管的压力

商用飞机领域的压力来自各个方面，既有经济方面的也有监管方面的，其中经济压力主要来自于航空公司客户。当然，每家航空公司都希望尽可能多、尽可能远且尽可能少花钱地运送乘客。由于这些压力的推动，产生了效率更高的发动机，设计也可能产生根本性的变化，例如上文讨论的翼身融合技术。另一个变化趋势是更高效的航线规划。根据 Skybrary (2014) 研究，一个概念方案是 4D 航迹导引。

飞机的 4D 航迹是在原有的三维空间基础上增加时间轴，从而称为四维导引。这就意味着事实上任何的延误都会导致航迹的扭曲失真，严重到飞行高度层改变或水平位置的改变。空中交通管制员的战术干预很少可以整体考虑对飞行航迹的这种影响因为前瞻时间过短（仅 20 分钟左右）。

4D 导引的优点包括：优化航空公司运营、降低成本及缩短时间、减少尾气排放并减少管制员的工作量。

其他压力，尤其是对发动机制造商的压力主要来自于减少碳排放量，以及

A recent trend in component procurement is for airline companies to procure or lease individual components themselves rather than using the supplier-provided components specified by the aircraft developer. Although the economic benefits for the aircraft operator may be substantial, there may be risk associated with this trend if the procured parts are not flight qualified. Parts may be valves, pumps, fans, and other parts. First of all, the environments for these parts may vary widely from airline to airline, from aircraft to aircraft and from aircraft zone to zone. The environment in an engine nacelle is considerably hotter than, for example, in the cargo bay. Hence, the responsibility is on the aircraft operator to assure that all the parts procured in these ways are flight qualified.

降低发动机噪声等方面。

部件采购趋势

部件采购的最新趋势是航空公司自己购买或者租借个别的部件，而不采用飞机研制方指定的供应商提供的部件。虽然经济利益对于飞机运营商非常重要，但是如果购买的部件不满足航空质量要求，航空公司将会面临更大的风险。这些零部件可能是各种阀、泵、风扇等其他零部件。最重要的一点是，这些零部件的工作环境对于不同的航空公司、不同的飞机和不同的飞机区域可能各不相同。例如，飞机发动机短舱内的环境与货舱相比要热很多。因此，飞机运营商有责任保证所有以这种方式采购的零部件均满足飞行要求。

3

Functional Analysis

… a motorcycle can be divided according to its components and according to its *functions*..The overall name of these interrelated structures, the genus of which the hierarchy of containment and structure of causation are just species, is *system*. The motorcycle is a system. (Robert Pirsig)

The International Council on Systems Engineering (INCOSE) Fellows (2006) state that a basic property of a system is its function. A function is a task, action, or activity performed to achieve a desired outcome. This principle applies to any system, including a motorcycle, as explained above by Pirsig (1974). This chapter will show how the functional architecture of the aircraft system becomes the basis for all aircraft SE analyses. It will show that the functional architecture is, to a given level, valid for all aircraft.

Functional analysis is a critical part of the SE process for commercial aircraft for three reasons:

- First, functional analysis is one of the primary techniques for establishing the completeness of performance requirements as discussed in Section 4.2. Functions are the primary prerequisite for the establishment of all performance requirements. This relationship between functions and performance requirements is valid for all systems, not just aircraft systems.
- Secondly, guidelines such as ARP 4754A (2010) for the certification of commercial aircraft call for a functional hazard assessment (FHA) based on a comprehensive functional analysis of the entire aircraft as shown in Section 10.2. The FHA examines all aircraft-level and subsystem-level functions and determines the safety criticality of these functions during aircraft operation.
- Finally, the functional architecture is the basis for the architecture of the entire aircraft. According to Rechtin (1991, p. 212), "Except for good and sufficient reasons, functional and physical structuring should match." This

第3章 功能分析

"摩托车可以根据其组成部分及功能进行分解。所有相关结构，恰属一类的内涵层次架构和有内在关系的结构，总称为系统。摩托车就是一个系统。"(Robert Pirsig)

国际系统工程协会 (INCOSE) 研究员 (2006) 指出，功能是系统的基本属性。功能是指为了获得期望的结果而执行的任务、行为或活动。如上面的 Pirsig(1974) 的介绍，这个概念适用于所有的系统，包括摩托车。本章将介绍飞机系统的功能架构如何成为所有飞机系统工程分析的基础。它将表明对于给定的层次，这种功能架构对于所有的飞机都是有效的。

功能分析是商用飞机系统工程过程的关键部分，原因有以下 3 条：

• 首先，如 4.2 节所述，功能分析是建立完整性能需求的主要技术手段之一。功能是建立所有性能需求的主要前提条件。功能和性能需求之间的关系对于所有的系统都是适用的，而不仅限于飞机系统。

• 其次，诸如 ARP4754A(2010) 之类的商用飞机合格审定指南要求以整架飞机全面的功能分析为基础进行功能危害性分析 (FHA)，如 10.2 节所示。FHA 将检查所有飞机级以及子系统级功能，并确定这些功能在飞机运行过程中的安全临界状态。

• 最后，功能架构是整架飞机架构的基础。根据 Rechtin(1997, 第 212 页)，"除非有合理且充分的理由，功能架构与物理架构应相互匹配"。这项规则既适用于系统的顶层架构也适用于单个子系统的架构。先行设计团队可以根据该规则来

rule applies both to the top-level architecture of the system and also to the architecture of individual subsystems. The Advanced Design organization can use this rule to create the architecture of the aircraft and of the individual subsystems.

The development of the functions of the aircraft system as discussed in Section 3.2 is one of the first steps in the aircraft system synthesis process, for it is around the system functions that the nature of the design begins to take shape. The final step in system synthesis occurs when the functions and requirements are converted into a design, as discussed in Chapter 7.

A sound practice of functional analysis is to define the limits of each function. These limits are normally defined by time and the scope of the activities included in the function. That is, the function has a definite beginning and end, and definite activities, all of which should be known. The importance of this concept will become more apparent for flight operations as discussed in Section 3.3.

Throughout this book, the names of functions will be capitalized. The names and organizations of functions are, to a certain extent, dependent on the interpretation of the analyst. So the reader should not interpret the functions and diagrams in this chapter to be definitive. It is the purpose of this chapter primarily to illustrate the significance and usefulness of functions and functional analysis.

When to perform functional analysis

Given the considerable effort that functional analysis may require, it is appropriate to ask: when should functional analysis be performed and when can it be skipped? Chapter 2 defines three types of commercial aircraft: new, derivative, and change-based. The most appropriate time to perform functional analysis is for new aircraft. This is because the aircraft is being created from scratch: its architecture, components, and requirements. For derivative and change-based aircraft the architecture and the components will, for the most part, already have been defined. For these cases, functional analysis can be avoided, or at a minimum modified only slightly. But, of course, the FHA will need to be done for all three types of aircraft. In addition, it must be remembered that the performance requirements may not have been defined for these cases. Or, perhaps the newly derived aircraft or changed aircraft may result in new performance requirements. For this reason the performance requirements should be

建立飞机及单独子系统的架构。

3.2 节介绍的飞机系统功能的研制是飞机系统综合过程的最初工作之一，因为围绕着飞机系统功能，设计的本质工作开始成形。飞机系统综合的最后步骤发生在功能和需求转化为设计之时，如第 7 章所述。

合理的功能分析做法是定义每一项功能的极限。这些极限通常由该功能所包含活动的时间及范围确定。也就是说，功能具有明确的起点和终点，以及确定的行为，所有的这些方面的内容均应被界定出来。这个概念的重要性在 3.3 节所述的飞行运行中体现得更为明显。

整本书中功能的名称都会以大写表示[1]。功能的名称及组织形式在一定的程度上依赖于分析人员的理解。对本章所列功能和图表的理解并不是唯一的。本章的主要目的是描述功能和功能分析的重要性和实用性。

何时执行功能分析

考虑到功能分析可能需要的巨大工作量，需要明确：应该何时进行功能分析，以及在什么情况下可以跳过功能分析。第 2 章定义了 3 种商用飞机：新研型、衍生型以及改装型。执行功能分析的最佳时机是新研飞机阶段，因为新研的飞机是从最初的草图开始的：其架构、部件及需求。对于衍生型和改装型飞机，飞机架构和部件大部分都已经确定。在这种情况下，可以不进行功能分析，或者仅稍微做一些更改。但是，上述 3 种类型的飞机同样需要进行 FHA 分析。此外，一定要注意在这些情况下飞机的性能需求可能尚未被定义。或者，也许新的衍生型飞机或者改装型飞机将产生新的性能需求。由于这些原因，关于这些部件的性能需求应该重新进行审视。关于性能需求及其开发过程的内容将在第 4 章详细介绍。

1　中文版以加粗字体表示。——译者注。

revisited for these components. See Chapter 4 for a complete discussion of performance requirements and their development.

The creation of a function

Before we discuss actual functions, we need to ask the questions: what is a function and how do you create one? A function is a description of what a system element does, but the system element is not named in the function. When you name the system element, you have a performance requirement. Without the system element, you just have a functional requirement.

The simplest form of a function is very simple; it is just a verb and a noun. Typical functions are *generate power* and *provide lift*. It is pretty obvious that a generator will generate power and a wing will provide lift, but that is not part of the function. Some practitioners try to avoid the verb *provide* in the name of the function, but logically there is nothing wrong with it.

The logic behind not naming the element is that this approach leaves the designer open to alternative ways to satisfy the function.

3.1 The SE Life-Cycle Functions

Table 3.1 compares the aircraft life-cycle functions with the traditional SE life- cycle functions from ANSI/EIA 632 (1999) and the life-cycle functions from the *FAA Systems Engineering Manual* (2014). The aircraft life-cycle functions have been organized to emphasize various aspects of the commercial aircraft life-cycle. These are the activities performed by the *developer* and the *users* of the aircraft from the moment of its conception to its disposal. That is to say, they are not the functions performed by the aircraft itself. Those are the functions of primary interest, but we will discuss them later in Section 3.3.

It is important to discuss the life-cycle functions because they have a significant impact on the aircraft itself. Furthermore, the life-cycle functions include the entire SE process, which is the subject of this book.

Figure 3.1 shows a typical SE life-cycle functional flow[1] for the development, manufacture, operation, and disposal of an aircraft system. This flow incorporates the aircraft life-cycle functions of Table 3.1. The table compares these phases with the

---

1   A rigorous functional flow diagram will have *and* and *or* gates. These have been eliminated for simplicity.

功能的产生

在讨论实际的功能之前，需要回答这样一个问题：什么是功能，如何产生一个功能？功能是关于系统元素可以做什么的描述，但是这个系统元素不能用功能来命名。当命名系统元素时，就有了性能需求。如果没有系统元素，就只有功能需求。

最简单的功能形式非常简单，就是一个动词和一个名词的组合。典型的功能例如产生能源或者提供升力。发动机提供能源，机翼提供升力，这些都是非常清楚的，但是它们不属于功能。一些工程师在命名功能时，试图避免使用动词"提供"，但是这在逻辑上并没有什么不妥。

不这样命名功能背后的逻辑是，这种方法让设计人员可以自由地选择满足特定功能的各种方案。

3.1 系统工程生命周期功能

表 3.1 对比了飞机生命周期功能，ANSI/EIA 632(1999) 所述的传统系统工程生命周期功能，以及《FAA 系统工程手册》(2014) 的生命周期功能。飞机生命周期功能重点关注商用飞机生命周期的各个方面，包含飞机的研制方和用户从概念设计开始到报废需要开展的活动。也就是说，它们不是飞机本身要执行的功能，是与主要利益相关的功能，我们将在后面 3.3 节中进行讨论。

关于生命周期功能的讨论非常重要，因为它们对于飞机本身影响很大。此外，生命周期功能包含整个系统工程过程，这也是本书的主题。

图 3.1 表达了关于飞机系统的研制、制造、运行及处置的典型系统工程生命周期功能流[1]。这个流程图纳入了表 3.1 中飞机生命周期的功能。该表将传统系统工程生命周期各个阶段与 FAA 的生命周期阶段进行了对比。这些对比表明生命

1　严格的功能流图将包含"与"和"或"门，简化起见，本图删除了这些内容。

traditional SE life-cycle phases and the FAA (2014, p. 5) life-cycle phases. This comparison shows that the different views of life-cycle phases only differ by scope.

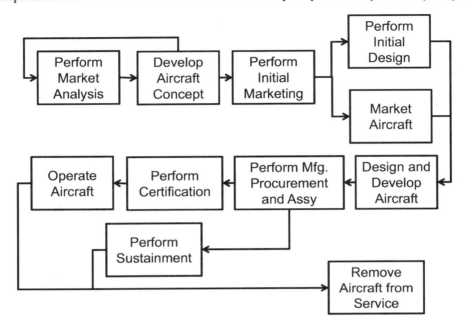

Figure 3.1 SE life-cycle functional flow for a commercial aircraft system

The important thing to remember about life-cycle functions is that they define the functioning of the entire aircraft system beyond the aircraft itself, and since this system is a *system*, all the pieces of it will relate to each other and not just constitute independent and unrelated components.

Table 3.1   Comparison of life-cycle functions

Traditional SE Life-Cycle Functions	Aircraft Life-Cycle Functions	FAA Life-Cycle Functions[*]
Development	Market analysis Perform initial marketing Perform initial design Market aircraft Perform design and development	Mission analysis Investment analysis

周期阶段的不同视角仅仅体现在范围上的不同。

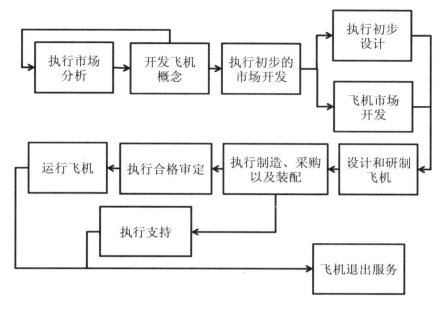

图 3.1　商用飞机系统的系统工程生命周期功能流

需要注意的是，生命周期功能定义了整个飞机系统的功能，而不仅仅是飞机本身的功能，因为这个系统是一个系统，所以各部分之间都将相互关联，而并非仅仅构成独立无关的部分。

表 3.1　生命周期功能对比

传统系统工程生命周期功能	飞机生命周期功能	FAA 生命周期功能
开发	市场分析 执行初步市场开发 执行初步设计 飞机市场开发 执行设计及开发	任务分析 投资分析
制造	执行制造，采购，以及装配	方案实施
验证	执行设计及开发 执行合格审定	

(Continued)

Traditional SE Life-Cycle Functions	Aircraft Life-Cycle Functions	FAA Life-Cycle Functions[*]
Manufacturing	Perform manufacturing, procurement, and assembly	Solution implementation
Verification	Perform design and development Perform certification	
Deployment	Operate aircraft	In-service management
Operations		
Support	Perform sustainment	
Training		
Disposal	Remove aircraft from service	Disposal

[*] *Source: FAA Systems Engineering Manual* (2014, p. 5).

The Perform Market Analysis function

The SE process begins, and the aircraft concept is born during this phase. This is a subfunction of the Development function in Table 3.1 and is where the basic top-level requirements are established. The process of identifying top-level requirements begins by identifying the types of customers. There are two kinds of customers: the aircraft market and specific customers. This function addresses only the general market as a customer. The basic questions to be answered are: How large is the market? How many seats are optimum? What are the airline route structures which will establish the range requirements? Are there any special requirements, such as short runways to consider, and so forth? However, the most important questions to be answered during this phase are economic ones: How many passengers are anticipated to fly in the near future? What will be the direct operating cost (DOC) of the aircraft the potential customers will expect as discussed in Section 8.6? What is the maximum price potential customer will be willing to pay for an aircraft? The answers to all of these questions will influence the design.

The Develop Aircraft Concept function

Within this function, another subfunction of the Development SE function, the systems engineer creates a concept, using the top-level requirements from the market analysis. The concept of this phase is what it is, namely, *a concept*. It is not a

56

传统系统工程生命周期功能	飞机生命周期功能	FAA 生命周期功能
部署	运行飞机	使用中的管理
运行		
支持	执行支持	
培训		
处置	飞机退出服务	处置

* 来源 :《FAA 系统工程手册》(2014，第 5 页)

执行市场分析功能

在这个阶段，系统工程过程开始启动，飞机的概念开始产生。这是表 3.1 中研制功能的一个子功能，基本的顶层需求也在这个阶段建立。识别顶层需求的最初环节是识别客户的类型。客户有两种类型 : 飞机市场和特定的客户。这个功能仅针对通用市场。需要回答的基本问题是 : 这个市场有多大？多少座级是最优的？确定航程要求的航空公司的航线结构是什么样的？是否存在任何特定的需求？例如短跑道的考虑等。然而，这个阶段需要回答的最重要的问题是经济上的 : 在不远的将来预计有多少乘客乘坐飞机？潜在客户期望 (8.6 节) 的直接运营成本 (DOC) 是多少？潜在客户最多愿意用多少钱来买一架飞机？对上述所有问题的答案将对设计产生影响。

开发飞机概念功能

该功能是**系统工程研制**功能的一个子功能，在这个功能内系统工程师利用从市场分析中获取的顶层需求建立一个概念。这个阶段的重点恰如其名 : 概念。它不是详细的设计，只是产生了工程草图。这个阶段将揭示一些参数，包括飞机预计总重、发动机数量等。但是详细的信息尚属未知。这个概念的目标是证明飞机能够被设计和制造出来，以满足市场的要求。概念开发包含建立最初的飞机架

detailed design. Only engineering sketches will be produced. This phase will reveal such parameters as the total estimated weight, the number of engines, and so on. But detailed information will not be known. The purpose of the concept is to verify that an aircraft can be designed and built to meet the market requirements. Concept development involves both the establishment of the initial aircraft architecture as discussed in Section 2.3 and the top-level aircraft system functions shown in Section 3.2. The concept of this phase is *subject to change* in the subsequent phases. After all, we have not yet determined specific customer expectations. The resulting aircraft may be very different indeed from the concept of this phase.

The concept developed in this phase may, of course, not even be a new aircraft as described in Section 2.2. It may be a derivative aircraft also described in Section 2.2, or it may be a change request-based design also discussed in Section 2.2. In any case, the adherence to SE principles will apply.

The Perform Initial Marketing function

Here is where we find out what specific customers' requirements are. Market analyses discussed in Section 3.1 will determine the need for aircraft in specific range and payload classes. For specific customers, the most effective way to perform this process, from an SE point of view, is to determine customer expectations from a functional point of view. That is, rather than emphasizing hardware options, we can stress the functions the customer wants to see. One way to do this is quality function deployment (QFD) described in Section 7.4. QFD is a very effective process for determining and prioritizing customer needs and presenting concepts to the customer which meet those needs. Based on the results of the real top-level requirements and an analysis to verify that the requirements are achievable, we are now ready to conduct a system requirements review (SRR) of the aircraft described in Section 12.4. The SRR provides concurrence from the customer(s) that the requirements are correct. Section 8.1 presents a list of typical top-level requirements which can be determined from market analyses and from specific customers.

The Perform Initial Design function

Based on the real requirements established with one or more prospective customers, we can now create a concept for a real aircraft. This step is also part of the

构 (2.3 节) 以及顶层的飞机系统功能 (3.2 节)。这个阶段的概念在随后的阶段可能发生变化，毕竟我们尚未确定具体客户的期望。最终的飞机可能与这一阶段定义的概念存在极大的不同。

当然这个阶段形成的概念甚至可能并不是新飞机 (2.2 节)。它可能是一个衍生型 (2.2 节)，或者它可能是一个基于要求改装的机型 (2.2 节)。在任何一种情况下，系统工程的概念都同样适用。

执行初步市场开发功能

在这个阶段我们知道了特定的客户需求是什么。市场分析 (3.1 节) 将决定航程及座级方面的需要。对于具体的客户而言，从系统工程的观点来看执行这个过程最有效率的方式，是从功能性角度来定义客户的期望。也就是说，相较于关注硬件选项，我们将重点放在客户期望看到的功能上。质量功能展开 (QFD)(7.4 节) 是一个确定客户的需求，并排列其优先次序，并将满足这些需求的概念展示给客户的有效过程。基于真正的顶层需求和分析得到的结论来证明需求是可实现的，现在我们已经可以进行系统需求评审 (SRR)(12.4 节) 了。SRR 提供了一个与客户共同工作的机会，以判断需求的正确性。本书 8.1 节介绍了典型顶层需求的列表，它可以根据市场分析和特定客户确定。

执行初步设计功能

基于与一个或者更多的预期客户确定的真正需求，现在我们可以建立一个真实飞机的概念设计。这个阶段也是表 3.1 中研制功能的一部分。我们将建立飞机的基本特性，但是仅仅是在顶层进行。我们现在可以进行系统设计评审 (SDR)(12.4 节) 了。这个评审也是与客户一起进行，主要是确认这个概念设计可以满足这些需求。SDR 的完成是下阶段工作开始的信号，即开始初步设计，从此由概念定义工作转换到设计工作。成功的 SDR 的输出是经批准的飞机系统规范。

Development function of Table 3.1. We will have established the basic characteristics of the aircraft but only at the top level. We are now ready to conduct a system design review (SDR) described in Section 12.4. This review, also with customer(s) present, confirms that the concept meets the requirements. The completion of the SDR signals the beginning of initial design and, hence, the transition from concept to design. The result of a successful SDR is an approved aircraft system specification.

One of the functions of an SDR is to identify the trade-offs that need to be performed before the next review, the preliminary design review (PDR) described in Section 12.4. System trade-offs are one of the main functions of the synthesis process described in Section 7.2. Here is where we evaluate the introduction of new technologies, such as composite materials. We have seen in Section 2.4 some of the possible technologies which may be introduced.

Chapter 12 discusses the various ways that these design reviews can be modified to adapt to the commercial aircraft domain without sacrificing quality.

The Market Aircraft function

In the commercial aircraft industry, marketing is much more than selling. Marketing is also part of the Development function of Table 3.1. From an SE point of view, the key marketing function is the determination of customer requirements. Now that we have settled on a basic aircraft design in the initial design phase above, the requirements for future marketing focus will be all change-based discussed in Section 2.2. That is, each customer will sometimes require extensive custom features on the fleet of aircraft. As we said before, these requirements have a significant impact on the aircraft design and therefore deserve as much attention and rigor as the aircraft-level requirements. With each custom feature comes the need for maintaining the aircraft performance, reliability, and safety, and meeting the host of other constraints. We can and should also perform QFD described in Section 7.4 with each of the new customers.

The Perform Design and Development function

In this phase we develop the detailed requirements described in Chapter 4 and conduct PDRs described in Section 12.4. We produce the detail drawings and review them at Critical Design Reviews (CDRs) also described in Section 12.5. During this function, verifications at the subsystem level are performed. Hence, this function includes

SDR 的一个功能是识别权衡方案，这些方案需要在开始下一次设计评审，即初步设计评审 (PDR)(12.4 节) 前执行。系统权衡分析是 7.2 节所述综合过程的主要工作之一。在这个阶段我们对引进的新技术 (进行评估)，例如复合材料。在 2.4 节中我们讨论了一些可以引入的技术。

第 12 章讨论了修订设计评审的不同方法，以保证在不牺牲质量的前提下应用于商用飞机领域。

飞机市场开发功能

在商用飞机工业领域，市场开发的工作内容远比销售工作要多。市场开发也是表 3.1 中研制过程的一部分。从系统工程的观点来看，主要的市场开发功能是确定客户的需求。现在我们已经在初步设计阶段定义了一个基本的飞机设计，对于未来市场开发的重点可能是 2.2 节所讨论的所有设计更改。也就是说，每一个客户都可能会时不时地提出大量关于机队定制特征的需求。正如我们之前所述，这些需求对于飞机设计有非常重大的影响，因此应该视其与飞机级需求一样重要和严格。鉴于每一个定制特征均来自于保持飞机性能、可靠性及安全性，并满足其他约束条件的需要，我们能够并且应该与每一个新的客户执行质量功能展开 (QFD)。

执行设计和开发功能

在此阶段开发第 4 章所述详细需求，并开展初步设计评审 (PDR)(12.4 节)。详细设计文件在这个阶段形成，并在关键设计评审 (CDR)(12.5 节) 中进行评审。子系统级的验证也在这个功能中进行。因此，这个功能包含了表 3.1 中的**研制及验证**功能。

这个阶段的关键点不仅仅是设计和研制飞机，而且要研制过程需要遵守的需求。这是系统工程过程的充分性，及其有能力确定超出飞机设计本身以外的系

both the Development and Verification functions of Table 3.1.

A key aspect of this phase is not only to design and develop the aircraft but also to develop the requirements for the manufacturing processes to follow. This is another example of the thoroughness of the SE process and its ability to determine the requirements for all aspects of the system far beyond the design of the aircraft itself.

As we will see in Section 10.2, the concept of organizational safety comes to the fore within this function. This concept requires that organizational aspects, such as aggressive schedules, do not increase the likelihood that the aircraft will have hazardous features.

The Perform Manufacturing, Procurement, and Assembly function

The Manufacturing function of Table 3.1 is an example of a phase of the SE process, other than Operate Aircraft, which can assign requirements to the aircraft itself. The most obvious example is the cost constraint imposed by the recurring (unit) cost of the aircraft as discussed in Section 8.6. A major driver of cost constraints is the cost of assembly. For example, this cost drives the number of parts to a minimum. Another example is the requirement to transport the aircraft to a central assembly facility in parts small enough to carry on a truck as discussed in Section 5.12. This requirement will force the design of the aircraft in segments rather than in large assemblies.

The Perform Certification function

After we assemble the complete test model, we flight test it and certify it. Flight test and certification are also part of the verification process described in Section 11.2.

The Operate Aircraft function

This function includes both the Deployment and Operations functions of Table 3.1. This is the phase that receives the lion's share of attention during the initial requirements definition phase of the aircraft. It is for this phase that we establish the top-level requirements based on the customer needs and determine them for the subsystems described in Sections 8.2 and Chapter 9. It is here that most of the requirements for the aircraft system are developed as discussed in Section 3.2. However, it is important not to forget the other phases.

统所有方面需求的又一例证。

正如 10.2 节将要讨论的, 组织安全的概念开始于本阶段前期。这个概念要求组织方面, 例如冒进的进度计划, 不得增加飞机有危害性特征的概率。

执行制造、采购及装配功能

除了运营飞机之外, 表 3.1 所述的制造功能也是系统工程过程阶段的一个例子, 该功能将需求分配到飞机本身。最明显的例子是由重复 (单位) 成本 (8.6 节) 带来的成本约束。成本约束的主要来源是装配成本。例如, 该成本推动零件数量最少化。另一个例子是如 5.12 节所述的需求, 即保证飞机零部件足够小, 以使其能通过卡车转运到装配厂房。该需求迫使飞机分段设计, 而不是按大组件设计。

执行合格审定功能

完成试验样机组装以后, 开始飞行试验并对其进行合格审定。飞行试验和合格审定也是验证过程 (11.2 节) 的一部分。

运行飞机功能

该功能涵盖了表 3.1 所述的**部署及运行**功能。这是飞机初始需求定义阶段中最受关注的一个阶段。正是在这个阶段, 我们基于客户需要建立顶层需求, 并针对子系统 (8.2 节及第 9 章) 确定这些需要。如 3.2 节所述, 大多数飞机系统需求针对此阶段进行开发。然而, 需要注意, 其他阶段的飞机系统需要也需要关注。

执行支持功能

支持是系统工程生命周期最为显著, 也最容易忽视的一个阶段。支持包含了表 3.1 中的**支持及培训**功能。维修工程师尝试将其对维修领域的关注体现在飞机设计中。但是, 传统的过程并非如 5.7 节所讨论的那么容易。只有系统工程过

The Perform Sustainment function

Sustainment is one of the most obvious, yet ignored, phases of the SE life-cycle. Sustainment includes both the Support and Training functions of Table 3.1. Maintainability engineers make valiant attempts to have their concerns addressed in the aircraft design. But traditional processes do not make that easy as discussed in Section 5.7. Only the SE process recommends the consideration of this phase as part of the functional analysis and the subsequent requirements development. Of course, maintainability is not the only sustainment function. There are also servicing and training, to name two. All of these provide aircraft requirements.

The Remove Aircraft from Service function

The need for disposal creates requirements for the aircraft. For example, regulatory documents prohibit the use of toxic materials. This function corresponds to the Disposal function of Table 3.1. This is another example of the impact of the life-cycle requirements on the aircraft.

3.2 Aircraft System-Level Functions

We are free to define the aircraft system as broadly as we need to. For example, we could define it to include the entire air transport system of the world, including all support systems and traffic control systems. For our purposes we only need to define it to include the aircraft, the pilot, and all the support equipment, training equipment, and facilities for the aircraft. We can call this the operational equipment, as opposed to business support equipment.

There are two reasons to identify the aircraft system functions and aircraft functions *at the same time*: First, the aircraft concept may require changes in some of the training, support, and facilities used with the aircraft. Certainly, with the introduction of major aircraft changes, such as the High-Speed Civil Transport (HSCT), this statement will be true. Secondly, the interfaces with these other non-aircraft elements may affect the design of the aircraft itself. Historically, aircraft designers have either ignored these aspects or postponed them until later. Most of the functions for the aircraft system, other than the aircraft itself, will be subfunctions of the Perform Sustainment function described in Section 3.1.

程才推荐将这个阶段的考虑作为功能分析，以及接下来的需求开发的部分工作。当然，支持功能不仅包括可维修，还包括服务和培训，这些方面都会产生飞机需求。

飞机退出服役功能

飞机处置的需要产生了对飞机的需求。例如，法规禁止使用有毒材料。这个功能就是表 3.1 中定义的**处置**功能。这也是生命周期需求对飞机影响的另一个例子。

3.2 飞机系统级功能

飞机系统的定义可以根据需要自由地定义。例如，可以定义飞机系统包含整个航空运输系统，包含所有的支持系统和空中交通管制系统。针对我们的目标，只需要定义它的范围，包含飞机、驾驶员、所有的支持设备、培训设备及与飞机相关的设施可以称之为运行设备，与之相对的是商务支持设备。

我们有两个理由说明为什么需要同时明确飞机系统的功能和飞机的功能：第一，飞机概念本身可能要求培训、支持及飞机所用设施方面的一些更改。而且，引入了重大的飞机更改，例如高速民用运输机 (HSCT)，也说明这是正确的。第二，与其他非飞机元素的接口也会影响飞机的设计。历史上，飞机的设计者要么忽略了这些方面，要么就是把它们放在最后才考虑。除了飞机本身以外，大多数的飞机系统功能都是**执行支持** (3.1 节) 功能的子功能。

培训功能

典型的培训功能包括**提供飞行机组培训**和**提供支持人员培训**。**提供飞行机组培训**功能可能会产生一些新的模拟器。更为重要的是，从飞机的角度出发，飞机本身可能需要一些特别的培训特性以方便执行培训功能。系统工程概念的一个

Training functions

Typical training functions include Provide Training for Flight Crews and Provide Training for Support Crews. The Provide Training for Flight Crews function may result in new simulators. More importantly, from an aircraft point of view, the aircraft itself may need special training features to facilitate the training function. It is a basic principle of SE that we can allocate requirements to any element. Hence, we can allocate the requirements resulting from the Provide Training for Pilots to the aircraft element. The same is true of any function.

Support functions

Typical support functions include Provide On-Aircraft Support and Provide Off-Aircraft Support. The support functions are probably the most important aircraft system functions except for the aircraft itself. They are very broad functions and cover many activities, including maintenance, servicing, and the replenishment of provisions for the crew and passengers. The maintenance functions will determine whether special maintenance tools or other equipment is needed, such as test equipment. These functions will influence the design of cargo loading equipment and other service equipment. If a more common environment exists in which the customer equipment is fixed, then we should consider these items as fixed external interfaces and design the aircraft to satisfy them. Consideration of the fixed interfaces should be part of the design process.

One of the most costly support functions of the airline is the acquisition, storage, and disposition of spares. The number of spares is a strong function of the airline's route structure. That is, the airline normally chooses to store spares at key *hub* facilities, which strongly affect the number of spares. Hence, the systems engineer should consider the cost of spares in the total system cost of the aircraft.

Facilities functions

Typical facilities functions include Provide Facilities for Storage and Maintenance of Aircraft and Provide Passenger Access to Aircraft. The facilities functions will determine whether new or modified facilities are needed to store and maintain the aircraft. Certainly, radically changed aircraft dimensions would influence these decisions. In addition, the system requirements may alter the passenger access tunnel, called the jetway. We will have to take into account the effect of the jetway on

基本原理就是：可以将需求分配给任何一个元素。因此，我们可以将**提供驾驶员培训**所产生的需求分配给飞机这个元素。这对于任何功能都同样适用。

支持功能

典型的支持功能包括**提供机上支持**和**机下支持**功能。支持功能可能是除了飞机本身以外最重要的飞机系统功能。它包含了非常宽泛的功能及很多的活动，例如维修、服务以及为驾驶员和乘客提供给养。维修功能将决定是否需要特殊的维修工具或者其他的设备，例如测试设备。这些功能将会影响货物装载设备的设计，以及其他的服务设备。如果在一个更加通用的环境里安装了客户设备，那么我们就要将这些设备视为固定的外部接口，并且在飞机设计过程中满足这些要求。固定接口的考虑应该成为设计过程的一部分。

航空公司最昂贵的支持功能之一是备件的采购、存储以及部署，备件的数量随航空公司航线结构的不同变化极大。也就是说，航空公司通常选择在关键的枢纽厂房存储备件，这将显著地影响备件的数量。因此，系统工程师应该在飞机整个系统成本中考虑备件的成本。

设施功能

典型的设施功能包含**为存储和维修飞机提供设施**，以及**为乘客上下飞机提供设施**。设施功能将决定是否需要新的或者改造的设施来存储和维修飞机。当然，飞机尺寸方面大的更改将影响这些功能。此外，系统需求也可能会改变乘客通道，一般称为登机通道。我们必须考虑登机通道对飞机设计的影响，因为登机通道和飞机有直接的接口关系。例如，外部物体，如皮托管不能布置在登机通道与飞机连接位置附近。

the aircraft itself, since the jetway interfaces directly with the aircraft. For example, external objects, such as pitot tubes cannot be located in the vicinity of the jetway attachment to the aircraft.

3.3   Aircraft-Level Functions

The concept of operations functional view

Finally, we get to the aircraft itself. The general functional description of how the aircraft operates is called the Concept of Operations (CONOPS), as shown in Figure 3.2 adapted from NASA (2001). This figure presents the functions of the aircraft from an operational point of view.

Future diagrams will employ the functional flow block diagram (FFBD) which is the preferred format of the *FAA Systems Engineering Manual* (2014, Section 4.4). However, many analysts prefer the IDEF0 (Integrated Definition for Function Modeling) methodology. The advantage of IDEF0 is that it accounts for the inputs and outputs of each function. This method will be shown also. There are other functional diagrammatic methods, such as the swim lane diagram also shown. The advantage of this method is that it emphasizes the time relationship among the components of the system.

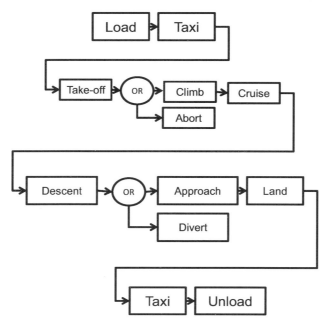

Figure 3.2   The commercial aircraft concept of operations

3.3 飞机级功能

运行概念的功能视角

最后，我们来到了飞机本身。关于飞机如何运行的通用功能性描述称为**运行概念** (CONOPS)，如图 3.2 所示，该图来自于 NASA 文件 (2001)。该图从运行角度说明了飞机的功能。

后面的图示将采用功能流框图的方式，这种方式是《FAA 系统工程手册》(2014，4.4 节) 首选的一种方式。然而，很多的分析更倾向于 IDEF0(功能建模) 方法。IDEF0 的优点是考虑到了每个功能的输入和输出。这种方法将在后面进行介绍。还有其他功能图的方法，例如泳道图等。这种方法的优点是强调了系统部件间的时序关系。

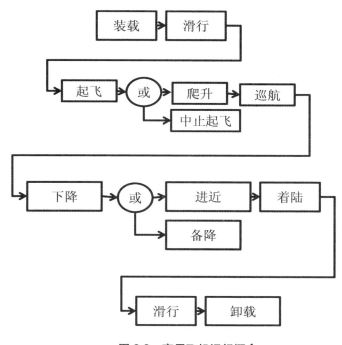

图 3.2　商用飞机运行概念

Table 3.2 compares the phases of operation with the aircraft-level functions. However, this table does not show all top-level functions. This view only shows this function from an operational phase point of view. In reality, the subfunctions of the CONOPS are multidimensional.

Table 3.2   Comparison of aircraft phases of operation and aircraft-level functions

Phases of operation	Aircraft-level functions
Load	Provide Ground Movement (pre-flight)
Taxi	
Take-off	Perform Flight Operations
Climb or Abort	
Cruise	
Descent	
Approach or Divert	
Land	
Taxi	Provide Ground Movement (post-flight)
Unload	

Each of the flight functions shown in Figure 3.2 should be matrixed against three other functional dimensions, as shown in Figure 3.3. The large number of functions implied by this chart, especially when expanded to the subsystem level, illustrates the monumental task required to assure the aircraft safety when the functional analyses suggested by the FAA in 4754A (2010) are developed.

All combinations of functions in this matrix do not necessarily exist. For example, the combination of the Perform Pre-Flight Operations function and the Provide Aerodynamic Performance function is not relevant because the aircraft does not fly during the pre-flight phase.

The concept of defining the functions precisely, as discussed at the beginning of this chapter, is especially important when considering the flight functions. The failure to define the limits may result in incorrect requirements and inadequate solutions.

Let's discuss each of the nodes of the Perform Air Transport Mission matrix:

表 3.2 将运行阶段与飞机级功能进行了对比。然而，这个表格没有示出所有的顶层功能。它只是从运行阶段的角度示出了这个功能。现实情况是运行概念的子功能是多维度的。

表 3.2　运行阶段及飞机级功能对比

运行阶段	飞机级功能
装载	提供地面运动（飞行前）
滑行	
起飞	执行飞行运行
爬升	
或	
中止起飞	
巡航	
下降	
进近	
或	
备降	
着陆	
滑行	提供地面运动（飞行后）
卸载	

图 3.2 中的每一个飞行功能均应与其他 3 个功能维度匹配，如图 3.3 所示。本图包含的大量功能，尤其是当扩展到子系统级时，描绘了在进行 FAA 推荐的 ARP 4754A(2010) 中的功能分析时，为保证飞机安全所需要的重要任务。

在这个矩阵中所有功能的组合未必都存在。例如，执行飞行前运行功能和提供空气动力学性能功能就是非相关的，因为飞机不会在飞行前阶段飞行。

正如在本章开头讨论的那样，精确定义功能概念对于考虑飞行功能是非常重要的。没有定义限制条件可能会产生错误的需求和不恰当的解决方案。

下面讨论执行航空运输任务功能矩阵中的各个节点：

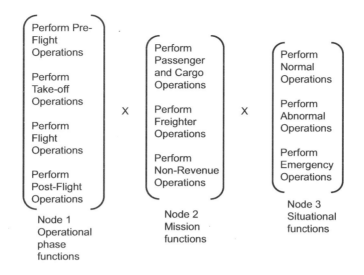

Figure 3.3   Matrix of the Perform Air Transport Mission function

Node 1: Operational Phase functions

We have broken the flight phases down into four basic phases shown in Figure 3.3. These are the same basic phases which define the Perform Air Transport Mission function of Figure 3.2 However, it may (and probably will) be necessary to break down the flight phases into even smaller phases during the analysis of specific subsystems. The flight phases which are important to the environmental control system (ECS), for example, will be different from the phases important to the flaps. The definition of the external environment associated with each phase is an essential aspect of the flight phase functions.

The Perform Pre-Flight Operations function

This function begins when the aircraft has begun to be prepared for flight. This function is related to the Perform Sustainment Function described in Section 3.1 of Figure 3.1 in that it consists of all the servicing and maintenance activities performed when the aircraft is either in a maintenance facility or on the airport ramp. The maintenance functions performed in these locations do affect the aircraft design. One such function would be Allow Removal and Replacement Access.

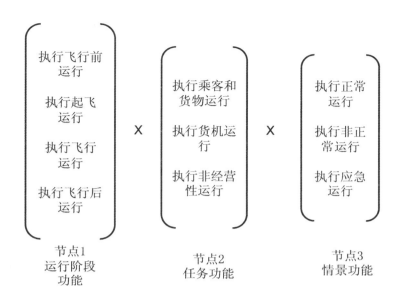

图 3.3　执行航空运输任务功能矩阵

节点 1：运行阶段功能

我们已经将飞行阶段分成 4 个基本的阶段, 如图 3.3 所示。它们与定义图 3.2 中的**执行航空运输任务**功能的基本阶段相同。然而, 在分析特定子系统时, 将 (以及可能将) 飞行阶段分解为更小的阶段或许是必要的。例如, 对于环控系统 (ECS) 非常重要的飞行阶段将有别于对襟翼重要的飞行阶段。规定每个阶段相关的外部环境是飞行阶段功能非常重要的内容。

执行飞行前运行功能

该功能开始于飞机开始准备飞行阶段。这个功能和 3.1 节介绍的**执行保障功**能有关, 它包含飞机在维修厂房或者在停机坪上开展的所有服务以及维修活动。在这些地方执行维修功能将影响飞机的设计。其中一个功能就是**允许移除和更换**。

The Perform Take-Off Preparations function

This function begins when the aircraft pushes away from the gate and ends when it receives clearance for take-off. Requirements pertaining to runway transit, aircraft navigation, and communications will be established during this phase.

The Provide Ground Movement function, shown in Figure 3.4, is a subordinate function to the Perform Take-Off Preparations function. In addition, most of the subfunctions of the Perform Flight Operations function described in Section 3.3 will apply. The primary subsystem to implement the requirements of this function is the landing gear and brakes subsystem described in Section 9.5. The landing gear subsystem will provide the carriage and steering while the aircraft is on the ground, and the braking subsystem will provide the braking.

The Perform Flight Operations function

This function begins when the aircraft receives clearance for take-off and ends when it pulls off the runway. Obviously, many subfunctions will occur between these times, for example, take-off, climb, cruise, descent, approach, and landing. Each of these subfunctions is important in aircraft sizing described in Section 8.2. Additional subfunctions can be defined which will lead to requirements definition for various subsystems. For example, the raising and lowering of flaps will define functions important to the flaps and other subsystems, such as hydraulics. We will expand the flight functions later in Section 3.3.

The Perform Post-Landing Operations function

This function begins when the aircraft turns off the runway and ends when it arrives at the gate. The requirements defined during this phase will be similar to those identified in the take-off preparations phase described in Section 3.3. This function will include the Provide Carriage, Braking, and Steering subfunction as shown in Figure 3.4 as well as most of the subfunctions of the Perform Flight Operations function described in Section 3. However, the possibility that pre-flight and post-flight functions may be different compels us to examine them separately.

执行起飞准备功能

该功能开始于飞机从航站楼登机口推出，到其接收到起飞放行许可结束。关于跑道穿越、飞机导航及通信方面的需求在这个阶段产生。

提供地面移动功能 (见图 3.4) 是**执行起飞准备**功能的一个下级功能。此外, 3.3 节中的**执行飞行运行功能**的大多数子功能将被执行。实现这项功能需求的主要子系统是起落架及刹车子系统 (9.5 节) 所述。起落架子系统将为地面上的飞机提供承载和转弯，刹车子系统将提供刹车。

执行飞行运行功能

该功能从开始于飞机接收到起飞放行许可，到其着陆后离开跑道时结束。显然，很多子功能将在这个时间段中作用，例如，起飞、爬升、巡航、下降、进近及着陆。每一项子功能对于确定飞机尺寸都非常重要 (8.2 节)。其他的子功能定义产生了不同子系统的需求。例如，升起和放下襟翼将定义对于襟翼及其他子系统重要的功能，例如液压的功能。3.3 节将对飞机功能进行详细讨论。

执行着陆后运行功能

该功能开始于飞机离开跑道，结束于到达航站楼登机门之时。在这个阶段定义的需求与飞行前准备阶段 (3.3 节) 识别出来的需求类似。该功能包含**提供承载、刹车和转弯**子功能 (见图 3.4)，还包括第 3 章讨论**执行飞行运行功能**的大多数子功能。然而, 飞行前与飞行后功能有可能不同, 因此, 必须分别对其进行检查。

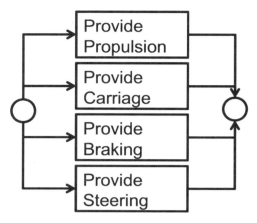

Figure 3.4   The Perform Ground Movement function

Node 2: Mission functions

The three mission functions shown in Figure 3.3 may apply separately or in combination.

The Perform Passenger and Cargo Operations function

This is the normal airline functional mode. It includes all the operations necessary to transport passengers and cargo between airline facilities.

The Perform Freighter Operations function

This function is well-known, for many airlines operate aircraft with the sole purpose of transporting freight. However, in the event the customer desires an aircraft with the capability of being converted from a passenger and cargo configuration to a freighter configuration, this function should be considered *in combination with* the Perform Passenger and Cargo Operations function. The reason for this requirement is that the convertibility requirement may impose special features on the aircraft that a pure passenger and cargo or a pure freighter version would not have.

The Perform Non-Revenue Operations function

All aircraft are required, from time to time, to operate in a non-revenue mode, that is to be flown from one point to the other with no passengers or freight. This

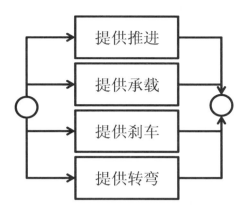

图 3.4　执行地面运动功能

节点 2：任务功能

图 3.3 所示的 3 项任务功能可以单独或者组合应用。

执行乘客及货物运行功能

这是航空公司正常的功能模式。它包括在两个航空公司设施之间运输乘客及货物的所有必要运行。

执行货机运行功能

该功能十分常见，很多的航空公司运营飞机只有一个目的，就是运输货物。然而，有时客户希望飞机具备从客机构型转化为货机构型的能力，此时这个功能应该与执行乘客及货物运行功能一起进行考虑。这个需求的理由是转化需求可能影响飞机的一些特性，这些特性是纯粹的客机或货机所没有的。

执行非经营性运行功能

所有飞机随时都需要以非经营模式运行，即在不搭载任何乘客或货物的情况下从一个地点飞到另外一个地点。这个功能考虑到此种运行模式产生的任何需求。

function accounts for any requirements which may result from this mode.

Node 3: Situational functions

This group of functions reflects the functional operations of the aircraft in a variety of situations related to the safety of the aircraft and the occupants.

The Perform Normal Operations function
This is the basic functional mode in which the aircraft operates as expected through all phases of flight.

The Perform Abnormal Operations function
This function is the basic operational mode in which the aircraft or some subsystem of the aircraft is not operating normally, but no danger to the passengers or crew exists. For example, if a generator malfunctions and a backup power system is providing sufficient power, we would say that the aircraft is performing the Perform Abnormal Operations function. This function is important in establishing the requirements for backup systems.

The Perform Emergency Operations function
This is the functional mode in which a possible catastrophic event is likely. It is important in establishing the requirements to survive the expected environment.

This function is also responsible for the protection of passengers in emergency conditions, such as the loss of cabin pressure, or in emergency landing operations. This function also leads to the requirement for the emergency equipment and features which should be carried on each aircraft.

Expansion of Flight Operations functions

We saw before in Section 3.3 that the Perform Flight Operations is just one of four operational phase functions. Figure 3.5 shows the subfunctions of the Perform Flight Operations function. These subfunctions are the heart of the architecting of a commercial aircraft. The key point to remember here is that functions are only functions. They are not subsystems. That is, we cannot assume that the Provide Environmental Control function only belongs to the Environmental Control Segment

节点3：情景功能

这组功能体现的是飞机在不同情况下的功能性运行，涉及飞机及机上人员的安全。

执行正常运行功能

这是一个基本的功能模式，在这个模式下飞机如预期情况进行所有飞行阶段的运行。

执行非正常运行功能

该功能是一个基本的运行模式，在这种情况下，飞机或者飞机其他子系统出现不正常的状况，但是对于乘客和机组没有危险。例如，如果一台发电机功能异常，并且备用电源系统可以提供充足的能源，我们可以说飞机在执行一个非正常运行功能。这项功能对于建立备用系统需求非常重要。

执行应急运行功能

该功能模式是指可能发生某一灾难性事件的模式。它对于建立在预期的环境下存活的需求非常重要。这项功能同样负责在紧急情况，例如座舱失压，或者紧急着陆运行情况下保护乘客。这项功能同样会产生对于每架飞机必备应急设备及其设计特征的需求。

飞行运行功能扩展

我们看到在3.3节之前，**执行飞行运行**只是4个运行阶段功能中的一个功能。图3.5中展示了**执行飞行运行功能**的子功能。这些子功能是构建一架商用飞机的核心。必须注意的是这些功能仅仅是功能，它们并不是子系统。也就是说，我们不能假定**提供环控**功能仅属于**环境控制部段**或者**提供乘客和机组起居设施**功能只

or that the Provide Passenger and Crew Accommodations function only belongs to the Interiors Segment. On the contrary, each of the following functions may be allocated to several of the aircraft segments.

It is this ability to trade off higher-level functions among several segments which allows optimization of the system (the aircraft) at the aircraft level rather than at the segment or subsystem level. Failure to optimize at the aircraft level may lead to optimization at the subsystem level and hence to a system which is not optimum.

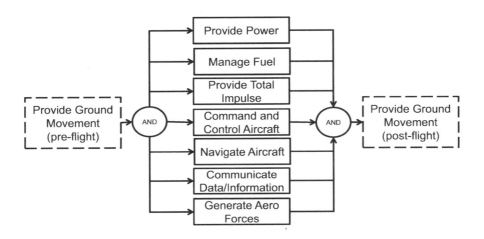

Figure 3.5   The Perform Flight Operations function

Although the flight functions listed in the following paragraphs may appear to pertain to individual subsystems, the SE process does not allow us to make that judgment at this time. They are still *top-level* functions. Not until Chapter 9 do we address the subsystems associated with the performance requirements of these functions.

The Perform Flight Operations function (IDEF0 view)

Originally developed as a software modeling tool, IDEF0 (Integrated Definition for Function Modeling) is widely used in SE as discussed by Buede (2000). The advantage of IDEF0 over the FFBD is that IDEF0 identifies the inputs and outputs of each function. FFBD focuses on the sequential occurrence of functions. These inputs and outputs form the basis for the cluster analysis to be shown at the end of this chapter. IDEF0 normally requires the assignment of these functions to system elements

属于**内饰部段**。恰恰相反，下面列出的每一个功能都可能会分配到飞机不同的部段。

正是这个在若干个部段之间权衡更高层级功能的能力使得可以在飞机层级优化系统 (飞机)，而不是在部件或者子系统层级进行优化。没有在飞机层级进行优化将导致须在子系统层级进行优化，从而使得系统并非最优。

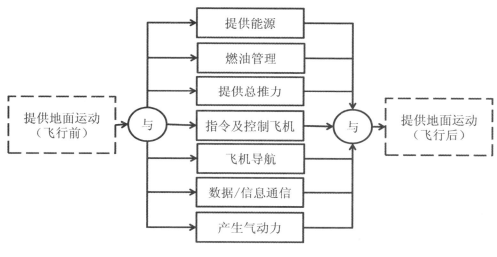

图 3.5 执行飞行运行功能

尽管下面各段列出的飞行功能可能在相关的子系统中出现，但系统工程过程不允许我们在此时作出判断，他们依然是顶层功能。一直到第 9 章，我们才开始说明与这些功能的性能需求相关的一些子系统。

执行飞行运行功能 (IDEF0 方法)

据 Buede(2000) 介绍，IDEF0(功能建模的综合定义) 在系统工程中广泛应用，该方法最初是作为一个软件建模工具而开发的。IDEF0 相对于 FFBD 的优势是它能识别每一个功能的输入和输出，而 FFBD 重点关注功能发生的顺序。这些输入

(components and subsystems), but for simplicity these elements are not shown in Figure 3.6. Figure 3.6 presents the same functions as in Figure 3.5 but in the IDEF0 format. The assignment of these functions to subsystems will be shown within the individual box describing each function.

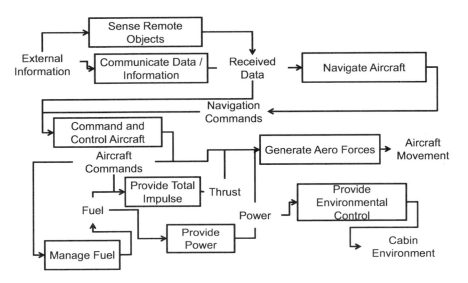

Figure 3.6   The Perform Flight Operations function (IDEF0 view)

The Generate Aero Forces function

We can break down the Generate Aero [aerodynamic] Forces into the following major subcategories: Provide Lift Performance, Provide Drag Performance, Provide Aerodynamic Stability, and Provide Aerodynamic Control, as shown in Figure 3.6. The primary aircraft segment allocated to perform the requirements associated with this function is the airframe segment described in Section 9.8. Of course, other segments, such as electrical described in Section 9.3 and mechanical described in Section 9.5, are strongly affected since they are required to move the control surfaces.

Of course, the lift and drag *performance* is the primary factor which allows the aircraft to achieve its speed (cruise, landing, take-off), range, cruise altitude, and other top-level performance requirements. These functions are achieved through the aerodynamic shaping of the wing and various other external components. Most manufacturers are employing winglets to enhance the lift and drag performance. Flaps provide augmented lift and drag for landing. The airframe segment described in Section

和输出形成了本章结尾提到的功能群分析的基础。IDEF0 通常需要将功能分配到系统元素 (部件及子系统),但为了简化起见,图 3.6 中没有示出这些元件。图 3.6 以 IDEF0 的方法展示了和图 3.5 中相同的功能。这些功能分配到一些子系统的方式将在描述每一功能的方框里示出。

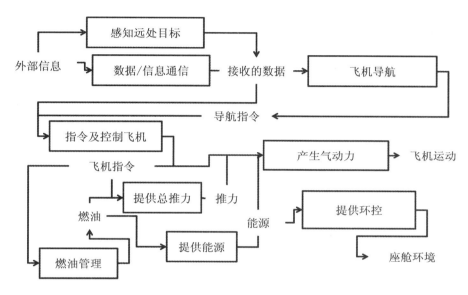

图 3.6 执行飞行运行功能 (IDEF0 方法)

产生气动力功能

产生气动力功能可分为如下几个主要的小类:**提供升力性能、提供阻力性能、提供气动稳定性、提供气动控制**,如图 3.6 所示。分配给执行与此功能相关需求的是 9.8 节所述的机体部段。当然,其他的部段,例如电气部段 (9.3 节) 及机械部段 (9.5 节) 都会显著地受到影响,因为要求它们使操纵面运动。

当然,升力和阻力性能是最主要的因素,它使飞机实现其速度 (巡航、着陆及起飞)、航程、巡航高度及其他顶层的性能需求。这些功能是通过机翼及其他外部组件的气动外形实现的。大多数的制造商采用翼梢小翼来提高升力及阻力性能。襟翼在着陆过程中提供增大的升力和阻力。9.8 节讨论的机体部段负责实现

9.8 bears the major responsibility for satisfying these functions.

The aircraft achieves aerodynamic stability through the lift and weight distribution if the *static margin* is positive; that is, that the lift forces act through a point (the center of pressure) a fixed distance behind the center of gravity. The tail surfaces (the empennage) provide the aft lift forces to assure that this is true.

Many aerodynamic surfaces provide control. These surfaces include the rudder (horizontal control), elevator (vertical control), and ailerons (roll control). In addition to these surfaces, the spoilers aid both the flaps and ailerons for landing and roll control.

The primary inputs to this function are the aircraft flight commands and power for moving the control surfaces.

The Provide Total Impulse function

This is the primary function for providing the propulsive force called thrust that propels the aircraft throughout its mission. The central subsystem for this function is the propulsion system.

The inputs for this function are fuel and aircraft commands to control the proper amount of thrust.

The basic subfunctions of Provide Total Impulse are Provide Forward Thrust and Provide Reverse Thrust. Subordinate functions will include Start Engines and Shut Down Engines.

Most modern commercial aircraft employ turbojet engines for thrust. However, there are a significant number of turboprop and reciprocating engine aircraft in service. We will see below and in the propulsion segment discussion discussed in Section 9.6 that the engines normally provide a number of functions other than providing thrust. These include Provide Bleed Air and Provide Mechanical Torque (for electrical generation). This example illustrates the generally held principle that a single subsystem can have performance requirements from multiple functions.

The Manage Fuel function

This function governs how much fuel is provided to the engines. In addition most aircraft have multiple fuel tanks, usually in the wings, to allow the fuel to be transferred from tank to tank in order to control the c.g. of the aircraft. The element of the aircraft that performs this function is the fuel management system, normally a separate software module within the aircraft.

Inputs to this function are from the aircraft control commands provided by both

这些功能。

如果静裕度是正的，飞机通过升力和重量分布可达到气动稳定。即，升力作用点（压力中心）在重心后面距离固定的一个位置。垂尾表面（尾翼）提供后部升力来确保这一点。

很多的气动面提供控制。这些气动面包含方向舵（水平控制）、升降舵（垂直控制）及副翼（横滚控制）。此外，扰流板帮助襟翼及副翼进行着陆和横滚控制。

这些功能的主要输入是使这些控制而运动的飞机指令及动力。

提供总推力功能

这是提供称为推力的推进力的一个首要的功能，它在飞机整个任务期间驱动飞机。这项功能的核心子系统是推进系统。

该功能的输入是燃油及控制合适推力的飞机指令。

提供总推力功能的基本子功能是**提供前进推力**和**提供反推力**。下级的功能将包含**起动发动机**和**关闭发动机**。

大多数的现代商用飞机采用涡轮喷气发动机提供推力。然而，有非常多的涡轮螺旋桨飞机和活塞式发动机在服役。推进部段及后面的讨论中看到发动机除了提供推力功能以外，通常还提供大量功能，包含**提供引气**以及**提供机械扭矩**（为产生电力）。这个例子说明了一个普遍的原理，即单独的子系统性能需求可能来源于多个功能。

燃油管理功能

该功能管理提供给发动机的燃油量。此外，大多数的飞机具备多个油箱，这些油箱通常在机翼中，其中的燃油可以从一个油箱传输到另外一个油箱，以控制飞机的重心。执行飞机这些功能的是燃油管理系统，该系统通常是飞机的一个独立软件模块。

the pilot and the automated control system.

The Command and Control Aircraft function

This function includes all commands that are necessary to move control systems or perform other actions to control the aircraft during flight including managing the fuel system and the propulsion system.

Two subsystems perform this function. One is the pilot and the other is the aircraft control system. That is, from an SE point of view, the pilot is a legitimate subsystem; otherwise the aircraft would not be a complete system. For two subsystems to perform this function there needs to be a complete set of rules to manage the interaction between the pilot and the automated control system. Billings (1997, pp. 237–260) has compiled a list of these rules. Among these he states, "To command effectively, the human operator must be involved."

Inputs to this function are first commands received directly from the navigation system and secondly information received directly by the pilot from information both inside the aircraft, the flight deck, and outside.

The Navigate Aircraft function

This function has several basic subfunctions: Determine Location of Aircraft, Determine Attitude of Aircraft, Determine Speed of Aircraft, Determine Direction of Aircraft, and Provide Flight Management. The majority of the requirements of these functions are the responsibility of the avionics segment described in Section 9.2.

Location can have many meanings. It includes the altitude, latitude, longitude, and the location relative to specific ground points. The attitude includes the pitch, yaw, and roll angles relative to an absolute coordinate system and to the air. The speed and direction can be relative to the air and to the ground. Speed also includes the rate of climb or descent. The Command and Control Aircraft is the function which provides commands to both the pilot and to the autopilot to control the aircraft and receives commands from this function.

The Communicate Data/Information function

There are two basic types of communications as expressed by the subfunctions: Provide External Communications and Provide Internal Communications. The requirements associated with these functions are the primary responsibility of the communications subsystem described in Section 9.2. External communications

这些功能的输入来自驾驶员及自动控制系统提供的飞机控制指令。

指令及控制飞机功能

这些功能包括在飞行过程中为使控制系统运动，或者执行其他动作以控制飞机的所有必要指令，包括燃油系统及推进系统。

执行该功能的子系统有两个，一个是驾驶员，另一个是飞机控制系统。从系统工程的观点来看，驾驶员是一个合理的子系统，否则飞机就不能成为一个完整的系统。对于执行这个功能的两个子系统，需要有一个完整的规则来管理驾驶员和自动控制系统之间的互动。Billings(1997，第 237 至 260 页) 编制了一份完整的规则列表。他表示"为了指令有效，驾驶员一定要参与进来"。

该功能的输入首先是直接从导航系统接收到的，其次是驾驶员从飞机内部，即驾驶舱或飞机外部获取的信息。

飞机导航功能

该功能有几个基本的子功能：**确定飞机位置、确定飞机姿态、确定飞机速度、确定飞机航向**及**提供飞行管理**。这些功能的主要需求由航电部段 (9.2 节) 满足。

位置的含义很广。它包含高度、经度、纬度及相对于地面具体某处的位置。姿态包含相对于绝对坐标系或者相对于气流的俯仰、偏航及横滚角度。速度和航向与气流相关，或者与地面相关。速度同样包含爬升或者下降的速率。**指令及控制飞机**功能既向驾驶员提供指令，也向自动驾驶仪提供指令，以控制飞机并接受来自该功能的指令。

通信数据 / 信息功能

该子功能包含两个基本类型的通信功能：**提供外部通信**和**提供内部通信**。与该功能相关的需求由通信子系统 (9.2 节) 满足。外部通信为驾驶员提供与塔台、

provide the means for the pilot to exchange information with the tower, with other aircraft, and with other ground nodes.

In addition, the pilot and other crew members must be able to talk to each other and to the passengers. Also, there must be a communications link between the flight deck and various parts of the aircraft for communication with the ground crew when the aircraft is on the ground.

This link supports the basic function of providing the pilots with sufficient information to fly the aircraft safely and monitoring the status of aircraft subsystems for malfunctions and any other hazardous condition. In addition, it provides the pilot with information outside the aircraft. Section 5.5 discusses the human factors aspect of establishing and implementing the requirements associated with this function. The indicating and recording subsystem described in Section 9.2 provides most of the information associated with aircraft status.

The Sense Remote Objects function

This function pertains to the sensing of exterior objects, such as other aircraft or the terrain. A radar system is the primary equipment to perform this function. The received information is provided to the navigation system.

The Provide Power function

All aircraft subsystems require power to operate. The most common power type is electrical, both alternating current (AC) and direct current (DC). In addition, the aircraft uses battery power for backup as well as the two other devices described below. In addition to electrical power, the aircraft uses hydraulic and pneumatic power. But, as Chapter 7 will show, these are only solutions. Therefore, any one of the three functions may not exist if it is not needed. The aircraft could operate on only one power source, if necessary. The optimum power source is the subject of trade-off.

Generators connected to the engines normally provide electrical power, In addition, devices called auxiliary power units (APUs) provide power in emergencies and when the engines are idle. Also for emergencies devices called ram air turbines (RATs) provide hydraulic power in emergencies. These devices were active in the Miracle on the Hudson incident described in Chapter 16.

其他飞机及地面站交换信息的手段。

此外，驾驶员和其他的机组成员必须可以相互通话，并且能与乘客通话。同样，当飞机停在地面的时候，在驾驶舱及飞机各部分之间必须有供使用的通信链路。

这种通信链路支持为驾驶员提供足够的信息，以保证安全地驾驶飞机、监控飞机各子系统的状态提供功能异常及任何其他危害性情况的信息。此外，它为驾驶员提供飞机之外的信息。5.5 节讨论了建立和实现与这一功能相关需求的人为因素方面的问题。指示记录子系统 (9.2 节) 负责提供大多数的飞机状态信息。

感知远处目标功能

该功能感知外部的物体，例如其他飞机或地形。雷达系统是执行这项功能的主要设备，将接收到的信息提供给导航系统。

提供能源功能

所有的飞机子系统都需要能源来运行。最常用的能源是电源，分为交流电源 (AC) 和直流电源 (DC)。此外，飞机使用蓄电池及下面介绍的两个设备来作为备用。除了电源，飞机还使用液压源和气源。但是，第 7 章所述的这些都只解决方案之一。因此，如果不需要，这三个功能中的任何一个都有可能不存在。必要时，飞机可能仅依靠一种能源运行。最佳的能源属于权衡分析的一个课题。

连接到发动机的发电机通常提供电源。此外，辅助动力装置 (APU) 在紧急情况及发动机慢车运行时提供能源。另外，称为冲压空气涡轮 (RAT) 的应急设备可在紧急情况下提供液压源。这些设备在第 16 章介绍的哈德逊河奇迹事件中都发挥了作用。

Other potential functions

The functions described above and in Figures 3.5 and 3.6 are the basic functions required to enable the aircraft to perform its operational mission. Other functions not described above have been suggested. Some systemists, people interested in studying and analyzing systems, argue that these functions are not valid functions since, in their opinion, they do not meet the criterion of a function which is to describe the exchange of energy between systems. They are included here both for completeness and because the reader may find them useful for identifying requirements and performing the other important activities associated with functions.

The Provide Environmental Control function

The primary environmental subfunctions the aircraft has to provide with respect to the passengers and the crew are: Provide Air Conditioning, Provide Pressurization, and Provide Oxygen. Provide Air Conditioning can be subdivided into Control Temperature, Control Humidity, and Control Air Quality. As we will see in Section 9.1, many subsystems will contribute to environmental control, in addition to the ECS itself.

In addition to controlling the environment of the passengers and crew, the aircraft must control its own environment. The primary subfunctions in this category are: Provide Ice Protection and Provide Rain Protection. Ice Protection includes both de-icing and anti-icing.

The primary input to this function is power. This can be electrical power, hydraulic power, or pneumatic power. The output is the energy to condition the aircraft, in the cabin and other parts as described above.

The Provide Passenger and Crew Accommodations function

This function includes a number of subfunctions, the primary ones of which include: Provide Passenger and Crew Space, Provide Seating, Provide Storage, Provide Lavatory Accommodations, Provide Galley Accommodations, and Provide Entertainment. The interiors segment described in Section 9.4 has the primary responsibility for meeting the requirements associated with these functions. The airframe segment, of course, provides the space for passengers and crew. There are other functions related to passenger life support and comfort. These include, for example, the pressurization, temperature control, and oxygen provided by the Provide

其他潜在功能

上面提及的，以及在图 3.5 和图 3.6 中介绍的功能是飞机能够执行其运行任务的基本功能。下面讨论上面没有提及的其他功能。一些系统工程师，以及对于研究和分析系统感兴趣的人认为这些功能并不是有效的功能，因为在他们看来，它们不满足功能的准则，即描述系统之间的能量交换。在这里提及它们是出于完整性的考虑，还因为读者可能发现，它们对于识别需求，以及执行其他与这些功能相关的重要活动非常关键。

提供环控功能

飞机必须为乘客和机组提供的主环境子功能为：**提供空调**、**提供增压**及**提供氧气**。**提供空调**又可细分为**控制温度**、**控制湿度**及**控制空气质量**。正如我们将在 9.1 节中介绍的，很多子系统将为环境控制提供贡献，并不仅仅局限于环控系统 (ECS) 本身。

此外，为了控制乘客和机组的环境，飞机必须控制其自身的环境。这一类的子功能是：**提供防冰除雨**功能。**防冰**功能包括**除冰**和**防冰**两种功能。

该功能的主要输入是能源。这可以是电源、液压源或者气源。输出是调节飞机座舱及上述其他部分环境的能量。

提供乘客及机组起居设施功能

该功能包含很多子功能，其中主要的一些功能是**提供乘客及机组空间**、**提供座位**、**提供存储空间**、**提供卫生间**、**提供厨房**及**提供机载娱乐**功能。9.4 节中描述的内饰部段负责满足这些功能的相关需求。当然机体部段为乘客和机组提供空间。还有其他与乘客的生命保障及舒适性相关的功能。这里面包含 3.3 节中讨论的**提供环控**功能提供的增压、温度控制以及供氧等功能。

Environmental Control function discussed in Section 3.3.

Emergency passenger provisions are established by the Perform Emergency Operations function described in Section 3.3, one of the situational functions.

The Provide Cargo Capability function

The primary subfunction associated with this function is Provide Cargo Space. Also included within this function is the Provide Cargo Loading subfunction. This subfunction includes the ability of the aircraft to allow for cargo ingress and egress. In addition, it can allow for autonomous cargo loading and handling systems described in Section 2.4.

Other functions associated with cargo can be found in other sections. For example, the Provide Environmental Control function described above provides for the environmental protection of the cargo. The Maintain Structural Integrity function described in Section 3.3 assures that the aircraft can sustain the cargo loads.

The Maintain Structural Integrity function

The two main subfunctions of the Maintain Structural Integrity function are Sustain Loads and Maintain Pressure, both of which are the primary responsibility of the airframe segment described in Section 9.8.

It is a common misperception that the aircraft structures do not have performance requirements since they "don't do anything." On the contrary, the ability to sustain loads is as valid a function as any other function. The input loads the airframe must satisfy come from a variety of sources: aerodynamic loads, inertial loads, and pressure loads.

This example illustrates the intuitive principle that all components should have at least one function. Although it may seem that the airframe segment described in Section 9.8 has the primary responsibility to fulfill this function, all segments also have to maintain structural integrity.

Structures may have other subfunctions, such as Allow Ingress and Allow Egress. These subfunctions apply to all openings and doors.

It is generally not necessary to examine all flight phases to determine the critical design conditions for structures since the Federal Aviation Regulations (FARs) will provide these. Design conditions will include crash loads, for example.

3.3 节中所述的**执行应急运行**功能提供乘客应急支持，属于一种具体的情景功能。

提供货运能力功能

与该功能相关的主要子功能是**提供货舱空间**，还包括**提供货物装载**子功能。这个子功能包含飞机的装货和卸货能力。此外，该功能可保证 2.4 节所述自主货物装载和搬运系统的运转。

与货舱相关的其他功能也出现在其他的章节中。例如，上面描述的**提供环控**功能为货舱提供环境保护。3.3 节中描述的**保持结构完整性**功能确保飞机可承受货物的载荷。

保持结构完整性功能

保持结构完整性功能的两个主要的子功能是**承受载荷**和**保持压力**，它们都由 9.8 节中描述的机体部件负责。

通常有一种错误的理解，认为飞机结构没有性能需求，因为它们"没有做任何事情"。然而，承受载荷的能力与其他功能本质上没有任何区别。机体必须满足的载荷有多个来源：气动载荷、惯性载荷以及压力载荷。

这个例子说明了一个直观的概念，即所有的部件均应至少具有一个功能。尽管机身部段 (9.8 节) 可能看起来主要负责满足这个功能，但是其他所有部段同样必须参与其中。

结构可能还有其他的子功能，例如**支持登机和离机**功能。这些子功能适用于所有的开口及舱门。

一般而言，没有必要检查所有的飞行阶段来决定结构的关键设计条件，因为联邦航空条例 (FAR) 将提供这些内容。例如，设计条件将包含坠撞载荷。

3.4 Functional Aspects of Safety

A key aspect of aircraft functional analysis is that it is used as the primary tool for identifying and classifying potential safety hazards. This analysis is a critical part of the design for certification as discussed in Section 10.2.

3.5 The Cluster Model

A methodology that provides more insight into a system is called cluster analysis. This methodology was first proposed by Lano (1979) and described by Hitchins (1993, pp. 135–147) and (2003, pp. 143–148, 449–457). The methodology has been incorporated into a software tool called CADRAT by Campbell (2013), which was envisioned by Hitchins.

The essence of the cluster model is that it optimizes the relationships between the elements of a system and presents clusters of entities with *functional affinity*. It is not necessary to know the identities of the elements of a system, for example, the subsystems or components. It is only necessary to know the relationships. The inputs and outputs of the functions in Figure 3.6 constitute the relationships. There is a relationship between *fuel* and *range* for example.

Figure 3.7 shows the output of the cluster model for the top-level entities of a commercial aircraft. These entities were extracted from the IDEF0 model shown in Figure 3.6. So what does the cluster model show? This figure shows that a commercial aircraft is a highly coupled system. If the system were uncoupled, the clusters would represent recognizable subsystems, such as the propulsion subsystem or the avionics subsystem. But as Figure 3.7 shows, the subsystems are apparent at the nodes of the matrix. But most importantly, this figure shows two dominant clusters that can be characterized in the following way:

The first and smaller cluster (from left to right) we call the energy cluster because it contains parameter related to power and thrust and to related functions such as aircraft movement and aircraft environment. This cluster shows that there is functional affinity among those parameters and to parameters they are connected to.

The second and larger cluster is dominated by those entities that pertain to the navigation and control of the aircraft and to related parameters such as information. Hence, there is functional affinity among these parameters as well.

It also shows that aircraft movement is related to parameters in both clusters.

3.4　安全性功能方面

飞机功能性分析的一个关键作用是识别潜在安全性危害，并对其进行分类。该分析是 10.2 节中描述的合格审定设计的关键部分。

3.5　群模型

有一种方法称为群分析，它提供了一种更深入系统内部的视角。这种方法由 Lano(1979) 首次提出，Hitchins(1993，第 135 至 147 页) 和 (2003，第 143 至 148 页，第 449 至 457 页) 对其进行了描述。该方法已被 Campbell(2013) 用软件工具 CADRAT 实现，该软件基于 Hitchins 的研究。

群模型的本质是优化系统元素之间的关系，并且利用实体群表明功能关系。没有必要了解系统元素 (例如,子系统或者部件) 的特性,而仅须了解其相互关系。图 3.6 中功能的输入和输出即构成这些相互的关系。例如，在燃油和航程之间存在的关系。

图 3.7 展示了商用飞机顶层实体的群模型的输出。这些实体是从图 3.6 所示的 IDEF0 模型中抽象出来的。群模型展示了什么？该图表明商用飞机是一个高度耦合的系统。假如系统是不耦合的，群模型将展示可辨识的子系统，例如推进子系统或航电子系统。但是如图 3.7 所示,这些子系统在矩阵的节点上是非常清晰的。但是最重要的是，该图展示的两个主要群组可以通过如下的方式进行说明：

第一个，也是相对较小的群 (从左往右)，我们可以称之为能量群，因为它包含与能量和推力有关的参数，以及其他相关功能，例如飞机运动和飞机环境功能。这个群表明了在这些参数与它们交联的参数之间存在功能上的紧密关系。

第二个，也是相对较大的群包含与飞机导航和控制及其他相关参数 (如信息) 有关的实体。因此，与这些参数之间也存在功能关系。

Since aircraft movement includes both forward movement and rotational (pitch, yaw, and roll) movement, this entity requires the combined involvement of many functions. This fact illustrates a certain degree of *loose coupling* in the aircraft system.

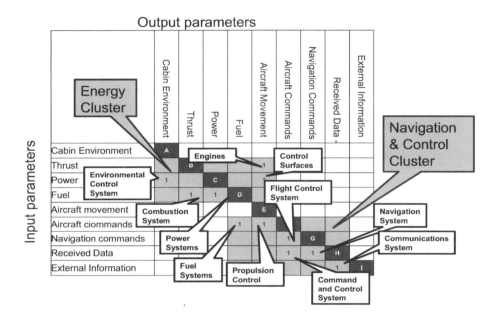

Figure 3.7   Cluster model

As noted above, this model shows that the aircraft is highly coupled, that is, that single entities play a role in many different functions. Two entities that stand out are power, thrust, and aircraft movement. This fact has relevance in the safety context; that is, the loss or malfunction of a single entity could mean the loss of several critical functions. There are many ways to mitigate this risk, many of which are used today. Two of them are as follows:

Functional redundancy

As pointed out in Chapter 16, *functional redundancy* means that a function is performed using two or more physically different means. The use of an APU as a *functionally redundant* source of electrical power is an example.

它还表明飞机的运动与两个群中的参数都有关系。因为飞机的运动既包含向前运动也包括旋转（俯仰、偏航以及横滚）运动，这个实体需要许多功能的组合。这个事实说明了飞机系统中一定程度的松耦合关系。

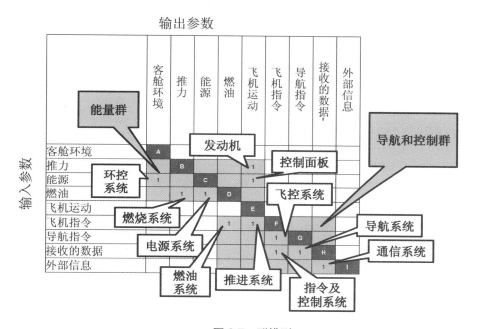

图 3.7　群模型

如上所述，该模型表明飞机是高度耦合的，也就是说单个实体在很多不同的功能中都会起作用。两个突出的实体是动力、推力及飞机的运动。这一事实与安全性方面同样相关；也就是说，一个实体的丧失或者功能异常可能意味着几个关键功能的丧失。有很多种下面正在使用的方法可以降低这种风险。以下列出其中的两种：

功能冗余

如第 16 章所述，功能冗余指的是一个功能通过两个或者更多的、物理上不同的方法实现。使用 APU 电源功能的就是功能冗余的一个例子。

Modularization

Called localized capacity in Chapter 16, *modularization* means that two or more independent means are used to perform a function. If one means fails, the other will operate independently. Functionally independent engines are an example.

In short, cluster analysis is a useful tool during the Advanced Design phase of development. It will reveal the shortcomings and strengths of any given design and will provide an understanding of the interrelationships among the aircraft parameters.

3.6   The Swim Lane Model

Another commonly used model is the swim lane model. Figure 3.8 shows an example of this model. The advantage of this model is that it shows a clear time dependency among the various elements of the system.

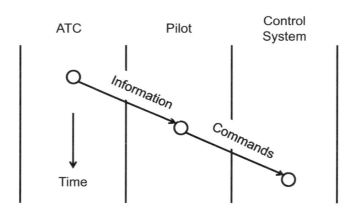

Figure 3.8   The swim lane model

模块化

第 16 章称为局部能力的模块化是指使用两个或者更多的独立手段实现某一个功能。如果一个失效了，其他的将独立运行。例如，功能上独立的发动机。

简而言之，群分析是前期研制阶段的一种有效工具。它将揭示任何给定设计的不足和优势，加深对飞机参数之间相互关系的理解。

3.6　泳道模型

另外一个普遍使用的模型是泳道模型。图 3.8 示出了该模型的一个例子。这个模型的优势是它给出了不同的系统元素之间清晰的时序关系。

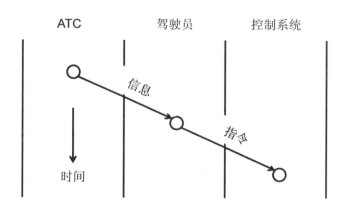

图 3.8　泳道模型

4

Requirements and Needs

The sources for requirements for the development of a new, derivative, or change- based aircraft design fall into two primary categories: regulatory and economic. Regulatory requirements are those which pertain to the safety of the aircraft and its occupants and are established by the Federal Aviation Administration (FAA) and documented in Federal Aviation Regulations (FARs). Economic requirements are those driven by the airline customer and pertain to the cost of purchasing, operating, maintaining, and servicing the aircraft. The SE process accommodates both categories of requirements.

Economic requirements, for the most part, emanate from customer needs, those desires to have an aircraft that economically profitable.

4.1   Requirements Definition

A requirement is a statement of required performance or design constraint to which a product must conform. One principle agreed on by most systems engineers is that a basic quality of a requirement is that it must be verifiable. The requirement is applied to the people, products, and processes, not to the engineer or the environment.

4.2   Requirements Types

Some sources, such as IEEE 1233 (1996), define as many as 25 requirements types. Many systems engineers use only three categories: functional requirements, performance requirements, and constraints. This method of categorizing requirements, used in this book, is simple and logical. The distinction between performance requirements and constraints is only important to the requirements *developer*. It helps him or her strive for the goal of requirements completeness. To the requirements *implementer* this distinction has no bearing on the task of converting the requirements into a design; he or she only has to make sure the design meets the requirement regardless of the requirement category.

第4章　需求与需要

研制新飞机、衍生型或改装型飞机的需求来源基本可归为监管及经济原因两大类。监管要求是指(美国)联邦航空局(FAA)在联邦航空条例(FAR)中规定的,关于飞机安全性的要求。源于经济原因的需求是指航空公司客户推动的需求,属于飞机采购、运行、维修及保养成本方面的需求。系统工程过程综合考虑上述两方面的需求。

其中,由客户提出的经济性需求大部分都是希望飞机具备良好的盈利性。

4.1　需求的定义

需求,是对产品必须符合的性能要求及设计约束的表述。大部分系统工程师公认的一点是,需求必须是可以验证的。需求适用于人、产品及过程,而不适用于工程师及环境。

4.2　需求的类型

IEEE 1233 (1996) 等材料定义的需求类型多达 25 种,但是许多系统工程师主要使用三类需求:功能性需求、性能需求及设计约束,本书采用该分类方法,因为该分类方法简单合理。性能需求与设计约束的区分仅对需求开发人员有意义,它有助于需求开发者达到需求完整性的目标。但对于需求实施人员,区分性能需

Functional requirements

Functional requirements are pretty simple; they are simply the functions as described in Chapter 3 but without any quantitative value for the function. When the quantitative value is added, the performance requirements are born, as described in the following section. Of all requirement types human factors requirements often prove to be of this type.

Performance requirements

A performance requirement is a measure of the extent to which a system performs a function. MIL-STD-961D (1995), for example, encompasses performance requirements within a category called *entity capability*. Performance requirements are a critical part of the certification process discussed in Section 4.9, Item 5. A basic concept is that a*ll performance requirements are traceable to functions*. Ideally, this concept applies even to humans within the system even though satisfying the human requirements is the most challenging of all as discussed in Section 5.5. This challenge is limited, of course, to those functions which apply directly to the satisfactory operation of the entire system. Even so, developing requirements for such critical functions as Maintain Vigilance consumes the energies of human factors specialists.

In the traditional non-SE environment, performance requirements are the most likely requirements to be neglected. A basic concept is that every aircraft element must have at least one performance requirement as described in Section 3.3. This concept is only logical because if it were not true, then elements would have no functions to perform and therefore no reason for existence.

Another concept is the traceability of performance requirements to higher-level requirements. This concept is a logical extension of the aircraft hierarchy in which the function of every element is to support the mission of the entire system. While it is not always practical to create complete top-to-bottom traceability of requirements, the concept assures that the origin of all requirements is known. It also assures that requirements get flowed down to lower levels of the aircraft hierarchy. Standard specification formats, such as the example shown in Appendix 2, often provide for the recording of requirements traceability.

求和设计约束并不影响其将需求转化为设计，需求实施人员只须确保设计满足需求，而无须关注需求的类别。

功能需求

功能性需求非常容易理解，它们是第 3 章描述的一些功能，但是并不包含该功能的任何定量数值。当添加了定量数值时，则产生了下文所述的性能需求。一般而言，在所有类型的需求中，人为因素需求属于这种类型。

性能需求

性能需求是系统执行某一功能程度的衡量指标。例如，MIL-STD-961D (1995) 就在称为实体能力的一类中包括了性能需求。性能需求合格审定过程 (4.9 节，第 5 条) 的关键部分。一个基本概念是，所有的性能需求均能追溯至相应的功能。理论上，这个概念甚至适用于某一系统范畴内的人，即使如 5.5 节所讨论的，满足人的需求在所有需求中最具挑战性。当然，这种挑战只局限于那些直接应用于系统符合要求的运行的功能。即使这样，开发例如**保持警觉**这样关键功能的需求，仍耗费人为因素专家很大的精力。

在未采用系统工程的传统环境下，性能需求往往是最容易被忽略的。一个基本概念是，每一飞机元件必须至少对应一条性能需求 (3.3 节)。这个概念是合乎逻辑的，因为如果飞机某一元件没有性能需求，即代表该元件不需要执行任何功能，那就意味着该元件没有存在的必要。

另一个概念是性能需求与上层需求的可追溯性。这个概念是对飞机架构的逻辑性延伸，在这个架构里，飞机每个元件所需执行的功能均是为了支持整个系统任务。虽然建立自上向下所有需求完整的可追溯性并非始终可行，但这个概念可以确保每条需求的来源得到明确，还可以确保高层级的需求根据飞机的架构向下传递至低层级。标准规范格式 (附录 2 例子所示) 往往要求记录需求的追溯性。

Performance requirements in an aircraft context

We saw before in Chapter 3 that the functional analysis creates a complete functional description of the aircraft, its subsystems, and its components. However, the process reveals nothing about the hardware descriptions. The allocation process creates a performance requirement for each function. (This is only one of three meanings of the word *allocate* (see Glossary).) At the top level, performance requirements will not exist for all functions. However, at a critical level, all functions will result in performance requirements.

At the top level, typical functions, such as Perform Transport Mission, will result in performance requirements, such as, "The aircraft shall be capable of transporting 250 passengers and 20,000 lb of cargo at least 7,000 nmi at a cruise Mach number of at least 0.8." This requirement is *top level*: that is, it does not depend on any design solutions. Requirements which do depend on design solutions are discussed below in Section 4.6. Other performance requirements, such as "Provide 11,000 lb of thrust," are derived from design solutions at the higher level of the system architecture.

We saw the Control Temperature function which was subordinate to the Provide Environmental Control function discussed in Section 3.3. The Control Temperature function is also top level since it does not depend on design solutions. It is necessary to capture all top-level performance requirements before proceeding to the derived requirements.

Performance requirements and interfaces

As we will see in Section 6.1, there is a strong link between interfaces and performance requirements. In brief, for every interface there are at least two performance requirements: For the delivering side of the interface, there will be a performance requirement to create and deliver the quantity being delivered, for example, electrical power. For the receiving side, there will be the requirement to *use* the received function, also, electrical power. For example, a fan will be required to create cooling using the received power.

1. *Emitted quantities*   Certain quantities are emitted from many of the aircraft components. These include noise, radiation, heat, and other quantities. These emitted quantities are often confused with the environments as discussed in Section 5.6. However, there does exist a performance requirement on each

飞机中的性能需求

在第 3 章提到，通过功能性分析可以获得对飞机、子系统及部件完整的功能描述，但这个过程丝毫未提及任何硬件方面的内容。需求分配过程将对每个功能建立性能需求（"分配"这个词在本书中有 3 个含义，此含义仅为（见"术语表"）其中之一）。在最顶层，不一定所有功能均有性能需求，但是在某个关键的层级上，所有功能均会有一组与功能相关的性能需求。

在飞机顶层，类似**执行运输任务**等典型功能会产生相应的性能需求，例如，"该飞机应能以 0.8 的最小巡航马赫数、运送 250 名乘客和 20 000lb 货物至少 7 000n mile"，这条需求是一项顶层需求，因为该需求不依赖于任何设计解决方案。依赖设计解决方案的需求将在 4.6 节中详细讨论。其他类似"提供 11 000lb 的推力"这样的性能需求，是根据系统架构上一层级的设计方案导出的。

在 3.3 节中我们提到，**控制温度**功能从属于**提供环控**功能，**控制温度**功能也是顶层功能，因为它也不依赖于设计解决方案。在导出需求之前，必须首先捕获所有顶层的性能需求。

性能需求与接口

在 6.1 节中将会提到，接口与性能需求联系紧密。简单来说，对于任何一个接口，至少有两条相关的性能需求：对于接口的输出端而言，将有产生并输出所提交的量的性能需求，例如电源。对于接口的接收端来说，会有使用接收到的功能的需求，同样也是电源。例如，需要一台风扇使用接收到的电源提供制冷。

1. 发射量　某些量会从飞机许多部件中发出，包括噪声、辐射、热等。这些发射量容易与 5.6 节所讨论的环境混淆。针对每一个部件都有性能需求，以限制其可能发出的这些量的数量。这些限制性的性能需求和其他性能需求一样，也是有效且可验证的。

component to *limit* the amount of each of these quantities the component may emit. These limiting performance requirements are just as valid and verifiable as any other performance requirements.

Although these emitted quantities are limited in their value, they do become the environment *for other components* in the immediate vicinity of the emitting component. The most striking example is electromagnetic interference (EMI). Electrical cables are limited in the EMI they may produce. Nevertheless, EMI is a real environment for electronic components. The system is protected from EMI in three ways: by limiting the output of EMI (a performance limit), by controlling the spacing between the cables and the electronic components, and by providing protective shielding for the electronic components.

2. *Delivered quantities and operations on incoming quantities* In general, the emitted quantities discussed above are generally *undesirable* quantities. However, many other quantities may be intentionally delivered from one component to another. These include electrical current, hydraulic fluid, and pneumatic air. These are *desirable* quantities. Thus, when an interface calls for an electrical current to be delivered from component *A* (an electrical cable) to component *B* (a fan), then two performance requirements are implied. First, it is a performance requirement for the cable to deliver the current to the fan. Secondly, it is a performance requirement for the fan to *use* the current to run. Both performance requirements are listed in the appropriate specifications.

Constraints and specialty requirements

Chapter 5 provides a more complete description of the concept of constraints. Suffice it to say that a constraint is any non-performance requirement; that is, any requirement that cannot be traced to a function. That is, it is a *global* requirement. Constraints can include mass properties, dimensions, environments, and design standards. MIL-STD-961D (1995) limits the use of the term *constraints* to such requirements as design standards. Most constraints are established within key engineering specialties, such as human factors, safety, production, and maintainability. The requirement for resilience as described in Chapter 16 also qualifies as a constraint.

In short, constraints are so numerous they often become the design drivers in spite of the fact that much emphasis in SE is on the development of functions, functional requirements, and the resulting performance requirements.

The specialty requirements pose a special challenge to the systems engineer. For

虽然发射量在数量上有限，但它们确实会影响发射部件附近的其他部件。电磁干扰 (以下简称 EMI) 就是一个典型的例子。必须对电缆可能产生的 EMI 进行限制。然而 EMI 是电子部件必须面对的环境。通过 3 种方式可以保护系统不受 EMI 干扰：限制 EMI 输出 (性能限制)，控制电子部件与线缆之间的间隔，以及在电子部件上加装防护罩。

2. 输出的量和获得量后的操作　通常，上述讨论的发射量均是不利的量。但有的时候将某量从一个部件传递到另一个部件是有意的，包括电流、液压油、气源等。这些都是有利的量。因此，当某种情况下，一个接口要求将电流从部件 A(电缆) 传递至部件 B(风扇) 时，就会产生两条性能需求，一条是关于电缆将电流传递至风扇的性能需求；另一条是定义风扇使用获得的电流运转的性能需求。上述两条需求均应列入相应的规范中。

约束及专业需求

第 5 章将更加完整地描述约束的概念。可以说，任何非性能需求都是约束，换句话说，只要不能追溯至相应功能的需求都是约束。约束是一个全局性的需求，可以包括质量、尺寸、环境及设计标准等。美军标 MIL-STD-961D (1995) 将"约束"这一术语的使用局限于设计标准之类的需求。大部分约束均在包括人为因素、安全性、生产及可维护性等的重要工程领域出现。第 16 章所描述的恢复力需求也定性为约束。

简单来说，虽然在系统工程领域一直强调功能开发、功能性需求及相应的性能需求，但是约束的数量之多使其成为设计过程中的驱动器。

专业需求大多数是定性的，这也给系统工程师带来了新的挑战。系统工程师的主要职责是将定性的特殊需求转换为可以被验证的定量需求，具体内容将在第 11 章详细介绍。表 4.1 提供了这种转换的一些例子。

the most part they are qualitative. The role of the systems engineer is to convert these qualitative requirements into verifiable requirements in accordance with Chapter 11. Table 4.1 provides some examples of this conversion:

Table 4.1   SE treatment of qualitative requirements

Specialty	Qualitative Requirement	Verifiable Requirement
Human Factors	The item shall be easy to reach.	The item shall be no more than three feet from the seat.
Maintainability	Provide access for replacement and repair of the item.	Provide a minimum of four inches for access on all sides of the item.

Design requirements

Design requirements are the design characteristics which are the product of the synthesis process as described in Section 7.7. They are the direct result of the performance requirements and constraints discussed above. Design requirements are the attributes of the item needed to meet the performance requirements and constraints. These could include, for example, physical dimensions or power required.

4.3   Requirements Development

A frequently asked question in SE is: When are all the requirements complete? The answer is that they are complete when the customers (including the regulatory agencies) say they are. SE provides some tools to assure that as many requirements have been captured as humanly possible. Here are a few steps.

The famous Vee model

One of the most often used ways of depicting the requirements development process is with the Vee model shown in Figure 4.1. Although the Vee model is a somewhat simplistic way of illustrating the process, it does bring out some of the important aspects of SE and especially how SE handles requirements. In short, the Vee model is a concise way of depicting the entire SE process. Following are some of those aspects illustrated by the Vee model:

表 4.1 系统工程对定性需求的处理

专业	定性需求	可验证的需求
人为因素	该项须易于达到	该项到座椅的距离不应超过 3in
可维修性	可更换或维修该项	该项各侧均有至少 4ft 的空间

设计需求

设计需求是综合过程 (7.7 节) 中产生的设计特征。它们是上述性能需求及约束的直接结果。设计需求是为了满足性能需求和约束所需项目的属性，例如，它们可能包括物理尺寸或所需功率等。

4.3 需求开发

在实施系统工程过程中，经常面临的一个问题是：什么时候所有的需求才是完整的？正确的答案是：当客户（包括监管当局）认为需求已经完整了的时候，需求才是完整的。系统工程提供了一些工具，以保证尽可能多地获取需求，主要包括以下几个步骤。

著名的 V 模型

描述需求开发过程最常用的就是图 4.1 所示的 V 模型。虽然 V 模型仅是对该过程的简化描述，但它却表明了系统工程的关键内容，尤其是系统工程如何处理需求。简而言之，V 模型是描述整个系统工程过程的一种非常精炼的手段。以下是 V 模型所阐明的系统工程的一些关键内容。

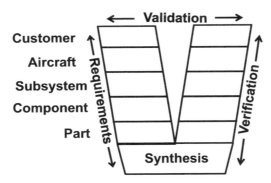

Figure 4.1 The Vee model

The Vee model overview

The left side of the Vee model represents the flow down (and up) of requirements as needed from level to level of the aircraft hierarchy. The right side depicts the flow up (and down) of the verification of the various requirements. Chapter 11 discusses verification. At the top of the Vee model the validation of customer needs is shown from left to right. Validation is also discussed in Chapter 11.

As will be explained below, the Vee model is only part of the picture. It represents the *reductionist* view of requirements development. That is, it only represents those requirements determined from the flow down from higher levels of the aircraft hierarchy. The broader view is called the *holistic* view which is discussed below. The holistic view accounts for requirements from other sources than flow down.

The aircraft abstract hierarchy

Chapter 2 explained how the aircraft system could be depicted in terms of an abstract hierarchy as in Figure 2.1 with the aircraft at the top and subsystems and so forth at lower levels. The Vee model shows that requirements are flowed down and allocated through the various levels of this hierarchy and that the verification tasks are flowed up through the same levels.

The time axis

The horizontal axis is time. Hence, the Vee model reflects the phased development inherent in integrated product development (IPD) discussed in Chapter 12.

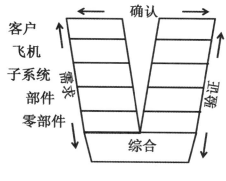

图 4.1 V 模型

V 模型概览

V 模型的左侧代表一些需求根据需要沿飞机架构自上而下（及自下而上）逐级传递的过程，右侧描述了各种需求自下而上（及自上而下）的逐级验证，第 11 章将详细讨论验证过程。在 V 模型的最顶端，是自左向右对客户需求的确认，确认工作也将会在第 11 章进行详细介绍。

正如下面将要介绍的，V 模型仅是该图形的一部分，它只代表了需求开发的简化视角。也就是说，它仅表达了沿飞机架构自上而下逐级确定的需求。更宽泛的视角称为整体视角，将在后面讨论。整体视角考虑到来自下行之外其他来源的需求。

飞机抽象层级架构

第 2 章介绍了如何以抽象层级架构表达飞机系统。如图 2.1 所示，顶层是飞机，子系统等在下层。V 模型表明各项需求通过这个自上而下逐级传递和分配，验证工作则通过同样的层级向上传递。

时间轴

V 模型的横轴代表时间。因此，V 模型反映了第 12 章讨论的集成产品研制 (IPD) 固有的阶段式开发。

Customer needs

Customer needs (also called customer requirements) are shown at the top of the Vee. As explained later in this chapter, the customer needs need to be converted into top-level aircraft (product) requirements before the flow down process can begin. It is important to distinguish customer needs from system, or product, requirements. Needs are often qualitative. It is not even known whether the needs can be achieved. It will be shown later in this chapter that a basic quality of a requirement in achievability. This is not so for needs; needs are not even verifiable since they are not product requirements. Working with the customer, the developer can conduct trade studies and other analyses to convert the needs into requirements. Section 7.4 will show that one way to convert needs into requirements is with quality function deployment (QFD) analysis.

Requirements flow down

Starting with the top-level aircraft product requirements (range, durability, and so forth), the Vee model shows the flow down all the way to the parts by allocation and requirements derivation as explained later in this chapter.

Synthesis

As explained in Chapter 7, synthesis occurs when the entire aircraft is integrated and assembled into a unified system. This occurs after all requirements have been defined.

Requirements flow up

In addition to requirements flow down as explained above, the Vee model also shows that requirements can flow up. This happens when it is desirable to use a commercial off-the-shelf product (COTS) to perform a certain function perhaps at the component level, such as a pump or a fan. When this happens, it is necessary to flow the requirements up to the aircraft level to confirm that it is not in conflict with any top-level requirements.

Verification flow up

Since verification must be performed at the same level as the requirement, the verification steps will be performed from bottom up as shown in the Vee model. For example, there will be verification tests (as required) for the parts, then components, and so forth. Most aircraft developers have laboratories in which complete subsystems

客户需要

客户需要 (也称为客户需求) 处于 V 模型的顶端。如本章稍后解释的，客户需要首先需要被转换为飞机 (产品) 顶层需求，然后才能开始向下传递。正确地区分客户需要与系统或产品需求是非常重要的。一般情况下，客户需要大多是定性的，有的时候甚至不知道这些需要能否满足。然而，正如后面章节将要介绍的，需求的一项基本特性就是可实现性，这一点和客户需要不同，有的时候客户需要甚至可能是不可验证的，因为它们并非产品需求。需求开发人员通过与客户共同工作，可以进行权衡分析或者其他分析方式，将客户的需要转化为需求。7.4 节将会介绍如何通过质量功能展开 (QFD) 将需要转化为需求。

需求向下传递

V 模型展示了通过分配与衍生将顶层飞机产品需求 (航程、耐久性等) 一直传递至底层部件的过程，后续章节将会介绍需求分配与需求衍生的内容。

综合

如第 7 章所述，综合在整架飞机集成并组装为一个统一的系统时即开始。在所有需求定义完毕以后才会进行综合。

需求向上传递

除了上面介绍的需求自上而下的传递过程之外，V 模型中也存在需求自下而上的传递过程。这种情况一般发生在采用泵或者风扇等货架产品 (COTS) 来实现某一功能的时候。在这种情况下，需要将底层需求自下而上传递至飞机级以便确认该底层需求不与任何顶层需求冲突。

逐级向上的验证

由于必须在需求的同一层级上进行验证，因此如 V 模型所示，验证工作需

can be tested, such as hydraulic systems, control systems, and so forth. Verification at the top level usually consists of flight tests.

Validation

As explained in Chapter 11, validation of customer needs must occur at the customer level. This can consist of flight demonstration tests (by the customer) or by buy-off of the lower-level tests. This is called system validation and should not be confused with requirements validation, which according to the FAA (2012, p. 54) is the assurance that the requirements are correct.

Reductionism vs. Holism

At the risk of introducing some somewhat esoteric terms, it is important to understand the difference between *reductionism* and *holism*. Many books have been written, for example Checkland (1999, p. 77), explaining why reductionism is undesirable and holism is desirable. In short, reductionism is vertical thinking, while holism is thinking about the whole, including the horizontal relationships between elements.

Hence, using the Vee model, as described above, alone without any consideration for the relationships across elements is reductionist and therefore incomplete. For example, you have a system A with two subsystems A_1 and A_2. Using the Vee model you could determine the requirements for A_1 and A_2, but you would know nothing about the relationship between A_1 and A_2.

Among SE sources, the one that addresses the holistic approach is Stevens et al. (1998, pp. 6, 206, 344). According to Stevens et al., the additional relationships that should be addressed include:

- the system architecture;
- the various disciplines involved (electrical, mechanical, and so forth);
- keeping requirements (including interface requirements), design, plans, and risks consistent;
- the various system development processes (engineering, integration, and so forth);
- requirements, costs, and timescales.

The *FAA Systems Engineering Manual* (2014) also notes the importance

要自下而上逐级开展。例如，首先根据需要开展零件级验证试验，此后开展部件级验证试验等。大多数飞机研制方均有对诸如液压系统、控制系统等整个子系统进行试验的实验室，顶层验证通常包括飞行试验。

确认

如第 11 章所述，必须在客户层面确认客户的需要。客户需要确认包括展示飞行试验（由客户实施）、低层级试验的验收等。该过程称为系统确认，切记不要将其与 FAA (2012，第 54 页) 定义的需求确认混淆，需求确认目的在于保证需求的正确性。

简化论 VS 整体论

在介绍略显深奥的术语之前，有必要了解简化论与整体论之间的差别。很多相关书籍分析了为何简化论是不利的，而整体论是有利的，比如 Checkland(1999，第 77 页) 著作所述。简单地说，简化论是垂直思考，而整体论则需要考虑整体，包括各元素之间的横向关系。

因此，采用如前文所述 V 模型，不考虑各元素之间的关系，这就是简化，因此是不完整的。例如，有一个系统 A，包括两个子系统 A_1 和 A_2，利用 V 模型可以确定 A_1 和 A_2 的需求，但是不能了解 A_1 与 A_2 之间的关系。

系统工程资料中，提出整体分析方法的是 Stevens 等人 (1998, 第 6, 206, 344 页)。根据 Stevens 等的理论，应予说明的其他关系包括：

· 系统架构；
· 涉及的相关学科（电气、机械等）；
· 保持需求（包括接口需求）、设计、计划及风险之间的协调；
· 各个系统研制过程（工程、集成等）；
· 需求、成本及进度。

of including the requirements for the operational phase as part of the holistic considerations.

So what do you do if you find that the reductionist and holistic requirements are in conflict? The answer is that you conduct a trade study. Section 4.8 discusses the principles for how to conduct requirements trade-offs.

Mission statement

The mission statement is the first step in the requirements development process. The mission statement is a simple statement of the purpose of the system being developed, the environment in which it will operate, and any special operational considerations which may be important. The value of the mission statement is to establish an understanding between the customer and the technical community about what the system is for. The mission statement will be the primary exhibit at the system requirements review (SRR) described in Chapter 12.

Mission statements apply at any level of the aircraft hierarchy. For example, at the aircraft level, the following simplified mission statement might be adequate:

> The purpose of the aircraft is to carry 250 passengers and 20,000 lb of cargo a total distance of 7000 nmi at a Mach number of 0.8. The aircraft will operate primarily in sandy desert climates.

At a subsystem level:

> The purpose of the subsystem is to protect the environmental control system from damage due to particulates in a sandy desert environment. The subsystem will remove at least 90 percent of the particulates from the air it receives.

Requirements allocation from functions

The allocation of functions to performance requirements is the basic completeness technique of SE as described in Section 4.2. This meaning of *allocation* is the second of three definitions (see Glossary). The principle of allocation of functions to performance requirements is part of the certification process.

《FAA 系统工程手册》(2014) 也强调了将运行阶段的需求纳入整体考虑的重要性。

在简化与整体需求之间存在冲突时应该如何处理？答案是进行权衡分析。4.8 节讨论了进行需求权衡分析的准则。

任务说明

任务说明是需求开发过程的第一步。任务说明是对如下内容的简单说明：系统研制的目的、系统运行环境及任何可能十分重要的特殊运行方面的考虑。任务说明的价值在于，在客户与技术团队之间就系统研制目的达成一致。任务说明是第 12 章所述系统需求评审 (SRR) 的主要评审对象。

任务说明可在飞机架构的任何层级应用。例如，在飞机级，如下的简化任务说明可能是充分的：

该飞机的用途是运载 250 名乘客及 20 000lb 货物，航程 7 000n mile，马赫数 0.8。飞机主要在沙漠气候下运行。

在子系统层级：

该子系统的用途是避免沙漠环境里的颗粒对环控系统造成损伤。该子系统将至少能去除其接收到的空气中 90% 以上的颗粒。

根据功能分配需求

将功能分配到性能需求是 4.2 节所述系统工程完整性的基本方法。这意味着分配是三种定义里的第二种（见"术语表"）。将功能分配至性能需求的原则是合格审定过程的一部分。

4.4 Requirements Sources

A basic practice of SE is to identify, justify, and record the source of all requirements. This process is called requirements validation. The principle of requirements validation is reinforced by the certification process which requires that all requirements be validated and recorded as shown in Table 10.1 as part of the certification data package. Furthermore, certification requires that a validation plan be submitted and that data be submitted to substantiate the validation. Section 4.10 explains requirements validation in more detail.

External requirements

External requirements are those which emanate either from the airline customer, regulatory agencies, such as the FAA or Joint Aviation Authorities (JAA), or industrial organizations.

Customer requirements

Unlike for the automobile industry, the requirements for an individual airline customer may result in a product which is highly focused on that customer. The process of capturing those requirements, negotiating with the customer, and incorporating the requirements into an aircraft can be a highly complex process.

Customer requirements are sometimes, and more accurately, called customer needs as described in the Vee model of Figure 4.1. This is because customer requirements are not necessarily achievable on first sight. The developer needs to work with the customer and develop a set of mutually agreeable requirements that can be flowed down to the aircraft level as shown in Figure 4.1. Requirements at the aircraft level are product requirements and must meet the criteria of *achievable* and *verifiable*.

The important aspect of capturing customer requirements is to be able to ascertain what the customer wants the product (either the aircraft or a subsystem) to do, that is, to understand the functionality of the need.

One common mistake is to accept buyer furnished equipment (BFE) unquestioningly without examining its requirements. There are some key questions which need to be answered: Does the equipment do what the customer expects it to? Will the equipment function in the environment of the host aircraft? Will the equipment interface with the aircraft in an acceptable manner? Will the equipment meet the

4.4 需求的来源

系统工程的基本做法是识别、证明并记录所有需求的来源。该过程称为需求确认。合格审定过程强调需求确认的原则，它要求对所有需求均须按照表 10.1 进行确认并记录，并将记录信息作为合格审定数据包的一部分。另外，合格审定过程要求提交确认计划，并提交证明该确认过程的数据作用。对需求确认的详细介绍见 4.10 节。

外部需求

外部需求是指那些来自航空公司客户、监管当局 [如 FAA 或联合航空局 (JAA)]，或行业性组织的需求。

客户需要

与汽车行业不同，个别航空公司客户的需求可能导致产品完全针对客户定制。捕获需求，与客户进行协商，并将这些需求纳入飞机之中，这是一个十分复杂的过程。

更准确地说，客户需求有时候是客户需要，如图 4.1 中的 V 模型所示。这是因为客户需求不一定一次就能达到。研制方需要与客户一起工作，开发出一系列双方都同意的、能向下传递到图 4.1 所示飞机级的需求。飞机级需求是产品需求，且必须满足可实现和可验证准则。

捕获需求最重要的方面是能够确定客户需要产品 (飞机或子系统) 完成什么任务，也就是说，理解需要的功能性。

无条件地接受买方提供设备 (BFE) 而不检查其需求，这是一种常见的错误。有一些核心问题需要回答：设备是否按照客户期望工作？设备在所安装的飞机环境下能否正常工作？设备能否按照合适的方式与飞机进行接口对接？设备是否满

safety requirements which have been imposed on the rest of the aircraft? All of these questions should be answered before the equipment is incorporated into the aircraft. The results should be briefed to the customer at the SRR as described in Section 12.4.

One method for capturing customer requirements is QFD. QFD is a structured methodology for conducting dialogue with a customer, prioritizing the customer's needs, identifying solutions to those needs, and scoring those solutions. QFD has been used successfully in the automobile industry and shows great promise for the aircraft industry.

Another type of customer requirements is *assumed* requirements. That is, during requirements development some requirements have to be assumed. For example, the number of cubic feet of storage area in the cabin is a requirement designated as assumed. The assumed value of the requirement is presented to the customer at the SRR for concurrence. If there is no concurrence, the requirement can be changed.

Regulatory requirements

Traditionally the regulatory requirements have been passed to the aircraft manufacturers through the FARs and have been verified in a certification plan submitted by the manufacturer. A major part of the certification plan is a Functional Hazard Assessment (FHA) which identifies hazard categories for identified components so that the proper designs and redundancies can be implemented.

The FAA, in cooperation with the Society of Automotive Engineers (SAE), has taken a major step towards incorporating the principles of SE into the certification process with the publication of the SAE ARP 4754A (2010). The guidelines of this document are not mandatory but recommended processes. It recommends that a thorough SE functional analysis be conducted for all levels of aircraft systems and that the hazard category be established for each function. It also recommends that combinations of functions be identified for potential hazards. It is expected that this process will result in increased aircraft safety. Ultimately, these recommended practices may be incorporated into an FAA advisory circular (AC) and eventually into a FAR.

Industry standards

Many standards are developed by industrial organizations, such as the SAE, the Institute of Electrical and Electronics Engineering (IEEE), and RTCA, Inc. The requirements validation should cite these sources. Examples of particular interest to the commercial aircraft industry include the SAE ARP 4754A and RTCA/DO-178B.

足飞机其他部分的安全性需求？在设备被集成到飞机上之前必须回答所有这些问题。结果须在系统需求评审 (SRR)(12.4 节) 上简要通报给客户。

捕获客户需求的方法之一是质量功能展开 (QFD)。QFD 是一种结构化方法，完成与客户对话、确定需要的优先级、提供针对这些需要的解决方案、对解决方案进行评级等工作。QFD 已经在汽车行业成功应用，并显示了在航空工业领域巨大的应用潜力。

另一种类型的客户需求是假定需求。即在需求开发过程中，假定一些需求。例如，座舱内存储空间的体积就是一种假定需求。需求的假定值在系统需求评审 (SRR) 上提交给客户，以达成统一意见。若意见不统一，则可对该需求进行更改。

监管要求

传统上，监管要求已通过 FAR 传递给飞机制造商，并通过制造商提交的合格审定计划进行验证。合格审定计划的主要内容是功能危害性评估 (FHA)，它判明特定部件的危害性类别，以确保实施恰当的设计及冗余度设置。

FAA 与汽车工程师协会 (SAE) 通过发布 SAE ARP 4754A(2010)，极大地推动了系统工程原则在合格审定过程中的应用。该文件并非强制性的，而是一系列推荐流程。该文件建议在飞机系统所有层级均进行全面的系统工程功能性分析，并针对各个功能划分危害性类别。该文件还建议对组合的功能是否存在潜在危害进行判断，期望通过该过程提高飞机安全性。最终，这些建议项将纳入 FAA 咨询通告 (AC)，并最终纳入联邦航空条例 (FAR)。

工业标准

(美国) 汽车工程师协会 (SAE)、电气电子工程师协会 (IEEE) 及航空无线电技术委员会 (RTCA) 等工业组织也制定了一系列标准。需求确认过程应援引这些标准。与商用飞机行业相关度较大的标准有 SAE ARP 4754A 及 RTCA/DO-178B 等。

Internal requirements

Many requirements originate from internal sources rather than from the airline customer. These requirements have many objectives: for example, to reduce cost, reduce weight, improve reliability, or simply to fix an item that does not work properly.

It is tempting to treat internal requirements less rigorously than, say, customer or regulatory requirements. Suffice it to say that internal requirements demand at least as much rigor as any external requirements. Just as much is at stake, for example, cost, safety, performance, reliability, to name a few of the more common ones. Before the redesign or fixing of a part or a subsystem, it should be established what the item was intended to accomplish, what environments it operated in, what other constraints were important, and what it had to interface with, both functionally and physically.

One innovation of recent years is the *quality circle*, a part of total quality management (TQM). In this process a team of employees attempts to resolve product or process problems in a group environment. This process has been regarded as highly successful. The combination of TQM with the SE rigor of defining requirements can result in value-added improvements. This approach can be applied to any of the requirements areas listed below.

Manufacturing initiated requirements

Many requirements originate in the manufacturing department of the aircraft company after the aircraft has been designed. The purpose of these requirements is usually to make the product easier to manufacture. Like most internal requirements, if the product had included manufacturing inputs during the initial aircraft development, these requirements would be unnecessary. However, they exist and should be included as a valid requirements category.

Product improvement

Design organizations will often initiate design improvements internally to improve a product. The purpose of these improvements may be to improve either weight or cost. Like manufacturing initiated requirements, these are after-the-fact requirements to correct oversights in the original design.

Design for manufacturing and assembly (DFMA)

内部需求

很多需求来自内部，而不是航空公司客户。这些需求针对很多目标：例如，降低成本、减轻重量、提高可靠性，或者是解决某个不能正常工作的设备问题。

人们在对待内部需求时，往往不像处理其他需求那么严格，如客户需求或监管要求。但可以这么说，内部需求至少需要保持与外部需求相同的严格程度。这不仅仅是因为它们许多都休戚相关，例如，常见的就有成本、安全性、性能和可靠性。在重新设计维修部件或子系统之前，应确定哪些需要完成，在哪些环境下运行，其他何种约束是重要的，以及与什么存在接口关系，无论是功能上还是物理上的关系。

质量圈是近年来出现的一种创新形式，它是全面质量管理 (TQM) 的一部分。在该过程中，由雇员组成的小组在团队环境中努力解决产品或过程问题。这种方法被证明十分有效。TQM 方法与系统工程对于需求的严格定义相结合，可以促进增值改进。这种方法可以应用于下列任何需求领域。

制造需求

飞机设计完成后，许多需求来自于飞机公司的制造部门。这些需求的目的通常是为了让产品更容易制造。和大多数内部需求类似，如果在飞机初始开发阶段就将制造输入纳入产品之中，那么这些需求 (制造需求) 可能就并非必要。然而，这些需求确实存在，而且应该归到有效的需求类别之中。

产品改进

设计团队通常在内部开展设计改进活动以改进产品。这些改进活动可能是为了减轻重量，也可能是为了降低成本。与制造需求类似，它们属于事后的需求以修正初始设计中的遗漏。

The basic purpose of DFMA is to reduce the recurring cost of an assembly. Other goals may be weight savings or improved reliability. These goals are normally accomplished by reducing the number of parts on a certain part of an aircraft. Other design improvements may accomplish the goals. When an assembly is identified as a potential area for improvement, a study is initiated to determine the cost savings, for example, of the redesigned assembly. If the non-recurring cost is returned in a given number of aircraft, say 20, then the project will be initiated.

Before the DFMA project can be initiated, both performance requirements and constraints for the original assembly should be determined and documented. The following questions are asked: What function did the original assembly have to perform? What were its performance requirements? What were its constraints? What environment did it have to withstand? What interfaces did it have with surrounding components? If the original requirements were not well documented, or perhaps not documented at all (after all, it was probably designed before SE was fully introduced into the organization), then the second question should be rephrased: What were its performance *capabilities*?

Product support initiated

The purpose of this type of change initiative is to improve the design of the aircraft so that it can be more easily maintained. For example, the structure surrounding a line replaceable unit (LRU) may be modified to allow a person to replace it more easily.

Technology focus initiated

This type of change is normally used to correct an operational deficiency. For example, an LRU may be replaced to provide a more reliable model.

4.5   Requirements Allocation to System Elements

This is the last major step in the requirements subprocess of the SE process. Here is where we decide what elements the requirements pertain to. At this point the shape of the element may not be known. It is only a *bucket* on the spec tree of Figure 2.1. Requirements may be allocated to multiple elements. For example, noise requirements are allocated to many elements. The allocation of requirements to elements is a basic requirement of certification.

面向制造及装配的设计 (DFMA)

面向制造及装配的设计 (DFMA) 的基本目标是减少装配过程中的重复成本。其他目标可能是减重或者提高可靠性。这些目标一般通过减少飞机特定部位的零件数量实现。其他设计改进也能完成该目标。当确定将一个装配过程作为潜在改进领域时，首先启动研究来确定成本节省量，例如，重新设计的装配过程可以降低的成本。如果非重复成本可在给定的飞机架数后收回，比如 20 架飞机，则该项目可启动。

在 DFMA 项目启动之前，应该确定针对原先装配的性能需求及约束，并归档记录。必须明确以下问题：最初的装配必须完成的功能是什么？其性能需求是什么？约束是什么？其必须承受的环境是怎样的？其与周边部件的接口是什么？如果没有很好地记录初始需求，或者根本没有记录 (毕竟，设计是在系统工程方法引入组织之前完成的)，那么第二个问题就变成：其性能能力如何？

源于产品支持的需求

这种类型更改的目的是改进飞机设计，以便维修过程更容易进行。例如，航线可更换单元 (LRU) 周围的结构可能进行更改，以更方便人员对其进行更换。

源于技术关注的需求

这种类型的更改通常用于修正运行缺陷，例如，改装某一 LRU，以便提供更可靠的型别。

4.5 需求分配至系统元件

这是系统工程过程中需求子过程的最后一个主要步骤，该步骤决定这些需求属于哪些系统元件。在该阶段，这个元件的形式可能尚不明确。它只是规范树 (图 2.1) 上的一片"篮子"。需求会分配到各个元件。例如，噪声需求即被分配到许多元件。将需求分配到各个元件是合格审定的基本要求。

4.6 Derived Requirements

Derived requirements are those requirements which depend on some feature of the solution to determine their values. For example, the value of engine thrust is a derived requirement determined from extensive trade-offs in the conceptual design process as discussed in Section 8.2. Most, but not all, subsystem requirements are, indeed, derived. Hence, derived requirements cannot be determined at the outset of the program. They are established gradually as the program progresses.

Another way to look at derived requirements is through the principle of requirements *harmonics*. Figure 4.2 illustrates this principle. In this figure the top-level requirement results in a solution, which, in turn, results in anther requirement (the derived requirement) which results in another solution. These alternating requirements and solutions are called requirements harmonics.

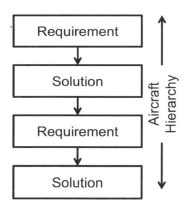

Figure 4.2   Requirements harmonics

Some sources, for example ARP 4754A (2010, p. 11), point out that derived requirements are not directly traceable to higher-level requirements since they may be traceable to design decisions. The key word here is *directly*; hence, since the design decision itself is traceable to higher-level requirements, the derived requirement will be indirectly traceable to the higher-level decisions. In some cases the derived requirement may be mathematically determined from higher-level requirements. For example, the requirement for the energy absorbed by a brake pad can be determined from the laws of physics only knowing the stopping distance, the weight of the aircraft, and the

4.6 衍生需求

衍生需求是指那些由对应方案的某些特性决定的需求。例如，发动机推力的大小是一种衍生需求，它是通过概念设计过程 (8.2 节) 中的全面权衡分析确定的。大部分情况 (当然不是全部情况) 下，子系统需求属于衍生需求。因此，衍生需求并不是在项目一开始就确定的，它们往往随着项目的进展而逐步建立。

衍生需求的另外一种研究角度是利用需求谐波准则。图 4.2 描述了这一准则的应用。图中，顶层需求产生解决方案，接下来该方案又生成另外的需求 (衍生需求)，该需求又产生其他的方案。这种需求与解决方案的交替即称为需求谐波。

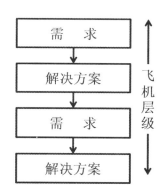

图 4.2　需求谐波

一些文献，如 ARP 4754A(2010，第 11 页)，认为衍生需求并非直接追溯到上层需求，因为它们可能可追溯到设计决策。此处的关键词是直接；因而，由于设计决策本身可追溯到一些上层级需求，因此衍生需求将间接追溯到这些上层决策。在某些情况下，衍生需求可能用数学方法由上层需求决定。例如，在只知道制动距离、飞机重量及着陆速度的条件下，关于刹车片吸收能量的需求可利用物理定律确定。在该计算过程中，涉及的设计决策只是刹车片的数量。另一个衍生需求是，燃油管路接头不得因累计产生电火花的电流量，该衍生需求取决于接头的表面面积。

landing speed. The only design decision involved in this calculation is the number of brake pads. Another derived requirement is the amount of current for which a fuel line coupling must be spark free due to a lightning strike. This derived requirement depends on the surface area of the coupling.

4.7   The Principle of Top-Down Allocation

This section will show how to allocate requirements from higher levels of the aircraft hierarchy as discussed in Section 2.3 to lower levels. For example, the analyst allocates requirements from the aircraft level to the subsystem level or from the subsystem level to the component level. This type of allocation corresponds to the third definition in the Glossary, namely, the breakdown of a top-level requirement into its subordinate components.

The principle of allocation assumes some degree of knowledge of the solution, that is, the design of the aircraft. Therefore, allocation will occur only after the appropriate level of aircraft design has been defined. For example, at the system design review (SDR) as discussed in Section 12.4 the aircraft architecture will have been defined, such as the one shown in Figure 2.1. Furthermore, at the SDR, an initial concept will have been developed. From this architecture and the initial concept we will be able to estimate the weight of each subsystem based on past experience. This estimate then becomes the *initial* weight allocation for each subsystem. We say *initial* because during the trade-offs discussed in Section 7.2 leading up to Preliminary Design Review (PDR) as described in Section 12.4 these allocations may, and probably will, change as a result of requirements trade-offs as discussed in Section 4.8.

Other allocated requirements, such as loads, shock, vibration, and noise, may require a greater degree of concept formulation to perform allocation. This requirement results from the fact that the allocation of these parameters requires the definition of the structure of the aircraft. In any event all allocation should be performed before PDR since all requirements are defined by PDR.

Requirements allocation is more than just a method of determining requirements at lower levels. It is also a method of program control. Some parameters, for example, weight and dispatch reliability, among others, are primary candidates for technical performance measures (TPMs) discussed in Section 12.7.

Weight

In commercial aircraft, weight is one of the most important design parameters.

4.7　自顶向下的分配原则

本节介绍如何将需求由飞机架构 (2.3 节) 的上层分配到下层。例如，分析人员将需求从飞机级分配到子系统级，或从子系统级分配到部件级。这种分配方式属于术语表中第三种定义，即，将顶层需求分解到其从属部件。

分配的原则基于对解决方案，即飞机设计某种程度上的了解。因此，分配只能在相应级别的飞机设计定义完成之后进行。例如，在系统设计评审 (SDR)(12.4 节) 阶段，此时，系统架构已经完成定义工作，如图 2.1 所示。另外，在 SDR 阶段，初始概念已完成开发。依据该架构及初始概念，可以根据以往经验估算各个子系统的重量。这个估算值又会对各子系统的初始重量分配。称之为初始，是因为在初步设计评审 (PDR)(12.4 节) 之前，权衡分析 (7.2 节) 过程中，这些重量分配过程很可能随需求权衡分析 (4.8 节) 结果而改变。

其他分配的需求，如载荷、冲击、振动及噪声，在实施分配之前可能需要更多地进行概念形成工作。该需求的产生是因为这些参数的分配需要基于飞机结构的定义。在任何情况下，所有分配过程均应在 PDR 之前进行，因为所有需求均在 PDR 之前定义。

需求分配绝不仅是一种在较低层级确定需求的方法，它也是一种项目控制方法。某些参数，例如重量及签派可靠性，与其他参数一起，均被视为技术性能指标 (TPM)(12.7 节) 的首要选项。

重量

对于商用飞机而言，重量是最重要的设计参数之一。100 lb 的重量就能对飞机运营及航程产生非常显著的影响。这就是为什么重量是技术性能指标 (TPM) 的首要指标之一，如上文所述。

One hundred pounds of weight can have a significant effect on the operating costs and on the range of the aircraft. That is why it is one of the primary TPMs, as described above.

Weight allocation from the aircraft level normally starts with the Manufacturer's Empty Weight (MEW) discussed in Section 5.2. MEW is the weight of all the aircraft structure and components, without any fuel, crew, passengers, or cargo.

Weight allocation is mathematically simple since weight is additive. As described above, the allocated weights result from an initial concept formulation. These allocations can then be adjusted as the final design takes shape. But, in any event, the objective is to keep the total weight of the aircraft (the *sum* of the allocated weights) below the initial constrained value.

Non-recurring (development) cost

Although aircraft manufacturers eventually amortize development costs over the unit costs of aircraft, they, nevertheless, set limits on development costs which they allocate to the aircraft components.

Recurring (unit) cost

We will see in Section 8.6 that recurring cost is a major component of the direct operating cost (DOC), which is a major design driver. Hence, allocating recurring cost to the aircraft elements is of primary importance in controlling the total aircraft cost. Successful cost control of each element would eliminate the need for cost reduction activities, such as design for manufacture and assembly (DFMA), as discussed in Section 4.4.

Direct Operating Costs (DOC)

Although DOC itself is not directly allocated to the aircraft elements, its components are described in Section 8.6. In addition, many derived requirements, such as weight, aerodynamic parameters, and fuel consumption, have direct impact on DOC. This dependence allows the management of DOC through these parameters.

Dispatch reliability

Dispatch reliability is one of the main drivers of airline indirect costs discussed

从飞机级开始的重量分配通常由制造空重 (MEW)(5.2 节) 开始。MEW 是指所有飞机结构及部件的重量，不包括燃油、机组、乘客或货物。重量分配从数学角度来说是比较简单的，因为重量是可以加法计算的。如前文所述，分配重量源自于初始概念形成。该分配过程在最终设计形成的时候会进行调整。但是在任何情况下，目标一定是保持飞机总重量 (分配重量之和) 低于初始约束值。

非重复（开发）成本

尽管飞机制造商最终是把研制成本分摊到单机成本，但是，他们一般对研制成本设定了限额，然后再将这些限额分配到飞机部件上。

重复（单位）成本

在 8.6 节我们提到，重复成本是直接运营成本 (DOC) 的主要组成部分，直接运营成本 (DOC) 是设计的主要驱动因素。因此，对于控制飞机总成本而言，将重复成本分配到飞机部件十分重要。对于各部件进行有效的成本控制，这将降低成本压缩工作的必要性，这些活动包括面向制造及装配的设计 (DFMA)(4.4 节)。

直接运营成本 (DOC)

尽管直接运营成本 (DOC) 本身并不直接分配到飞机各元件，8.6 节仍对其组成部分进行了讨论。另外，许多衍生需求，如重量、空气动力学参数及燃油消耗，都对直接运营成本 (DOC) 产生直接影响。这种依赖关系使得我们可以利用这些参数来管理直接运营成本 (DOC)。

签派可靠性

签派可靠性是航空公司非直接成本 (8.6 节) 的主要影响因素之一。因此，飞机制造商及航空公司期望将其保持在尽可能高的水平，通常在 0.99 左右。签派

in Section 8.6. Hence, manufacturers and airlines like to keep it as high as possible, normally in the range of 0.99. Allocation of dispatch reliability is not as straightforward as additive parameters, such as weight. Reliabilities must be multiplied to determine total reliability. Appendix 1 provides a simplified explanation of how to allocate reliability.

Maximum Allowable Probability (MAP) of failure

Like dispatch reliability discussed in Section 4.7, the maximum allowable probability (MAP) of failure can be allocated to all aircraft elements. This allocation is an integral part of the safety analysis discussed in Section 10.2.

Internal noise (sound levels)

The internal noise is the noise that each aircraft element emits, and not the noise environment it must withstand discussed in Section 5.9. Noise allocation is dependent on two factors: frequency and the location at which it is measured. Frequency is straightforward: the noise should be allocated at each frequency band of concern. Location is somewhat more complex. For the simple case of a sound emitter in the cabin, such as an air conditioning duct, the noise the duct can emit is limited by the location of the passengers with respect to the duct. When the noise allocation is made, specific cabin locations will have to be specified.

For noise emitters outside the cabin, the noise allocation will depend to a great extent on the insulation capabilities of the cabin walls, that is, on the design solution of the aircraft itself. Hence, this allocation cannot be made until the cabin wall concept is formulated. Hence, it is a derived requirement.

External noise

External noise, also frequency dependent, and primarily from the engines, is primarily limited by regulatory requirements pertaining to airport noise limitations. Allocation of noise to other components will be limited by their effect on ground crews.

Electrical loads

Electrical load analysis will determine the loads delivered to various components

可靠性的分配并不像重量等累加型参数那样直截了当。可靠性必须相乘以确定总的可靠性。附录 1 提供了可靠性分配的简化分析方法。

最大允许故障概率 (MAP)

与签派可靠性 (4.7 节) 类似，故障最大允许概率 (MAP) 可分配到所有飞机元件。该分配过程是安全性评估 (10.2 节) 的组成部分。

内部噪声（声级）

内部噪声是指飞机每个元件发出的噪声，而不是飞机必须承受的环境噪声 (5.9 节)。噪声分配取决于两个因素：频率及测量点的位置。频率比较容易理解：噪声应分配在关注的每一个频带上。位置因素相对而言比较复杂，以飞机座舱中噪声源，例如空调管道作为一个简单的例子，该管道发出的噪声取决于乘客相对于管道的位置。当进行噪声分配时，必须规定具体的座舱位置。

对于座舱外的噪声源，其噪声分配很大程度上取决于舱壁的隔音效果，即取决于飞机本身的设计方案，从而只能在舱壁概念形成后进行这种分配。因此，它属于衍生需求。

外部噪声

外部噪声也与频率相关，其来源主要是发动机，且主要由与机场噪声限制有关的规章要求限制。噪声向其他部件的分配取决于它们对于地勤人员的影响。

电气负载

电气负载分析决定输出到飞机各部件的负载。该分析过程可用来调整发电机的尺寸 (对发动机输出产生影响)，或限制输出到各部件的电力。

of the aircraft. This analysis can be used either to resize the generator (and hence place a penalty on the engine output) or to limit the electrical power delivered to various components.

Air distribution

For the most part, air distribution is the responsibility of a single subsystem, the environmental control subsystem (ECS). Nevertheless, the ECS needs to allocate the air distribution to the various compartments and areas: passenger cabin, flight deck, lavatories, galley, cargo, and other special areas.

Fuel consumption

Fuel consumption can be directly allocated to four factors: propulsive thrust, electrical power, hydraulic power, and air bleed for pneumatics. The degree to which each one of these consumes fuel is strongly dependent on the demands for each of these quantities. Minimum fuel consumption results from trading off these power drains. As discussed above, fuel consumption has a direct effect on DOC.

Emissions

Toxic emissions are a major concern in the design of modern aircraft. These are primarily the responsibility of the power plant and the APU.

Maintenance cost

Total maintenance cost is a component of DOC discussed in Section 8.6. Maintenance cost is a prime candidate for allocation to the various aircraft elements. The cost allocation will depend to a great extent on the element type, for example, electrical, electronic, mechanical, or structural.

Loads, shock, vibration

More than most other allocated parameters, loads, shock, and vibration are strongly dependent on the aircraft design. Hence, analyses will determine how these

空气分配

对大多数零部件，单个子系统，即环控子系统 (ECS) 负责空气分配。然而，环控子系统 (ECS) 需将空气分配到各舱室及区域：客舱、驾驶舱、盥洗室、厨房、货舱及其他特定区域。

燃油消耗

燃油消耗可直接分配到 4 个因素：推进推力、电源、液压源及用于气源的引气。这些油耗很大程度上取决于对这些量的需求。最小燃油消耗源自于对于这些能源消耗的权衡分析。如前文所述，燃油消耗直接影响直接运营成本 (DOC)。

排放

有毒物排放是现代飞机设计的一项主要考虑因素。它们主要由动力装置及辅助动力装置 (APU) 产生。

维修成本

总维修成本是直接运营成本 (DOC)(8.6 节) 的一部分。维修成本是分配到飞机各元件的首要因素。该成本的分配很大程度上取决于该元件的类型，例如，电气、电子、机械或结构元件。

载荷，冲击，振动

相对于大多数其他分配参数，载荷、冲击及振动对于飞机设计的依赖程度更高。因此，分析过程将确定如何将这些参数由结构传递到飞机各部分。该分析将确定它们如何影响不同的部件。该分配过程进而可用于限制源载荷，或者通过结构设计来限制载荷的传递。

parameters are transmitted through the structure to various parts of the aircraft. These analyses will determine how they affect different components. This allocation can then be used either to limit the source loads or to design the structure to limit the transmission of the loads.

Durability

A driving requirement in the commercial aircraft domain is durability. The question is how many take-offs and landings can the aircraft make before cracks begin to appear in the structure? This requirement is particularly important in the regional jet market. The reason this requirement is important is because regional jets make far more take-offs and landings than larger jets. Another reason this is important is because metal fatigue is notoriously difficult to predict. For composite structures the issue is delamination, which is also difficult to predict. Either way, the only path to having a durable aircraft is to use the best data and the best prediction methods available and then add a good safety margin.

The allocation method for durability is simple: it is that every component on the aircraft has to meet the same requirement as the whole aircraft. Of course, structural components will receive the most attention for this requirement.

4.8   Requirements Trade-Offs

Chapter 7 discusses design trade-offs; this is the most common understanding of trade-off. Requirements trade-offs are less familiar. For example, at the aircraft level, there are trade-offs to minimize the take-off weight (W_{to}) and the DOC. Another derived requirement, the drag coefficient of the aircraft, CD, results from minimizing these parameters. This process is called a requirements trade-off.

A second aspect of requirements trade-offs pertains to allocated requirements, as discussed above. We saw how to establish certain requirements by top-down allocation that is dividing a top-level parameter, such as dispatch reliability, into parts and allocating them to different elements as described in Section 4.7. The first cut at this allocation is just a guess based on estimates of past reliabilities of those elements. In subsequent passes you can refine those estimates by trade-offs. This process answers the question: What allocation of dispatch reliabilities will result in the highest total dispatch reliability? Varying the allocation may result in a higher level of dispatch reliability. This procedure is another requirements trade-off.

耐久性

商用飞机领域的一项重要需求是耐久性。这个问题是：在结构出现裂纹之前，飞机可以完成多少次起飞和着陆？该需求在支线喷气飞机市场尤为重要，因为相对于大型喷气机，支线喷气机要完成更多次的起降。另一个方面是因为对金属疲劳的预测极为困难。对复合材料结构而言，问题在于分层，它也难以预测。无论哪种结构，设计耐久性优秀的飞机的唯一方法是依据最好的历史数据，以及利用现有最好的预测方法，并且给予合理的安全余量。

针对耐久性的分配较为简单：即飞机每个部件都必须满足和整机同样的需求。当然，结构部件对这一需求最为关注。

4.8 需求权衡

第7章讨论了设计权衡分析，这是对于权衡的最常见理解。人们对需求的权衡不太熟悉。例如，在飞机级，应采用使起飞重量 (W_{to}) 及直接运营成本 (DOC) 最小的权衡。另一项衍生需求，即飞机的阻力系数 CD，源自于这些参数的最小化。该过程称为需求权衡。

需求权衡的第二个方面与前文所述需求分配相关。我们了解到如何通过自顶向下的分配，即将签派可靠性之类的顶层需求分解成几部分，并以 4.7 节所述方式将其分配到各个元件来建立特定需求。分配的第一步是基于这些元件既往可靠性进行估算。之后可通过权衡对这些估计结果进行细化。该过程要回答以下问题：哪一种签派可靠性的分配可使总签派可靠性最高？尝试不同的分配可能产生相对高的签派可靠性方案。该过程是另一种形式的需求权衡。

Requirements trade-offs and examples

Amazingly SE literature provides very little guidance and very few examples of how to perform requirements trade-offs Figure 4.3 shows in a conceptual way the reason that vertically flowed down requirements may be in conflict with requirements from other sources. In this diagram requirements are shown flowing down from a top-level system called System A to two lower-level subsystems called Subsystem A_1 and Subsystem A_2.

As was explained above, requirements can be flowed down from A to A_1 and A_2 by various means including allocation, derivation, and straight flow down. In addition, both A_1 and A_2 may be the recipients of requirements from other sources including interfaces, production, operations, and differing requirements from different stakeholders. Some requirements conflict simply because they were incorrectly allocated. Since these requirements are not flowed down, this figure shows them as horizontal entries. Hence, the flowed down requirements, shown as vertical and the horizontal requirements, may be in conflict since they come from different sources; there is a need for a trade-off between them. The nature of the trade-offs will be determined by the situation and the nature of the requirements.

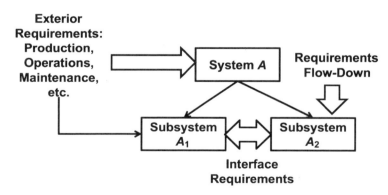

Figure 4.3   Sources of conflicting requirements

Linear addition

In some cases requirements from different sources may be added together. For example, the energy the brake pads need to absorb can be derived by a simple calculation which depends on primarily the weight of the aircraft, the landing speed, and the length required for stopping. Let's call that energy E_1. However, additional

需求权衡及示例

令人意外的是,系统工程文献中关于如何开展需求权衡的例子非常少。图4.3通过一种概念性的方法展示了垂直分解的需求与其他来源的需求相冲突的原因。图中来自顶层系统(称为系统A)的需求自上而下传递到两个下级的子系统,即子系统 A_1 和子系统 A_2。

如前文所述,需求可以通过不同的方法由 A 自上而下传递到 A_1,A_2,包括:分配、衍生及直接向下分解。另外,A_1 和 A_2 都可能是其他需求的接收方,这些需求的来源包括接口、生产、运行,且需求随利益攸关方的不同而不同。某些需求冲突仅仅是因为它们没有被正确地分配。由于这些需求并不是向下分解,图中它们以水平条目存在。向下分解的需求,在图中显示为垂直及水平需求,它们可能相互冲突,因为它们来自不同源头;有必要在它们之间进行权衡。权衡的性质视情况及需求的性质决定。

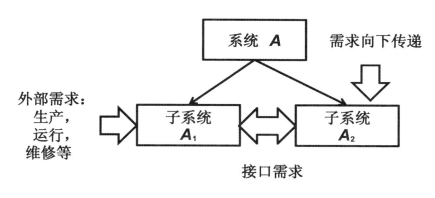

图 4.3　冲突需求的来源

线性相加

在某些情况下,不同来源的需求可以相加。例如,刹车片需要吸收的能量可以通过简单的计算得出,该能量主要取决于飞机重量、着陆速度及刹车距离,这里将该能量称为 E_1。然而,在潮湿跑道上着陆时该能量会增加,这里称之为

energy may be required for stopping on wet runways. Let's call that energy E_2. In addition, the customer may want to carry an additional amount of cargo. This will result in an additional amount of energy E_3. Hence, it may be assumed that these energy requirements may be added together resulting in a total energy $E_1 + E_2 + E_3$. This linear assumption may not always be valid, so the analyst will need to examine it carefully.

Modeling

In other cases the combination of requirements is so complex the only way to combine them is through a computer model of the system. Once again the example selected pertains to an aircraft brake system and the complex phenomenon of friction-induced instabilities as described by Hamzeh et al. (1999).

Requirements weighting

In some cases the correct requirement may be determined by examining how the conflicting requirements would affect the design based on a metric such as cost or weight. This technique would most likely involve a preliminary design based on the conflicting requirements. Such techniques, such as parametric cost analysis (PCA) could be used.

The QFD method described in Section 7.4 uses the weighting method. This method is particularly applicable when customer requirements are in conflict.

Absolute limits

Sometimes requirements are not tradable. Take maintenance requirements for example. The space required to repair or replace a component is limited. If this requirement is in conflict with another requirement, the space requirement will probably prevail. In general, though, when two requirements are in conflict and one of them is more restrictive than the other, the more restrictive one will prevail. Of all requirement types safety will most likely prove to be the most restrictive.

Reallocation

One of the primary methods of flowing down requirements is allocation. Many requirements are allocated, such as maintenance labor-hours, reliability, and weight. One way to resolve a conflict is to reallocate the requirements among the components they were allocated to.

E_2。另外，客户可能需要运输额外的货物量，这会导致额外的能量 E_3。因此，可以假设这些能量需求通过相加得到总能量 $E_1+E_2+E_3$。这种线性的假设并非总是可行，因此，分析人员必须仔细检查。

建模

在其他情况下，需求的结合十分复杂，唯一可行的结合方法是通过计算机对系统进行建模。这里可再次以飞机刹车系统，以及 Hamzeh 等 (1999) 所述的由摩擦导致不稳定的复杂现象为例。

需求加权

在某些情况下，可以基于成本及重量等指标，通过检查需求冲突如何影响设计来确定正确的需求。该方法极有可能涉及基于冲突需求的初步设计。可采用诸如参量成本分析 (PCA) 的一些方法。

质量功能展开 (QFD)(7.4 节) 方法应用了加权方法，该方法在客户需求发生冲突时更为实用。

绝对限制

某些需求不能进行权衡分析，以维修需求为例，维修及更换部件的空间受到限制，如果该需求与其他需求冲突，将以空间需求为主。一般来说，当两个需求发生冲突，且其中某一个限制性更强时，将以限制性强的需求为主。在所有需求类型中，安全性需求被证明是最具有限制性的。

重新分配

向下传递需求的主要方法之一是分配。很多需求通过分配进行处理，如维

4.9 Requirements Categories for Certification

The categories of requirements data required for certification are in agreement with the traditional SE categories but are broken out as shown below (1–8), as summarized from ARP 4754A (2010). The requirements list is comprehensive, showing that certification is concerned about the complete requirements formulation of an aircraft.

1. *Safety requirements* For certification purposes, safety requirements are identified separately because of their primary importance in the certification process. Safety is discussed more thoroughly in Section 10.2.

2. *Functional requirements* For certification, the term functional *requirements* refers to all requirements except safety requirements. These include customer requirements, operational, performance, installation and physical, maintainability, and interface requirements.

3. *Customer requirements* Customer requirements include both market-driven and specific customer requirements as described in Section 4.4.

4. *Operational requirements* These are the requirements associated with the interface between either the flight crew or the maintenance crew and the aircraft.

5. *Performance requirements* For the purposes of certification, performance requirements are in agreement with the SE definition discussed in Section 4.2.

6. *Physical and installation requirements* Many of the physical and installation aspects, for example, mounting provisions, will also appear as interface considerations, as shown below in Item 8. Others, for example, access, will be part of maintainability requirements as described in the following section.

7. *Maintainability requirements* Maintainability in a certification context is compatible with the SE view of maintainability discussed in Section 5.7. Although certification focuses on safety related maintainability, this category of requirements also includes economic related maintainability discussed in Section 8.6.

8. *Interface requirements* This category concerns the functional and physical interfaces which are part of the standard SE methodology as described in Sections 6.1 and 6.2.

修工时、可靠性及重量。在已分配了需求的部件之间重新分配需求是解决冲突的方法之一。

4.9 合格审定需求类别

如 ARP 4754A(2010) 所述，针对合格审定的需求数据类别与传统的系统工程类别一致，但是分为如下几种 (1-8)。需求列表是全面的，表明合格审定过程关注飞机需求构建的完整过程。

1. 安全性需求　就合格审定而言，安全性需求是单独进行定义的，因为安全性在合格审定过程中十分重要。在 10.2 节将更加详尽地讨论安全性。

2. 功能性需求　就合格审定而言，功能性需求包括除安全性需求以外的全部需求。包括客户需求、运行需求、性能需求、安装需求及物理需求、维修性需求及接口需求。

3. 客户需求　客户需求包括 4.4 节所述市场驱动需求及特定客户需求。

4. 运行需求　这些需求是飞行机组人员或维修人员与飞机之间接口相关的需求。

5. 性能需求　就合格审定而言，性能需求与 4.2 所述系统工程定义一致。

6. 物理及安装需求　很多物理及安装需求，也会以下文中第 8 项中接口需求的形式出现，如安装措施。其他需求，如可达性，将作为可维修性需求的一部分，在下文中进行讨论。

7. 可维修性需求　合格审定中的可维修性与系统工程中的可维修性相兼容。尽管合格审定着重于与安全性相关的维修，但这一类需求也包括与经济性相关的可维修性需求 (8.6 节)。

8. 接口需求　这一类关注功能及物理接口，它们是 6.1 及 6.2 节所述标准系统工程方法的一部分。

4.10   Requirement Validation

Finally, before we enter these requirements into a specification, we have to ask ourselves: Are these valid requirements? This process is called requirements validation, as distinct from system validation discussed in Chapter 11. There is a standard list of tests for requirements validity. ARP 4754A (2010, p. 54) provides such a list. This list elaborates on that list and provides some guidance of how to achieve some of the subtler aspects of requirements validity.

Verifiable

Above all, requirements must be verifiable by one of the four methods described in Chapter 11, namely, test, demonstration, analysis, or inspection. To meet this criterion the requirement must be physically possible.

Not vague or unambiguous

Vague and ambiguous requirements are some of the more common problems with requirements. For example the requirement, *the [item] shall be as efficient as possible,* is vague. This problem calls for a trade-off to determine exactly how efficient the item should be. The trade-off options in Section 4.8 should provide a solution to vagueness.

Explicitly stated

By *explicit* it is meant that the requirement must state exactly what system or subsystem must meet the requirement and what the system or subsystem must be required to do. The template for a requirements statement is often put this way:

The [system or subsystem] shall [perform the specified action] in the [specified environment].

There are several things to notice about this template:

- First, by *system or subsystem* the requirement may apply to any level of the system architecture.
- By *shall* the requirement is mandatory.
- By *specified action* the requirement performs some specified quantitative goal. This item should also include specific tolerances on the performance of the action.
- By *specified environment* the environment may be specified within the

4.10　需求确认

最后，在我们将需求加入规范之前，必须明确这样一个问题：这些是有效的需求吗？该过程称为需求确认，且不同于第 11 章所述的系统确认。针对需求确认有一系列标准检查单。ARP 4754A(2010，第 54 页) 提供了这样的一份清单。该清单描述详尽，提供了一些关于如何完成需求确认的工作指南。

可验证

首先，需求必须能通过第 11 章所述的四种方法之一进行验证，即试验、演示、分析或检查。要满足这些准则，需求必须实际可行。

避免不明确或含糊不清

不明确与含糊不清是需求中比较普遍的问题。例如，该 [项] 应尽可能高效这条需求就是不明确的。该问题需要进行权衡分析，以准确定义该项应该如何高效。4.8 节中的权衡选项提供了针对模糊的解决方案。

明确表述

明确意味着需求必须准确说明哪些系统或者子系统必须满足这些需求，以及这些系统及子系统必须完成的工作。需求表述的模板通常是如下形式：

[系统或子系统] 应在 [规定环境中][完成规定动作]

关于该模板须注意一些事项：

· 首先，"系统或子系统"是指需求可能应用于系统架构的任何层级

· "应"是指需求是强制性的。

· "规定动作"是指需求要完成一些定量化的目标。该项还应包括该规定动作执行的具体容差

statement, or it may reference an environment specification, such as a specification documenting the environment in various zones of the aircraft.

The use of this template will demonstrate that the requirement is identifiable as a requirement.

Achievable

If the implementation of a requirement requires the rewriting of the laws of physics, then this requirement needs to be rewritten. However, achievability may be limited by more practical aspects, such as the lack of space, time, or technological maturity of a component.

Justified by supporting data

This means that the appropriate analyses have been conducted with correct data. This test may apply to requirements that have mathematically derived from a higher level in the system hierarchy. Requirements with error of fact will not meet this criterion.

Not in conflict

Requirements may be in conflict if they emanate from different sources. For example, as discussed above, the requirements developed from flow down may be in conflict with an interface requirement. The solution to resolving conflicting requirements is though a trade-off as described in Section 4.8.

Necessary

It is not always easy to tell when a requirement is not necessary. The way to tell whether a requirement is necessary or not is to see what deleting it does to the design. If the design does not change when the requirement is deleted, the requirement is not necessary and can be deleted.

Not redundant

Like unnecessary requirements redundant requirements are not always easy to recognize if they are worded differently. However, the test is the same; if a redundant requirement can be deleted without changing the design, it should be deleted also.

"规定环境"是指在表达中可以规定环境要求，或者可参考环境规范，例如记录飞机各区域环境的规范文件。

使用该模板是该需求可以作为一项需求的辨识标志。

可实现

如果一条需求的执行需要改写物理定律，那么这条需求就需要更改。然而，可实现性会受限于更多的实际因素，如缺乏空间、时间或部件的技术成熟度。

经由支持数据论证

这意味着已用正确的数据进行了相应的分析。该测试可应用于那些来自系统架构较高层级经数学推导而来的需求。存在事实错误的需求将不满足该标准。

无冲突

来自不同源头的需求可能会发生冲突。例如，在前文中，自上而下分解得到的需求可能与接口需求相冲突。解决该冲突的方法是进行4.8节所述的权衡分析。

必要

分辨需求在什么时候并非必要，这并非易事。分析一条需求是否必要的方法是看删除该条需求后对设计有何影响。如果删除该需求后设计并无改变，则该需求不是必要的，可以删除。

非冗余

与非必要需求类似，如果表达方式不同，则不易辨别冗余需求。但是，测试的方法是相同的；如果某一冗余的需求可以被删除，且不改变设计，则该条应被删除。

How well, not what

For example the statement *a fan shall be installed* is not a valid requirement. It is a statement of a task. It does not say how well the fan should work or how much air it should move.

Understandable by supplier

A common problem is that supplier requirements are often written in terms a supplier cannot understand usually due to the physics of the situation. For example, the requirements may be written in terms of decibels, while the supplier only understands vibration. The requirement should be rewritten in term of vibration.

Valid source

The requirement should come from a valid source. Valid sources include the customer, regulatory agencies, and rigorous derivation from higher-level requirements.

Part of a complete set

The requirement when included in the larger set of requirements should form a complete set of requirements.

Complete and valid safety set

Are all requirements necessary for a safety analysis complete and valid?

Allow for flexibility and expansion

As described in Section 5.13, the requirements for new designs often allow for the design of derivative aircraft as defined in Section 2.2.

4.11 Avoiding Requirement Creep

Having read the previous sections of this chapter, readers may feel overwhelmed. The problem the readers will counter is the sheer number requirements. Every specification will have, for example, at least one performance requirement. On top of this there may be hundreds of unnecessary requirements in the tool. Deciding how to decide may be a more difficult task. Here is a simple scheme for deciding: Assign three people to the task of deciding which requirements or categories of requirements are in or out of the tool. The following three people are suggested:

"好"的程度，而非"什么"

例如，"应安装风扇"这条说明就不是一条有效需求，它是任务的说明，它没有说风扇应如何正常运转，或者它应让多少空气流动。

被供应商理解

一种常见的问题是，由于物理含义的原因，供应商需求术语经常不能被供应商理解。例如，需求可能以分贝表示，而供应商只能理解振动，因此，该需求应按照振动重新编写。

有效来源

需求应来自有效来源。有效来源包括客户、监管当局及由上层需求严格衍生而来的需求。

完整集合的一部分

当包含于较大一整套需求之中时，需求应构成一套完整的需求集合。

完整及有效安全性集合

安全性分析必需的所有需求是否完整且有效？

考虑灵活性及扩展

如 5.13 节所述，针对新设计的需求通常须考虑到衍生型飞机 (2.2 节) 的设计。

4.11 避免需求蔓延

阅读完本章前面部分后，读者可能会倍感压力。问题在于读者会遇到大量的需求，例如，每条规范至少对应一条性能需求，工具最顶端中可能存在数以百计的非必要需求。决定如何取舍将是一项艰难的工作。这里提供一种简单的方法用于决策。 决策工作可以由三类人员完成，该决策过程确定哪些或者哪类需求包含在工具中，哪些不包含在工具中。如下为建议的三类人员：

The Chief Engineer or Project Manager

More than anyone else, the chief engineer should have a comprehensive understanding of the technical aspects of a project and the specifications associated with the project.

A common practice for the Chief Engineer or Project Manager to do is to prepare a document with a name like Project Description. This is a very preliminary description of the project before any design work has begun. One of the features of this document is to describe the project in terms of the disciplines required. This document should aid all participants on the project, especially technical specialists in determining whether they have any role on the project or not.

The Technical Specialist

Certainly, if the component being procured has any software in it, the judgment of the software engineer will be necessary. The identity of any technical specialty will depend, to a great extent, on the nature of the component. If there are any electrical parts in the component, both the electrical engineer and the EMI specialist will be required. More importantly, if these components are *not* present, then the requirements for them can be omitted. This is what we are trying to accomplish: the elimination of requirements that are not needed and would clutter up our tool and specifications.

If any advanced technologies are involved, such as composite materials, the participation of specialists in those technologies would be essential. These specialists could then indentify associated specialists, such as production engineers, who could identify the production requirements associated with composite materials.

The Systems Engineer

Although systems engineers are normally associated with the *process* of SE, their broad familiarity with the many aspects of the design, across disciplines, make them particularly able to identify the particular requirements that are either necessary or unnecessary.

They are also adept at the methodologies for resolving conflicts between requirements, as described above.

总工程师或项目经理

总工程师对项目技术方面，以及与该项目相关的规范的理解应远比其他任何人全面。

一个常见的做法是，总工程师或项目经理编制名为**项目描述**的文件。该文件是在任何设计工作开始之前对项目的初步描述。该文件的特征之一是，以相关专业要求的形式描述项目。该文件应有助于该项目的所有参与者，尤其是技术专家确定其在该项目中应当发挥的作用。

技术专家

当然，如果采购的部件中包含软件，那么软件工程师的判断就十分必要。技术专家的身份很大程度上取决于部件的属性。如果部件中有任何电气零部件，就同时需要电气工程师及电磁干扰专家参与。更重要的是，如果没有这些部件，就会忽略它们的需求。这就是我们试图完成的：去除那些不需要的，或者会扰乱我们的工具及规范的需求。

如果涉及任何先进的技术，如复合材料，则该方面的技术专家的参与就十分重要。这些专家进而确定相关专家，如生产工程师，他们能确定与复合材料相关的生产需求。

系统工程师

尽管系统工程师通常与系统工程过程相关，但他们熟知设计中跨学科的许多问题，这使得他们能较好地辨别必要及非必要的需求。

如前文所述，他们对于解决需求之间冲突的方法也较为熟悉。

Summary of requirements creep adaptation

As will be seen in Chapter 13 "Adapting the Systems Engineering Process to the Commercial Aircraft Domain," adaptation is all about risk. That is, you want to gain the maximum effectiveness while at the same time minimizing the risk of deleting steps or processes. Requirements screening is one of those processes in which this principle is paramount.

Therefore question that the three requirements screeners, above, have to keep in mind is: can this specialty be ignored at this phase without causing any undue risks? They are the only ones who can make this judgment; however, it is a necessary step.

小结

正如第 13 章"在商用飞机领域应用系统工程"中将会介绍的，应用均与风险有关。就是说，在最大程度地降低删减步骤及过程的风险，同时保证最高效能。需求筛选是这些过程之一，在该过程中这些准则十分重要。

因此，以上三类需求筛查人员必须谨记的问题是：在该阶段，该专业能否在不产生不恰当风险的前提下被忽略吗？只有他们才能做出这些判断；无论如何，这是必要的步骤。

5

Constraints and Specialty Requirements

The term *constraints* refers to all non-performance requirements, that is, requirements which cannot be determined from functions. Whereas performance requirements establish how well a system or subsystem should perform, constraints define the limits on that performance. As we have noted, definitions differ among systems engineers. For example, MIL-STD-961D (1995) uses the term *constraints* for a limited set of design constraints. The present definition, though, is preferred for clarity and simplicity.

Specialty requirements are those requirements set by the various engineering specialties. These include, but are not necessarily limited to, human factors, reliability, maintainability, safety, environments, mass properties, and software. Because of their importance in the certification process, safety and software are discussed in Chapter 10. Other specialties include tooling, manufacturing, and facilities. Constraints and specialty requirements are discussed together here because many constraints arise from the engineering specialties. However, specialty requirements may be either performance requirements or constraints.

5.1 Regulatory Requirements

There are two primary sources of regulatory requirements: First are the civil aviation agencies, the US Federal Aviation Administration (FAA) and the Joint Aviation Authorities (JAA), the latter representing other, primarily European, countries. Other countries, such as Canada and Russia, also have their own aviation regulatory requirements. The primary interest of all regulatory agencies is the safety of the aircraft, crew, and passengers as discussed in Chapter 10. These agencies regulate the entire aircraft manufacture and operation system, including manufacturers, airlines, airports, and air traffic control.

The second source of regulatory requirements is the environmental requirements established by such agencies as the Occupational Safety and Health Administration (OSHA) and the Food and Drug Administration (FDA). These agencies are responsible

第5章　约束及专业需求

"约束"是指所有非性能需求，即那些不能根据功能确定的需求。性能需求决定系统及子系统性能的优劣，约束则定义了性能方面的极限。如前文所述，不同系统工程师对于名词定义有不同的理解。例如，MIL-STD-961D(1995) 将"约束"定义为设计约束的有限集合。出于清晰及简洁的考虑，本书采用该定义。

专业需求是指由各工程专家设定的需求。专业需求包括但不限于人为因素、可靠性、可维修性、安全性、环境、质量及软件。考虑安全性及软件对于合格审定过程中的重要性，第 10 章将专门讨论。其他专业还包括工装、制造及设施。由于许多约束来自于各工程专业，因此把约束与专业需求一并讨论。然而，专业需求可能是性能需求，或者约束。

5.1　监管要求

监管要求主要有两个来源：第一个是民用航空管理机构，如美国联邦航空局 (FAA) 及 (欧洲) 联合航空局 (JAA)，后者主要代表欧洲国家。其他国家，如加拿大及俄罗斯，也有各自的航空监管要求。监管当局主要关注飞机、机组和乘客的安全，如第 10 章所述。监管当局对整架飞机制造及运营体系进行监管，包括制造、航空公司、机场及空中交通管制。

监管要求的第二个来源是由职业安全与卫生管理局 (OSDA) 及食品药品监督管理局 (FDA) 等机构提出的环境要求。这些机构负责控制有毒物排放及其他可

for controlling toxic emissions as well as other hazardous materials which might be a threat to the passengers, crew, and the public as well as sanitary construction of airplane water systems and galleys.

Like all requirements, regulatory requirements are subject to verification in accordance with the principles of verification discussed in Section 11.3 of which certification is a major part.

The FAA (2012, p. 4) has provided a list of example regulatory documents. They are as follows:

- FAR 25.1301 "General Requirements for Intended Function." This document pertains to the installation and function of individual items of equipment. It states that each item must:
  - be of a kind and design and appropriate to its function;
  - be labeled as to its identification, function, or operating limitations, or any applicable combination of these factors;
  - be installed according to its limitations specified for that equipment; and
  - function properly when installed.
- FAR 25.1309 "Equipment Systems and Installation." The importance of this document is that it establishes the relationship between the severity of a failure and the probability of its occurrence.
- AC 20-152 "Invokes RTCA DO-254 Design Assurance Guidance for Airborne Electronic Hardware." This is a document providing guidance for the development of airborne electronic hardware.
- AC 20-115B "Invokes RTCA DO-178B Software Guidance." This document is used for guidance to determine if the software will perform reliably in an airborne environment.
- AC 20-174 "Invokes ARP 4754A (2010) Guidelines for Development of Civil Aircraft and Systems." This document is used as guidance for the certification of aircraft systems, meaning subsystems in the context of this book.
- ARP 4761 "Guidelines and Methods for Conducting the Safety Assessment Process on Civil Airborne Systems." This document is used for guidance in determining the compliance of aircraft systems with FAR 25.1309 (described above).

能对乘客、机组及公众及飞机水系统和厨房构成威胁的危害性材料。

与所有需求类似，监管要求需根据 11.3 节所述的验证原则进行验证，它是合格审定工作的主要内容。

FAA(2012，第 4 页) 提供了一份监管文件示例清单，包括：

- FAR 25.1301 **"预期功能通用要求"**。该文档描述各单项设备的安装及功能。每项设备均必须：

  —为适用其功能的种类和设计 ；

  —按其名称、功能，或使用限制，或这些因素任何适用的组合 ；

  —根据该设备规定的条件进行安装 ；

  —安装后工作正常。

- FAR 25.1309 **"设备系统及安装"**。该文件的重要性在于其建立了故障的严重程度及其发生概率之间的联系。

- AC 20-152 **"援引《RTCA DO-254 机载电子设备设计保证指南中应用》。"** 该文件对航空机载电子设备研制提供指导。

- AC 20-115B **"援引《RTCA DO-178 软件指南》。"** 该文件用作确定软件在机载环境里能否可靠运行的指南。

- AC 20-174 **"援引《ARP 4754A 民用飞机及系统研制指南》(2010)。"** 该文件用作飞机系统的合格审定指南，在本书中它属于子系统。

- ARP 4761《民用航空机载系统安全性评估过程指导方针及方法》。该文件被用作飞机系统符合 (上述)FAR 25.1309 要求的指南。

5.2   Mass Properties

The term *mass properties* refers to all parameters pertaining to the aircraft's weight and its distribution, such as weight, center of gravity (c.g.), and moment of inertia.

Weight

Weight is a requirement which generates much attention in the commercial aircraft industry. During preliminary design a weight goal is established for all future work discussed in Section 8.2. The weight of the aircraft is then broken down and allocated to the major subsystems shown in Figure 2.1. In this way, weight becomes a derived performance requirement since the weight is a critical factor in determining the range of the aircraft.

Weight may also be constrained by airport factors, such as the strength of the runway. Many airports limit the weight of the aircraft which may land. This type of weight limit is a true constraint.

Weight can be limited by transportation factors. For example, if a segment of the aircraft, for example, the fuselage or wing must be transported by rail, road, or air from a supplier location, it is often necessary to break the aircraft into segments for transportation as discussed in Section 5.12. The size of each segment may be limited by weight. This weight should be specified in the design of the segment. Military specifications, such as MIL-STD-961D (1995), often have special sections for transportation limits. However, if these limitations, such as weight, are taken into account in the weight limits, this section is not necessary.

The weight of the passengers and cargo is a performance requirement as discussed in Section 4.2. From this requirement and other performance requirements (range, speed, and so on), we calculate the total weight of the aircraft as described in Section 8.2. However, weight as a constraint sets the boundaries on the aircraft performance. As shown above, weight is also a factor in DOC and hence is used to manage DOC.

Weight can be expressed in different ways. Each method constrains the design in a separate way. Following are the principal weight parameters:

Manufacturer's empty weight (MEW)
MEW is the primary weight parameter used for design. It includes all system and

5.2 质量属性

术语"质量特性"指描述飞机重量及其分布的所有参数,例如重量、重心 (c.g.) 及转动惯量。

重量

重量是商用飞机工业中关注较多的一项需求。如 8.2 节所述,在初步设计阶段就需设定一个重量目标,以便开展后续的工作。然后将飞机重量分解并分配至如图 2.1 所示的主要子系统。由于重量是决定航程的关键因素,因此就成为一种衍生的性能需求。

重量亦会受机场因素的限制,如飞机跑道强度。许多机场对着陆飞机的重量进行限制。这种类型的重量限制是一种真实的约束。

重量亦受限于运输因素。例如,机身或机翼等飞机部段必须通过铁路、公路或者航空从供应商处运走,通常需要以 5.12 节所述方式将飞机分解成若干部段以便运输。每个部段的尺寸也会受限于重量。部段设计时必须考虑重量。军用规范,如 MIL-STD-961D(1995),通常会设立单独章节来说明运输限制。然而,如果在重量限制方面已考虑这些限制条件,那么这些章节就并非必需。

乘客及货物重量属于性能需求 (4.2 节)。依据该需求及其他性能需求 (航程、速度等),可以按照 8.2 节所述方法计算出飞机总重量。然而,重量作为一种约束,可以确定飞机性能的边界。如上所述,重量也是直接运营成本 (DOC) 的一个因素,并可用于控制直接运营成本 (DOC)。

重量可以通过不同方式表达。每种方法均通过单独方式约束设计方案。以下为主要重量参数 :

subsystem equipment items on the aircraft. It does not include the crew, fuel, cargo, luggage, expendable items, such as food, or crew items. MEW is the basic weight which can be broken down and allocated to the various subsystems.

Maximum take-off weight (MTOW)

MTOW is important from a performance point of view since it is a parameter in determining the take-off distance. It is also the parameter used to assure that the aircraft stays within the runway strength limits, as mentioned above.

Other weights

Other weights, whose significance is obvious by their names, are as follows:

- taxi weight
- landing weight
- zero fuel weight
- maximum jacking weight
- maximum towing weight.

Each type of weight should be specified and managed to control its particular aspect of aircraft design.

Center of gravity (c.g.)

The c.g. of the aircraft is one of two principal parameters which together determine the stability of the aircraft. The other parameter is the center of pressure (c.p.). The distance between the c.p. and the c.g. is called the static margin. If the c.p. is behind the c.g., the static margin will be positive, and the aircraft will be stable. If the static margin is negative, the aircraft will be unstable. It is our purpose here to show that the value of the c.g. is a parameter which can be established as a requirements constraint.

The center of gravity is the point through which the weight vector for the whole aircraft passes. The c.g. is determined by the weight distribution of the individual components. Thus, if the c.g. for the whole aircraft is set at a predetermined location, then the location of the individual components will be limited. If the location of components cannot be changed to meet the c.g. requirements, then ballast must be

制造空重 (MEW)

制造空重 (MEW) 是设计时的主要重量参数。MEW 涵盖飞机所有系统及子系统设备。该参数不包括机组、燃油、货物、行李、消耗品，如食物、或机组用品。MEW 是一个基本重量，可分解并分配至各个子系统。

最大起飞重量 (MTOW)

从性能角度而言，MTOW 是一个非常重要的参数，因为 MTOW 决定了起飞距离。MTOW 也作为保证飞机处于跑道强度极限内的参数，如前文所述。

其他重量

其他一些可以顾名思义的重量 (指标)，依据其命名可以直观看到其重要性，如下所示：

- 滑行重量

- 着陆重量

- 零燃油重量

- 最大顶起重量

- 最大牵引重量。

应规定并管理每种重量，以控制飞机设计的特定方面。

重心 (c.g.)

飞机重心是共同决定飞机稳定性的两个主要参数之一。另一个参数是压力中心 (c.p.)。重心及压力中心的距离称为静稳定性裕度。如果压力中心在重心之后，则静稳定性裕度为正值，飞机处于稳定状态。如果静稳定性裕度为负值，飞机处于不稳定状态。上述讨论意在说明重心值可作为需求约束的参数。

飞机重心是整架飞机重量矢量经过的点。重心由各个部件重量分布决定。因

added. Of course, ballast is always considered a last resort. A properly designed aircraft will need no ballast.

Moment of inertia

The moment of inertia is important in determining the forces required to roll or pitch the aircraft. For example, engines mounted nearer the fuselage will result in a smaller roll moment of inertia with a resulting requirement for less aileron torque to roll the aircraft.

5.3 Dimensions

Dimensions are any sort of dimensional limits which may limit the size of the aircraft or its subsystems. At the aircraft level, the wing span, for example, may be limited by airport facilities, such as the width of gates. At the subsystem level, the size of the equipment may be limited by the surrounding equipment or aircraft structure.

Like weight, dimensions may be limited by rail, road, or air transportation requirements. Dimensions also include dimensional tolerances, which is one requirement for the interchangeability of parts listed in Appendix 2.

5.4 Reliability

In the aircraft industry there are many requirements parameters included in the general category of reliability. Each of these may be specified in the requirements analysis. However, only two of these are considered important top-level requirements. These are dispatch reliability and operational reliability. The others are derived requirements: that is, they depend on some aspect of the solution for their value as discussed in Section 4.6.

Dispatch reliability is the pre-flight probability that the equipment will perform as specified within 15 minutes after being called upon to do so. Dispatch reliability is driven by the minimum equipment list (MEL), that is, the list of equipment which can be inoperable and still fly the aircraft safely.

The aircraft manufacturer establishes a master minimum equipment list (MMEL) from which the FAA reviews and approves. The airline selects a subset called the minimum equipment list (MEL). Hence, the airline may choose to delay a departure based on criteria more stringent than the FAA and tailored to the airline's airplane

此，如果重心被限定在预先确定的位置，则单个部件的位置将被限制。如果部件分布不能改变以满足重心需求，则必须增加配重。当然，增加配重总是作为最后考虑的解决方案，一架设计合理的飞机应无需配重。

转动惯量

转动惯量是决定飞机横滚及俯仰所需力量的重要参数。例如，如果发动机安装位置靠近机身，飞机转动惯量就相对较小，飞机横滚时所需副翼扭矩就较小。

5.3 尺寸

尺寸是限制飞机或其子系统体积大小的空间限制。在飞机级，例如翼展可能会受限于机场设施，如航站楼登机口宽度。在子系统级，飞机设备的尺寸受周边设备及飞机结构的限制。

和重量一样，尺寸也会受铁路、公路，或航空运输要求的限制。尺寸也包括尺寸公差，该公差是附录 2 所列部件的互换性相关的需求。

5.4 可靠性

在航空工业领域，一般的可靠性通用类别里包含很多需求参数。这些需求参数在进行需求分析时可单独予以规定。然而，(这些需求中) 仅有两种被视作重要的顶层需求，即签派可靠性及运行可靠性。其他需求则属于衍生需求：即它们的值依赖于设计方案的某些方面，如 4.6 节所述。

签派可靠性是飞行前，设备按指定要求在 15 分钟内运行的概率。签派可靠性由最低设备清单 (MFL) 驱动，也就是说，该清单里的设备不能正常运转时飞机依然能安全飞行。

飞机制造商制订出主最低设备清单 (MMEL)，经 FAA 评审并批准。航空公

configuration and operation.

Operational reliability is the probability that the aircraft completes its mission. Both dispatch and operational reliabilities are parameters highly valued by the airline customers as discussed in Section 8.5.

The following reliability parameters are also used in the design of the aircraft. However, they all are derived requirements:

1. *LRU MTBF*   The line replaceable unit (LRU) mean time between failures (MTBF) is the average time in LRU hours between confirmed LRU failures.
2. *LRU MCBF*   The LRU mean cycles between failures (MCBF) is the number of operating cycles between confirmed LRU failures.
3. *LRU MCBUR*   The LRU mean cycles between unscheduled removals (MCBUR) is the average LRU operating cycles between unscheduled LRU removals.
4. *LRU MTBUR*   The LRU mean time between unscheduled removals (MTBUR) is the average time between LRU unscheduled removals.

Although most engineers understand the SE principle of allocation for parameters which are additive, such as weight, it is much more difficult to understand the principle of allocation when it is applied to reliability. Some reliabilities are additive; others are multiplicative. Appendix 1 provides a description of the allocation of multiplicative parameters, such as reliability, and shows some examples.

5.5   Human Factors

The EIA 632 (1999) definition of SE states that the purpose of SE is to evolve "*people, product, and process*" solutions. The establishment of requirements for people is one of the most difficult areas of SE. This difficulty is due to the vast complexity of human behavior and the inability of analysts to understand and quantify all of the independent variables that influence behavior. Chapanis (1996) examines human factors in an SE context and shows that, in spite of these difficulties, SE does indeed provide a useful set of tools for the establishment of human factors requirements.

Two views of humans in the system

People requirements can be viewed in two ways: as part of a system or as an

司选择一部分建立最低设备清单 (MEL)。因此, 航空公司可选择比 FAA 更严格的, 且根据该航空公司飞机构型及运行定制的准则, 选择推迟起飞。

运行可靠性是指飞机完成任务的概率。如 8.5 节所述, 签派可靠性及运行可靠性都是航空公司客户高度重视的参数。

以下可靠性参数在飞机设计中亦有使用。然而, 这些参数均属于衍生需求 :

1. LRU MTBF 航线可更换单元 (LRU) 平均无故障间隔时间 (MTBF) 指两次确认 LRU 故障之间的平均时间。

2. LRU MCBF 航线可更换单元 (LRU) 平均故障间隔周期 (MCBF) 指确认 LRU 故障之间的操作循环次数。

3. LRU MCBUR 航线可更换单元 (LRU) 平均非计划拆换间隔 (MCBUR) 指 LRU 非计划拆换间的平均 LRU 操作循环。

4. LRU MTBF 航线可更换单元 (LRU) 平均非计划拆换间隔时间 (MTBBUR) 指 LRU 非计划拆换间的平均时间。

虽然大多数工程师能理解系统工程累加型参数的分配原则, 如重量的分配, 但是在应用于可靠性时, 理解这个分配原则就困难得多。一部分可靠性参数是相加的, 其他的则是相乘的。附录 1 提供了一些关于相乘参数 (如可靠性) 的分配描述, 并列举一些示例。

5.5 人为因素

根据 EIA 632(1999) 对系统工程的定义, 系统工程的目的在于提出关于"人、产品、过程"的解决方案。关于人的需求的建立是系统工程难度最大的领域。这种困难来自于人类行为的高度复杂性, 以及无法理解所有影响行为的独立变量并对其进行定量分析。Chapanis(1996) 从系统工程角度对人为因素进行了研究, 结果显示, 尽管存在各种难度, 但是系统工程的确提供了一套有效的工具来建立人为因素需求。

interface with an aircraft component.

First, if people are viewed as part of the system, then SE can lay a limited number of requirements on the people. These requirements usually concern the number, training, and sometimes physical characteristics of the flight and maintenance crews. Chapanis (1996), however, cautions against relying too much on the model of the human as a system component because of the large number of variables which may affect human behavior. Nevertheless, Satchell (1993) cites an FAA study which showed that 33 percent of aircraft incidents (were due to "deviation from basic operating procedures." This is why the Commercial Aviation Safety Team (CAST) (2011) has placed such a high priority on improving operational procedures. Furthermore, many of these deviations resulted from training deficiencies. Hence, assigning requirements to the human, such as training requirements, is an essential part of the requirements process.

Secondly, people can be considered to be an interface with the aircraft components. It is this second view in which human factors can establish requirements for the components. Even this view of human factors is not without difficulties. FAR requirements will say, for example, that the fire extinguisher must be "readily available." It is the role of the human factors specialist to transform this statement into a verifiable requirement in accordance with SE principles. The final requirement may be, for example, that the fire extinguisher must be "no more than six feet from the oxygen mask." Of course, this requirement will have to be established by a human factors analysis and verified by examination as discussed in Section 11.2. Finally, it must be pointed out that the true systemist would hold that the view of the human as a system component is the correct one since the system is not *complete* without the human.

The following sections will discuss the five human interface types: flight crew, cabin crew, maintenance, ground service crew, and passengers. For each of these types, the human factors specialist will transform the human factors considerations into valid, verifiable SE criteria:

Anthropometry and strength

Anthropometry and strength pertain to how the system should allow humans to *fit* into it. Humans must be able to reach, stand, or sit where appropriate. Flight deck and other controls should allow movement within human strength capabilities. In most cases, dimensions should consider the fifth to the ninety-fifth percentile male and female population. All emergency equipment and exits, lights, and communications should meet the anthropometry and strength requirements which would enable the

166

关于系统中的人的两种观点

人的需求可以从两种角度理解：作为系统的一部分，或者充当与飞机部件的接口。

首先，如果人被视为系统的一部分，那么系统工程只需涉及有限数量的人的需求。这些需求通常涉及人数、培训，有时是飞行及维修机组的身体特征。然而，Chapanis(1996) 针对过多依赖将人作为系统组成部分的模型发出警示，因为影响人的行为的变量数量极大。但是，Satchell (1993) 引述 FAA 一项研究表明 33% 的飞机事故源于"对基本操作程序的偏离"。这就是商用航空安全小组 (CAST)(2011) 如此重视改进操作程序的原因。另外，许多偏离是因为培训的不足。因此，人的需求，例如培训需求分配，是需求管理过程的重要工作。

其次，人可被视作与飞机各部件之间的接口。这是第二种观点，该观点认为人为因素能建立关于部件的需求。这种关于人为因素的观点同样面临困难。例如，FAR 要求灭火器必须"随时可用"。人为因素专家的职责是按照系统工程的要求，把这些意图转化为可验证的需求。比如说，最终的需求可能是这样的："灭火器离氧气面罩的距离不得超过 6ft"。当然，这种需求必须通过人为因素分析制订，并经 11.2 节所示方式进行验证检查。最后，必须指出的是，一个真正的系统学家会坚持，将人作为一个系统部件的观点才是正确的，因为离开人的系统是不完整的。

以下部分将讨论五种类型的人的接口：飞行机组、乘务组、维修人员、地勤人员及乘客。针对每种类型，人为因素专家把人为因素方面的考虑转化为有效且可验证的系统工程准则。

人体测量学及力量

人体测量学及力量定义人如何适应并融入系统。人必须能到达、站立或者

humans to use them effectively. For example, "the door handle shall allow operation with a force of 35 lb" is a clearly verifiable requirement.

Human factors analysis is critical in determining the reach and posture requirements for passengers, cabin crew, maintenance, and ground servicing. Three-dimensional computer models of both the human body and also the components to be maintained have been helpful in establishing these requirements. Finally, the study of human factors treats all aspects of passenger needs for controls, convenience, and comfort.

Vision

Flight deck vision requirements are concerned with the ability of the system to provide flight crew ability to see inside and outside the flight deck in accordance with FAR standards. In addition, the cabin crew should have a clear view of equipment and passengers both inside and outside the aircraft. Maintenance and service personnel should be able to see what they are working on. Passengers should be able to see relevant controls, indications, and safety equipment.

Ingress and egress

The human factors specialist has the responsibility for setting the requirements for flight crew ingress and egress during normal, abnormal, and emergency conditions. The flight crew should have egress through the window or flight deck door. Cabin crew should have egress through all normal and emergency exits. Maintenance and service personnel should be able to access all equipment and to reach service ports and stations. Passengers should be able to have ingress with carry-on luggage (under normal conditions) and have egress quickly (during abnormal and emergency conditions). These requirements should be specified in terms of the actual minimum physical dimensions and verified by examination.

Sound

The establishment of sound level requirements for human communication and stress management as discussed in Section 5.5 is an example of the necessity and value

坐在适当的地方。驾驶舱及其他控制装置应该允许在人力范围内的运动。大多数情况下，尺寸应考虑第 5 至 95 百分位的男女人群。所有应急设备及撤离通道、照明灯及通信设施必须满足人体测量学及力量需求，以便使人能有效地使用这些设备。例如，"舱门能用 35lb 的力操作"就是一条明确可验证的需求。

人为因素在确定乘客、客舱机组、维修及地勤的可达性及姿势需求时十分关键。对人体及需维修部件进行三维建模能帮助确定这些需求。最后，人为因素研究解决乘客在控制装置、便捷性及舒适性等所有方面的需要。

视景

驾驶舱视景需求关注系统按照联邦航空条例 (FAR) 标准要求，赋予飞行机组观察驾驶舱内外情况的能力。另外，客舱机组应能清晰观察到飞机内部和外部的设备及乘客。维修及地勤人员应能观察到工作区域。乘客应能观察到相关控制、指示及安全设备。

登机和离机

人为因素专家负责制订正常情况、非正常情况及紧急情况下飞行机组登机和离机的需求。飞行机组应能通过窗户及驾驶舱舱门撤离。乘务组应能通过正常及紧急出口撤离。维修及地勤人员应能够到达所有设备并抵达服务端口及站点。乘客应能携带随身行李登机 (正常情况下) 并能快速撤离 (非正常及紧急情况下)。这些需求应以实际的最小物理尺寸规定，并通过检查进行验证。

声音

针对人员通信及压力管理建立声级需求 (5.5 节)，这是体现系统工程方法必要性及价值的一个很好示例。该工作要求音量方面的限制通过各子系统分配到所有潜在声源，如风扇、空调导管、泵、发动机等。

of the SE approach. This task requires that limits on sound levels be allocated across many subsystems to all potential sound sources, such as fans, air conditioning ducts, pumps, engines, and so on.

All sound information, such as alerts and communications, should be within human range and sufficiently above background noise levels to be understood. This information is used to establish the noise level limits discussed in Section 5.9. Maintenance and service crews should be able to communicate with others as necessary. Passengers should be able to communicate with each other and with the cabin crew. Overall, the aircraft environment should be free of annoying noise or pulse patterns.

Touch

Touch requirements pertain to the physical interaction of the human with the equipment. Touch temperatures should be below the value suitable for the material: that is, the temperature at which a material can be touched by a human is different for different materials. All rough or sharp edges should be eliminated. Controls (handles, knobs, and so on) should be similar for commonality and dissimilar enough to reduce confusion. Handle directions and emergency equipment should be intuitive. In some cases, in the flight deck, controls should conform in shape to the associated aircraft control surface. Equipment should be designed to allow maintenance and service personnel to work by feel. Passengers should be free from the dangers of sharp edges and be able to distinguish controls.

Cognitive considerations

Cognitive requirements—general

In general, cognitive requirements pertain to the way humans receive and act on information. These requirements are most important in the flight deck. Information should be provided to humans in such a way that they can understand and handle it with an error rate below some specified level. For the cabin and flight crews, memory requirements should be minimal, emergency operations should be intuitive, and workload and error must be minimized. Cognitive requirements also pertain to the training and psychological requirements applied to cabin crew members to enable them to maintain control of passengers in a variety of situations, such as during an emergency evacuation.

170

所有声音信息，如警告及通话内容，均应在人的可承受范围内，并且足以超过背景噪声，使人可以理解。该信息用来确定噪声声级极限(5.9节)。必要时，维修和地勤人员应能与其他人通话。乘客应能够相互通话，或者与乘务组通话。总而言之，飞机环境应无恼人的噪声及脉冲干扰存在。

触觉

触觉需求关注人与设备的物理交互。接触温度应低于材料适宜温度值：即，对于不同材料，人可接触的温度是不同的。应去除所有粗糙及锋利的边缘。控制装置(把手、旋钮等)应具有相似的共性，且足够差异化以减少混淆。把手的方向及应急设备应直观明了和简便易用。在某些情况下，驾驶舱控制装置的形状应与相关的飞机控制面一致。设备应设计成使维修及地勤人员能依靠感觉工作。乘客应无接触锋利边缘的危险，且能辨别不同控制装置。

认知考虑

认知需求 - 概述

一般来说，认知需求关注人员接收信息并做出反应的方式。这些需求在驾驶舱最为重要。信息必须以可理解的、并能以低于规定误差率进行处理的方式提供给人。对于乘务及飞行机组来说，需要记忆的内容应尽量少，应急运行简便易用，工作负荷及误差也应尽可能少。认知需求亦关注适用于乘务及机组成员的培训及心理学需求，以使其能在各种条件下仍能管控乘客，如应急撤离时。

对于所有人(飞行机组、乘务组、维修人员、地勤人员及乘客)，各种设备均应简便而且直观易用。

认知需求 - 驾驶舱

驾驶舱是认知人为因素需求的重点。可通过人为因素分析确定一些环节的

For all personnel (flight crew, cabin crew, maintenance personnel, service personnel, and passengers), equipment should be simple and intuitive.

Cognitive requirements—flight deck

The flight deck is the focal point of cognitive human factors requirements. Human factors analysis determines the requirements for flight deck information processing, the levels of automation, situational awareness, resource management, and procedures. Normal, abnormal, and emergency conditions should be analyzed in accordance with the situational functions described in Section 3.3. Particular attention should be paid to avoiding design induced errors and maintaining pilot vigilance. Vigilance is the ability of observers to maintain their attention and be alert to stimuli over prolonged periods of time. One area of particular focus is the design of cockpits which will enhance the pilot's situational awareness so that he or she will not become dependent on the aircraft's automated systems but will remain aware of hazardous situations. Human factors specialists endeavor to determine the root cause of accidents and establish requirements which will prevent their recurrence.

Although most aircraft accidents are attributed to human error, this fact does not imply that pilots are always culpable or even poorly trained. It implies that an aircraft needs to be designed to allow the pilots to handle the situations in which accidents are most likely to occur. For this reason, the role of the human factors specialist is important and the use of SE principles will aid in the task of establishing the requirements to meet this challenge.

Satchell (1993), for example, provides an analysis of flight deck requirements pertaining to resource management to avoid human error. Satchell suggests several functional requirements for the alerting system. For example, the alerting system "must measure the vigilance of the operator." This step is in clear agreement with the SE functional analysis process. However, the exact requirements for this measurement and the criteria for verification rest with the human factors specialist. He or she should use the human factors tool box of professional expertise, analysis, historical data, and simulation to establish these criteria.

From an SE point of view, the Navigate Aircraft function discussed in Section 3.3 can be allocated either to the flight crew or to the equipment. Section 9.2 discusses the allocation of this function to the avionics segment and the synthesis of the avionics equipment. From a human point of view, the subfunctions are more complex. In addition to the direct functions, such as Provide Control Commands, some additional

需求，包括驾驶舱信息处理、自动化水平、情境感知、资源管理及程序。应依据 3.3 节所述场景功能描述分析正常、非正常及紧急情况。应特别注意防止设计诱导的错误并保持驾驶员警觉。警觉是指观察者保持注意力专注，并在较长时间内维持对激励保持警觉的能力。驾驶舱的设计是一个重点关注的环节，通过设计提高驾驶员的情境感知，使其保持对危害性情况的警觉，而不是依赖飞机自动系统。人为因素专家努力找到导致事故的根本原因，并建立需求以避免事故再次发生。

尽管大多数飞机事故可以归因于人为错误，但这并不意味着驾驶员处理不当或者培训不足。它意味着飞机设计应能使得驾驶员在最可能发生事故的情况下能掌控局面。基于这个原因，人为因素专家的角色非常重要，同时系统工程准则的应用可以帮助我们完成建立需求应对这个挑战的任务。

例如，Satchell (1993) 提出一项关于驾驶舱需求的分析，该分析专注于通过资源管理避免人为错误。Satchell 给出一些关于预警系统的功能建议。例如，预警系统"必须测试操作者的警觉性"。该步骤与系统工程功能分析过程明显一致。然而，关于测试过程的准确需求及验证准则留给了人为因素专家。他或者她应该利用包含专业技术、分析、历史记录及模拟功能的人为因素工具箱来建立这些准则。

从系统工程角度来看，3.3 节所讨论的**飞机导航**功能要么可以分配给飞行机组，要么可以分配给设备。9.2 节分析了如何将该功能分配至航电部件，以及如何集成航电设备。从人的角度来说，子功能更为复杂。除了直接功能如**提供控制指令**之外，下文将对其他一些附加的子功能进行分析。

一个针对驾驶舱的潜在子功能是**提供边缘化心理调节**。边缘化心理是一种复杂的心理状态，它是由驾驶员的角色从直接接触和控制飞机转变为系统监控者而导致的。Satchell 分析了边缘化心理的成因、主要的累加机理制及调节方法。边缘化心理是导致警觉下降的原因之一。

173

subfunctions are discussed below.

A potential subfunction to be treated for the flight deck is Provide Peripheralization Modulation. Peripheralization is a complex psychological state which results from a shift in the pilot role from direct contact and control of the aircraft to one of system monitor. Satchell discusses the causes of peripheralization, primarily increased automation, and ways to modulate it. Peripheralization is a contributor to the loss of vigilance.

Another possible subfunction is Manage Stress. Stress is an emotional state, either detrimental or beneficial, which may affect flight crew performance. Stress may be caused by various factors, called stressors, such as temperature and vibration. Stress may be caused by outside factors, such as personality and cultural traits. Some stress factors may be static, others dynamic. Some stress may be beneficial; that is, it may provide a certain degree of alertness. Stress may result in arousal, that is, a state of alertness following termination of the stressor or perturbing event.

An obvious subfunction is Manage Workload. Shafer (1987) defines workload as the number of things an operator has to do within any particular time period modified by their level of difficulty. Workload assessment is a major human factors effort. A major requirement is to reduce the workload during periods of high activity, such as flying at low altitudes, and during emergencies. Human factors specialists have attempted to quantify workload difficulty with a measure called the Bedford scale. This scale helps human factors specialists create workload requirements which approach the SE goal of verifiability.

Human-centered Automation

Billings (1997) discusses the concept of human-centered automation. The premise of this concept is that humans must be the focal point of all requirements in the design of automated flight deck and air traffic control (ATC) systems. These requirements will augment the requirements for the avionics segment described in Section 9.2. The result of Billings' study is a set of requirements for both the flight deck and ATC systems. A typical requirement is that "designers should keep human operators involved in an operation by requiring of them meaningful and relevant tasks, regardless of the level of management being utilized by them." This type of requirement may not be easy to verify or to synthesize in accordance with SE principles. However, its articulation is a major step towards assuring that human requirements become the focal point of flight deck and ATC design. Chapter 6 on Interfaces discusses Billings' rules more extensively.

另外一种可能的子功能是**压力管理**。压力是一种情绪状态，或者有害，或者有益，该状态会影响飞行机组的工作表现。压力可由称为压力源的多种因素造成，例如温度及振动。压力可能源于外在因素，如性格及文化特点。一些压力因素是静态的，另一些则是动态的。某些压力有时候是有益的，例如，会在一定程度上提高警觉。压力可能导致兴奋状态，即压力及干扰事件终止后出现的一种警觉状态。

工作负荷管理是一种典型的子功能。Shafer (1987) 给出了工作负荷的定义，即，依据困难程度修正的，任一特定时间内操作者必须完成的工作数量。工作负荷评估是人为因素分析工作的主要内容。一项主要需求是降低繁忙工作时，如在低空飞行期间，以及紧急状况下的工作负荷。人为因素专家试图用称为 Bedford 标尺的尺度量化工作负荷的困难程度。该标尺有助于人为因素专家产生接近系统工程可验证性目标的工作负荷需求。

以人为本的自动化

Billings(1997) 探讨了以人为本的自动化的概念。此概念的前提是，人必须是自动控制驾驶舱及空中交通管制 (ATC) 系统所有设计需求的核心。该需求将增加 9.2 节所述航电部分需求。Billings 的研究结论是一系列针对驾驶舱及 ATC 的需求。需求的一个典型的例子是"无论人类操作者所使用的管理强度如何，设计者应通过提供相关工作，保持其参与操作。"这种类型的需求不容易进行验证，或者不易按照系统工程的准则进行集成。然而，这种表达是朝着保证人的需求成为驾驶舱及 ATC 设计核心的方向迈出的重要一步。第 6 章中"接口"将对 Billings 规则进行更深入的讨论。

设备安全性

人为因素分析对于设备安全性至关重要。人为因素安全性分析用于解决设

Equipment safety

Human factors analysis contributes greatly to the equipment safety analyses. Human factors safety analysis addresses the potential safety hazards which equipment poses to humans. Areas addressed include headstrikes, decompression, seat design for crash loads, corners, edges, and walking surfaces. For maintenance and service personnel, other considerations, such as fuel sparks and electrical shocks, are considered.

Conclusions

No other specialty offers such difficulties and opportunities for improvement in aircraft design as human factors. Difficulties arise from both the complexity of human behavior and the large number of variables involved. SE offers an integrated and methodical approach to addressing human factors within an aircraft context.

5.6 Environments

This category includes all environments which every component of the aircraft must endure; such as temperature, pressure, shock, vibration, and so on. The systems engineer should assure that all requirements are satisfied under the appropriate environments *and* combinations of environments. This category does not include any phenomena which the item might emit. For specifications, it is not necessary to specify the environment for each performance requirement. It is only necessary to have a single requirement which says that "the system shall meet its performance requirements during and after exposure to the following environments." Or, alternatively, the requirement might say that "the system shall meet its performance requirements *following* (as opposed to during) exposure to the following environments." Hence, it is important to specify the exact operational requirement relative to the environment.

It is up to the systems engineer to make sure that the requirements are verified in the appropriate environments. Although environmental analysis is an area which has always been important in the design of aircraft, SE brings to the table a methodology which recommends a much more rigorous incorporation of environments into the requirements process. That is, it recommends a multidimensional evaluation of environments: by operational phase, aircraft zone, the degree of abnormality of operation, and geographical zone.

In all cases it is important to specify quantitatively the range of environmental

备对人造成的潜在安全性危害。研究领域包括头部撞击、失压、针对坠撞载荷的座椅设计、转角区域、边缘及行走表面。此外，还需针对维修及服务人员考虑燃油火花和电击。

结论

人为因素给完善飞机设计带来的困难和机遇远胜其他专业领域。困难来自于人类行为的复杂性及相关变量的众多。系统工程提供了一种完整且清晰的方法来处理飞机设计中的人为因素问题。

5.6 环境

本类别包括飞机各部件必须承受的全部环境因素，如温度、压力、冲击、振动等。系统工程师须确保在适当环境及环境组合下满足所有需求。本类别不包括任何放射现象。对规范而言，针对每条性能需求规定其环境并无必要。仅需有一条需求，即"无论是正在或者曾经暴露于如下环境之中，系统均应满足其性能需求。"或者，换言之，该条需求可以表述为"在暴露于如下环境后（相对于暴露期间），系统应满足其性能需求。"因此，有必要根据环境规定准确的运行需求。

确保所有需求都可在适当的环境下验证，这是系统工程师的职责。尽管环境分析始终是飞机设计的一项重要工作，但系统工程提供了一种分析方法，该方法将环境以更严格的方式纳入需求过程。即，该方法引入了多维度的环境评估方法：依据运行阶段、飞机区域、运行的异常程度及地理位置进行环境评估。

任何情况下，定量规定环境条件都十分重要。例如，温度的度数、每立方英尺沙尘中的颗粒数量等。这些需求大部分都可在标准规范中查到。

conditions, for example, temperature in degrees, sand in particles per cubic foot, and so on. Most of these requirements will be found in standard specifications.

Environments by phase

The critical feature of environments is that they constitute the conditions under which *all* performance requirements should be specified and tested. For example, as we have seen in Section 3.3, the environments should be established for each phase of aircraft operation, including maintenance and highway transport. For structural design, the FARs specify the conditions under which the critical environments occur.

Environments by zone

It is important to specify environments *by aircraft zone*: for example, in the passenger compartment, cargo compartment, wing, wheel-well compartment, and so on. Many environments are very zone dependent because they are attenuated by the aircraft itself. Hence, it will be impossible accurately to estimate the environments in those zones until after an initial concept is formulated. An example of such a requirement identified after concept development is noise which is greatly attenuated by the body of the aircraft. Zonal environments are derived since they depend on aircraft design for their values. Hence, zonal environments could be identified no sooner than the system design review (SDR) discussed in Section 12.4 and no later than preliminary design review (PDR) also in Section 12.4.

Environments by normal, abnormal, and emergency operation

Included in these functions are the normal, abnormal, and emergency aircraft operations. The functions force the requirements analysis to examine all environmental conditions any element of the aircraft may encounter. Cabin depressurization is an example.

Environments by geographical region

Environments for specific customers can often be more severe than for most other customers. In that case, it should be assured that the aircraft will withstand the

按阶段划分环境

环境的关键特点在于其构成了所有性能需求定义及试验的条件。例如,如 3.3 节所示,对飞机运行的各个阶段均应规定其环境,包括维修及高速公路上的运输。对于结构设计来说,FAR 规定了关键环境出现的条件。

按区域划分环境

按飞机区域规定环境也十分重要。例如,在客舱、货舱、机翼、轮舱等区域。很多环境因素与区域相关度很高,因为飞机本身会对其起到减弱作用。因此,在初步概念建立之前,很难对这些区域的环境进行精确评估。在概念开发后进行需求定义的一个例子是噪声,机体会对噪声起到极大的减弱作用。区域环境因其数值依赖于飞机设计,所以是衍生出来的。因此,区域环境在系统设计评审 (SDR) (12.4 节) 之后、初步设计评审 (PDR)(12.4 节) 之前予以明确。

按正常、非正常及应急运行划分环境

飞机正常、非正常及应急运行均包含在这些功能之中。这些功能迫使需求分析检查飞机任何元件可能遇到的所有环境条件,座舱减压就是其中一个例子。

按地理位置划分环境

对于一些特定客户而言,他们所面对的环境比其他大多数客户更恶劣。在这种情况下,比如说,应保证飞机能应对沙漠中长时间热渗透的环境,或者北极的长时间低温环境。同样,沙漠环境还会带来其他恶劣影响,如提高了沙尘的等级。地理环境属于飞机架构的顶层需求,因为它不依赖于设计。因而,进行系统设计评审 (SDR) 时,可能会介绍地理环境定义。

environments of long heat soaks in the desert, for example, or for long cold exposure in the Arctic. Similarly, the desert environment will have other severe aspects, such as increased levels of sand and dust. Geographical environments are requirements at the top level of the aircraft hierarchy since they do not depend on the design. Hence, geographical environmental definition would be presented at the SDR.

Environment types

Following is a list of environments which should be specified for an aircraft:

Temperature

For exterior components, the temperature environment will be determined by the altitude temperature profile and aerodynamic heating. In addition, on the aircraft there is considerable *induced* heating, for example, for components in the vicinity of the engine. Temperatures on the ground will be determined by the *heat soak*, that is, for aircraft which have been sitting on the runway for hours on a very hot day.

Pressure

For exterior components, the pressure will normally be determined by the atmospheric pressure and the aerodynamic pressure profile on the aircraft. For internal components, the pressure will be either in the pressure shell, or outside. Among the emergency operations is the decompression condition. During decompression, not only should pressure be specified, but also the pressure *rate*. The airframe's ability to maintain the cabin pressure is considered a performance requirement rather than a constraint.

Electromagnetic interference (EMI) and high-intensity radio fields (HIRF)

EMI is always an *induced* environment, that is, it is an environment produced by some other aircraft component, namely, electrical components. In addition, EMI can be produced by on-board equipment carried by passengers, such as portable computers. Electronic equipment is the primary type of component susceptible to EMI. Hence, the EMI environment is the environment for which each piece of electronic equipment should be designed. In addition, the equipment should be located no nearer than specified distances from electrical cables.

HIRF is similar to EMI except that it is an externally generated, normally ground-based, electromagnetic environment the aircraft may be subjected to.

环境类型

以下为应对飞机规定的环境清单：

温度

对于外部部件而言，其温度环境取决于海拔温度分布及气动加热。另外，飞机上存在明显的感应加热效应，例如，发动机附近的部件。地面上的飞机温度取决于热浸效应，一般出现在炎热天气里在跑道上停放数小时的飞机上。

压力

对于外部部件而言，正常情况下压力取决于大气压力及飞机的气动压力分布。内部部件的压力则是压力壳体内部或外部的压力。应急运行的情况之一是失压条件。在失压期间，不仅应规定压力，还应规定压力变化速率。飞机机体维持座舱压力的能力被视为性能需求，而不是一种约束。

电磁干扰 (EMI) 及高强度辐射场 (HIRF)

电磁干扰 (EMI) 始终是一种感应环境，它是其他一些飞机部件，即电子部件产生的环境。另外，乘客随身携带的机上设备，如便携式计算机也会产生电磁干扰 (EMI)。电子设备是最容易受到电磁干扰 (EMI) 干扰的设备。因此，所有电子设备都应该在设计时就考虑电磁干扰 (EMI) 环境。另外，这些设备离电子线缆的距离不应小于规定数值。

高强度辐射场 (HIRF) 类似于电磁干扰 (EMI)，只是它是飞机外部产生的，通常是基于地面的电磁环境。

冲击、振动及载荷系数

冲击是由较大冲撞，比如硬着陆或者地面冲撞的结果。规定冲击环境是因为

Shock, vibration, and load factors

Shock is the result of major impacts, such as hard landings or ground impact. Shock environments will be specified *as they are transmitted* by the aircraft structure. Similarly vibration can result from engine or auxiliary power unit (APU) operation or the result of aerodynamic vortices. Like shock, vibration will be transmitted by the aircraft structure. Thus, for both shock and vibration, the environment for each aircraft component will depend on its location in the aircraft and the phase of flight.

Shock and vibration can also occur during transportation to the final assembly location, during final assembly, and on rough ground taxiways and runways. Hence, these phases should also be considered in the determination of the shock and vibration environments.

Load factors are those loads which are not of a fast, impulsive nature. These are the g-loads experienced by the aircraft in flight and upon landing. Although shock, vibration, and other load factors may be applied to any component, they are the primary environments against which all structures are designed.

Lightning

All equipment is normally subject to the requirement to withstand the effects of single- and multiple-stroke lightning. These effects are most critical for fuel lines, electrical conduits, and other lines that pass through wing tanks. Line connectors that are not totally conductive are subject to sparking and may result in the inflammation of the vapors in the fuel tank. Newer aircraft suppress this effect by inserting nitrogen enriched air (NEA) into the vapor region.

The lightning requirement for individual connectors is normally of the form <The connector shall be spark-free when subjected to a current of TBD amps.> Of course, TBD (to be determined) will vary from connector to connector throughout the aircraft. The value of TBD can be determined by *deriving* it from the basic input charge for the whole aircraft, which may be, for example, 200,000 amps. This is an example of a derived requirement as discussed in Chapter 4.

This requirement can be verified by passing a current through a connector in a laboratory and observing a resulting spark, if any. If there is no spark, the requirement will have been met. This is an example of verification by test as described in Chapter 11.

Sand and dust

Normally, sand and dust are only encountered when the aircraft is on the ground.

它们的传递是通过飞机结构实现的。同样，振动是由发动机、辅助动力装置 (APU) 或气动涡流产生。与冲击类似，振动也会通过飞机结构传递。因此，对于冲击及振动，飞机每个部件的环境取决于其在飞机上的位置及飞行阶段。

在部件运输到总装地点途中、总装期间、在粗糙的滑行道或跑道上滑行时，也会产生冲击及振动。因此，在确定冲击和振动环境时也应考虑这些阶段。

载荷因子是指那些非快速、非冲击型的载荷。该载荷是指飞机在飞行及着陆时承受的重力载荷。尽管冲击、振动及其他载荷系数适用于所有部件，但它们是所有结构设计主要考虑的环境因素。

雷电

所有设备通常均须满足抵御单次或多次雷击影响这一需求。这些影响对于燃油管路、导线管道及穿过机翼油箱的其他管路最为重要。不完全导电的管路接头容易产生电火花，并导致油箱内燃油气化物燃烧。新一代飞机通过在蒸气区域注入富氮气体 (NEA) 的方法来抑制这种影响。

针对单个连接器的雷电防护需求一般采用以下格式 < 在某待定安培值的电流通过时，该连接器应无电火花。> 当然，这个待定安培值对于整架飞机各个连接器而言各不相同。该待定电流值可以根据全机的基本输入电荷导出，例如，有可能是 200 000 A。这是衍生需求 (第 4 章) 的一个例子。

该需求可以在试验室内进行验证，方法是使电流通过连接器并观察有无电火花。如果没有出现电火花，则证明已满足该需求。这是通过试验对需求进行验证 (第 11 章) 的一个例子。

沙尘

正常情况下，只有在飞机处于地面时才会遭遇沙尘。在起动、滑行或起飞时，沙尘只会影响发动机运行、飞机外表面及吸入的空气。在地面维修、存储或装载

During run-up, taxiing, or take-off, sand and dust are only likely to affect engine operation, the exterior of the aircraft, and the intake air. In a ground maintenance, storage, or loading condition, it is important to specify the sand and dust as they might affect cargo loading equipment. Hence, every phase of the flight profile is important. Additionally, if the support equipment is part of the aircraft *system* as discussed in Section 2.3, then the effect of the sand and dust on that equipment should be considered also. Furthermore, the effect on interior systems should be considered when the aircraft is on the ground with the doors open.

Sand and dust may enter the aircraft air conditioning system through the bleed air system especially when the aircraft is operating in a particularly severe environment, such as in a desert climate. In these instances extra filters are often added to the bleed air system.

Fungus

Aircraft materials should be free from damage by various types of fungus since many carriers operate in regions of the world where fungi are abundant. Lists of fungi are available through standard sources.

Corrosive environment

This pertains to any sort of toxic environment which the aircraft materials may come in contact with, including human perspiration, while the aircraft is being flown or maintained. More severe toxic environments include acids or hydraulic fluids. This category also includes salt spray. Corrosion protection is an FAA item for continued attention.

Solar radiation

Solar radiation is primarily an environment which is important when the aircraft is on the ground. Solar radiation includes both the spectral distribution and level of solar radiation the aircraft will be subjected to. It not only affects the exterior of the aircraft but also is the input heat load for calculating the maximum interior temperatures on the ground.

Humidity

Most aircraft components are designed to a humidity range from 0 to 100 percent.

条件下，规定沙尘条件十分重要，因为它们可能影响货物装载设备。因此，沙尘对于飞行包线的各个阶段都十分重要。此外，如果支持设备被视为飞机系统 (2.3 节) 的一部分，那么还应考虑沙尘对该设备的影响。当飞机停留在地面且舱门打开时，可能还应考虑沙尘对内饰系统的影响。

沙子和灰尘可能通过引气系统进入空调系统，尤其是当飞机在恶劣环境下运行时，例如在沙漠气候下，在这些情况下，往往需要在引气系统中增加额外的过滤设施。

霉菌

飞机材料不得因各种霉菌造成损坏，因为许多飞机在世界各地区运行，其中存在大量霉菌。霉菌清单可以通过标准数据库获取。

腐蚀环境

这涉及飞机在飞行或者维修时，飞机材料可能接触到的任何腐蚀性环境，包括人体汗液。其他更具腐蚀性的环境还包括酸或液压油。该类别环境还包括盐雾。腐蚀防护是 FAA 的持续关注项。

太阳辐射

当飞机停放在地面上时，太阳辐射是一个重要的环境因素。太阳辐射需求应包括飞机承受的太阳辐射谱分布及太阳辐射水平。太阳辐射不仅影响飞机的外部，也是计算飞机在地面上的最高内部温度时的热载荷输入。

湿度

大多数飞机部件设计时考虑的湿度范围是从 0 到 100%。

降水

降水需求规定了飞机必须承受的雨、雪、冰雹、雾及霰环境。降雪需求包

Precipitation

The precipitation requirements specify the environments for rain, snow, hail, frost, and sleet which the aircraft must withstand. Snow requirements include snow loads on the aircraft.

Foreign object debris (FOD)

FOD is the environment of objects which may be on the runways of various airports and may damage the aircraft, especially the engines.

Noise

This is the noise each aircraft component must *endure* as opposed to the noise it emits as explained in Section 5.9. This environment would be particularly important to components within the engine nacelle, for example.

Volcanic ash

The attention to volcanic ash has increased since the eruption of the Icelandic volcano Eyjafjallajökull in 2010. At a meeting of the International Volcanic Ash Task Force (IVATF) in 2010 the FAA presented a list of ways to mitigate the effects of volcanic ash. According to EasyJet (2014) following the eruption of 2010 development has begun on a device called AVOID (airborne volcanic object identifier and detector) that would detect volcanic ash and allow the pilot to avoid the volcanic ash area. Implementation of this device would be an example of employing the *drift correction* resilience principle described in Chapter 16.

5.7 Maintainability

The concept of maintainability includes both the constraints on the item to be maintained (the aircraft) and the requirements for the maintenance equipment (ground support equipment). Hence, maintainability spans the elements of the aircraft system.

SE brings two concepts to maintainability which promise to improve the quality of the aircraft system. First is the concept of verifiable requirements. Maintainability has traditionally been thought of as a discipline not amenable to verifiable requirements. We have seen that the role of the systems engineer is to convert the qualitative requirements into verifiable requirements. Secondly, SE helps build maintainability into the aircraft from the inception of the program.

括飞机上雪的荷载。

机场跑道异物 (FOD)

机场跑道异物 (FOD) 是可能出现在各类机场跑道上，会造成飞机损伤，尤其发动机损伤的外来物环境。

噪声

此处是指每架飞机的部件必须承受的噪声，而不是 5.9 节所述的，由它们本身发出的噪声。例如，噪声对于发动机短舱内的部件尤其重要。

火山灰

由于冰岛 Eyjafjallajökull 火山盖在 2010 年的爆发，对火山灰的关注有所增加。在 2010 年国际火山灰工作团队 (IVATF) 的一次会议上，美国联邦航空局 (FAA) 提出一系列减轻火山灰影响的方法。2010 年火山爆发以后，易捷航空 (2014) 开始了称为 AVOID(机载火山灰识别及探测装置) 的开发，该设备可对火山灰进行检测，并使驾驶员能避开火山灰分布区域。该装置的实施将是第 16 章中所述应用漂移修正恢复力准则的一个例子。

5.7 可维修性

可维修性的概念既包括对被维修对象 (飞机) 的约束，也包括对维修设备 (地面支持设备) 的需求。因此，可维修性概念覆盖了飞机系统的各个元件。系统工程将两个概念引入维修性范畴，这两个概念有望改进飞机系统的质量。首先是可验证需求的概念。可维修性传统上不被认为是可以按可验证需求处理的学科。我们已经知道，系统工程师的作用是把定性的需求转化为可验证的需求。其次，系统工程有助于从项目启动伊始就将可维修性纳入考虑范围。

Quantitative maintainability requirements

Many maintainability requirements are quantitative and are therefore verifiable.

1. *Maintainability cost*   The principal requirements parameter for maintainability is maintenance cost per 1 000 flight hours (MN\$/1 000FH). This parameter can be converted into maintenance man-hours per 1 000 flight hours (MMH/1 000FH). Material cost per 1 000 flight hours (MT\$/1 000FH) can also be separately specified.

2. *Fault isolation*   Specific and verifiable time limits can be established for both on-aircraft and off-aircraft fault isolation times. On-aircraft repair time is measured by the mean time to repair (MTTR). A basic maintainability requirement for both safety and economic failures is that "all failures shall be evident" or not affect safety, economics, or operations (for any subsystem or for the aircraft as a whole). *Evident* can either mean that there is an indication to the crew during the course of their normal duties (alert, light, and so on) or that it would be evident to an observer, such as to a person doing an aircraft walk-around during turnover.

Qualitative maintainability requirements

Qualitative maintainability covers many areas: accessibility, interchangeability, off-the-shelf components, overhaul, component wear, standard tools, handling equipment, support equipment, erroneous installation, contamination, identification tags and labels, repair procedures, corrosion. It is necessary to express all qualitative requirements in verifiable terms.

Safety-related maintainability

Safety-related maintainability is of primary importance. Many of the items discussed above, such as fault isolation requirements, are safety related. These items will be part of the certification data package.

Maintenance types

定量可维修性要求

许多可维修性需求都是可量化且可验证的。

1. 维修成本　可维修性需求的主要参数是每 1 000 飞行小时的维修成本 (MN$/1 000FH)。此参数可以转化为每 1 000 飞行小时维修工时 (MMH/1 000FH)。每 1 000 飞行小时材料成本 (MT$/1 000FH) 也可以单独规定。

2. 故障隔离　可对机上及离机故障隔离时间规定具体、可验证的时间限制。机上修理时间通过平均修理时间 (MTTR) 来衡量。针对安全性和经济性的一个基本可维修性需求是"所有故障均应是显而易见的",或不影响安全性、经济性或运行 (对于任何子系统或整架飞机)。明显表示在机组正常执勤过程中向机组发出指示 (警示、灯等),或指易于被观察者察觉,例如在飞机转场期间对飞机进行巡检的人员。

定性可维修性需求

定性可维修性涵盖了许多领域：可达性、互换性、货架产品、翻修、部件磨损、标准工具、搬运设备、配套设备、错误安装、污染、识别标记及标牌、修理程序及腐蚀。必须以可验证的术语表述所有的定性需求。

与安全相关的可维修性

与安全相关的可维修性最为重要。这类可维修性的很多内容在上文已经讨论过,例如故障隔离需求就是安全相关需求。这些可维修性是合格审定数据的一部分。

维修类型

在以下 6 种维修情况下会产生维修成本:过夜检查、A 检、B 检 (很少使用),

Maintenance costs are incurred during six types of maintenance: overnight checks, A checks, B checks (rarely used), C checks, unscheduled maintenance, and fixed interval checks. The overnight check is done when the aircraft is in service at any airport and consists of routine tasks, such as fluid checks. The A checks occur at predetermined intervals and consist of preventive maintenance tasks. B checks and C checks are at incrementally longer intervals than A checks. Each level of maintenance type is progressively deeper in its diagnostic inspections. Although it is a basic goal of maintainability to minimize the scheduled maintenance, it is never a practice to *require* scheduled maintenance. Unscheduled maintenance is driven by the MTBUR.

Fixed interval maintenance

Fixed interval maintenance is also undesirable and is maintenance mandated at intervals different from the regular scheduled maintenance. It is a maintenance cost driver and results from FAA mandated certification maintenance requirements (CMRs). A CMR is a required periodic task established during the design certification of the aircraft as an operating limit. It is intended to detect hidden faults, that is, safety-significant latent failures which would, in combination with one or more other specific failures or events, result in hazardous or catastrophic failure conditions.

The way to reduce the number of CMRs is to do a thorough fault tree analysis and to pass the system safety analysis (SSA). These steps may result in requirements for increased redundancy and reliability and a reduction in non-evident faults.

Accessibility

We saw above that accessibility was listed as a qualitative requirement needed to be turned into a verifiable requirement. There are two ways to do this: The first is to establish a time limit on the removal and replacement of an LRU. These limits can easily be verified by demonstration. The second is to establish, during the design phase, dimensional criteria for accessibility: for example, the space between the structure and an LRU should be at least TBD inches for removal and replacement. This requirement can be verified by examination.

Trade-offs between accessibility and reliability are a necessity. For example, if the MTBF of a component exceeds the planned life of the aircraft, then its accessibility is not so important. On the other hand, components with low or unknown reliabilities

C 检、计划外维修及定期检查。过夜检查一般是针对在任一机场使用中的飞机，且包括若干日常工作，例如液体检查。A 检按照预定时间间隔进行，包括预防性维修。B 检和 C 检的时间间隔相对 A 检而言逐步增加。可维修性的每个级别在诊断检查方面逐步加深。尽管尽量减少定期维修是维修的一个基本目标，但是定期维修并不是常规要求。计划外维修由平均非计划拆换间隔时间 (MTBUR) 驱动。

固定间隔维修

固定间隔维修也并非有利的维修类型，它指的是在不同于常规定期维修的时间间隔下强制的维修。固定间隔维修会导致维修成本增加，其来源是 FAA 强制性审定维修要求 (CMR)。CMR 是在设计合格审定过程中作为使用限制规定的要求定期完成的任务。其目的是检测潜在缺陷，即潜在的重大安全性故障，当它们与一个或多个其他特定失效或事件一起时，就可能导致危害性或灾难性失效状态。

减少 CMR 数量的方法是进行全面的故障树分析，并通过系统安全性分析 (SSA)。这些步骤可能会带来增加冗余度和可靠性的需求，以及减少不明显的失效的需求。

可达性

通过前文已知，可达性需求被视为定性的需求，且必须被转化为可验证需求。有两种方法可以做到这一点：首先是为某一 LRU 设定拆换和更换时限。这些时限可以很容易地通过演示验证。第二是在设计阶段，规定可达性的尺寸准则：例如，规定结构和某一 LRU 之间的空间应至少保持若干英寸，以便拆卸和更换。该需求可以通过检查进行验证。

在可达性和可靠性之间进行权衡十分必要。例如，如果一个部件的平均无故障时间 (MTBF) 超过飞机的计划寿命，则其可达性便不再重要。另一方面，对于可靠性低或可靠性未知的部件，其可达性要求则极其重要。

191

need to be extremely accessible.

Spares

The cost of spares at the time of purchase of the aircraft is a major factor in the purchase of aircraft. Reduction in the number and cost of spares would significantly aid in the sale of aircraft. The following items drive the number and cost of spares. If these items are addressed during the requirements phase, then SE can be a major help in the reduction of the number and cost of spares:

1. *MEL*   A goal of maintainability is to put all aircraft components on the MEL. A longer MEL will reduce the number of spares. The best way of getting an item on the MEL is to build redundancy into the system. This can be done by creating an up-front requirement for redundancy. Of course, all items cannot be redundant. Trade-offs of redundancy against cost and reliability are necessary. It is also possible to make a subsystem fault tolerant by assuring that minor failures do not render a larger system inoperative.

2. *MTBF, MTBUR, and MTTR*   It is also possible to specify the required values for these parameters, which can be built into the item. Another maintainability goal is to make MTBUR equal to MTBF, that is, to remove all LRUs only at the scheduled rate. There are two ways to do this: design foolproof fault detection procedures, and make built-in test equipment (BITE) 100 percent reliable.

3. *Non-interchangeability*   It is possible to specify the interchangeability of an item, either with other aircraft models or with other items on the same aircraft. This type of requirement is a standard part of specifications (Appendix 2). Interchangeability does not necessarily mean that interchangeable items are identical or made by the same supplier. Interchangeability can be achieved by assuring that their interfaces, both functional and physical, are identical (Chapter 6).

4. *Sole-sourcing*   This is an economic factor which is part of the supplier management process. It means the awarding of a contract to a supplier without a competition. Competition among potential suppliers will assure both lower aircraft costs and lower spares costs. The manufacturer should consider the possible consequences of sole-sourcing as part of the risk management process discussed in Chapter 15.

备件

购买飞机时，备件的成本是飞机采购的一个主要因素。减少备件的数量和成本将显著促进飞机销售。下列项目影响备件的数量和成本。如果这些项目在需求阶段处理，那么系统工程将在减少备件数目和降低备件成本方面产生极大作用：

1. 最低设备清单 (MEL)　可维修性的一个目标是在最低设备清单 (MEL) 中包含所有飞机部件。较长的 MEL 可以减少备件的数量。将某个项目列入最低设备清单 (MEL) 的最好方法是在系统中保持冗余度。这可以通过创建冗余度的前期需求来完成。当然，并非所有的项目都会有冗余，应针对成本及可靠性权衡冗余。另外，也可以通过保证小故障不引起较大系统不工作，以使一子系统成为容错的子系统。

2. 平均无故障时间 (MTBF)、平均非计划拆换间隔时间 (MTBUR) 及平均修理时间 (MTTR)　规定这些参数的要求值是可行的，这些数值将被纳入该设备中。另一个可维修性目标是使平均非计划拆卸间隔时间 (MTBUR) 等于平均无故障时间 (MTBF)，即，仅按照计划速率拆换所有航线可更换单元 (LRU)。有两种方法可以做到这一点：设计简单的故障检测程序，并确保机内检测装置 (BITE)100％可靠。

3. 非互换性　可以利用其他飞机模型或相同飞机上的其他设备规定某一设备的互换性。这种类型的需求是规范的标准部分 (附录 2)。互换性并不一定意味着可互换项完全相同或由同一供应商制造。互换性可以通过确保其接口即功能接口与物理接口两者的同一性 (第 6 章) 来实现。

4. 独家采购　这是部分供应商管理过程中的经济因素。这意味着给予供应商一份排除竞争的合同。潜在供应商之间的竞争将确保较低的飞机成本和备件成本。制造商应将独家采购可能带来的后果纳入风险管理 (第 15

5. *Number of line stations and transit time*   The aircraft manufacturer has little control over the number of airline hubs and the distance between them. Nevertheless, they can be considered to be part of the aircraft system in a larger sense. The airline should consider the effect of these factors on spares costs when establishing their line maintenance within their route structure.

Building maintainability into the aircraft

Maintainability is an area in which integrated product development (IPD) multifunctional teams are most valuable. This value results from the fact that an organization's maintainability department may be organizationally and physically separated from design engineering. The IPD multifunctional teams encourage the maintainability engineers to be an integral part of the requirements and design process described in Section 12.3. In this way maintainability requirements can be incorporated into the design before concepts are formulated. That is, in the SE process the design is not evaluated for maintainability after it has been formulated, but rather the maintainability requirements are incorporated with all other requirements.

5.8   Design Standards

This is another category which would include things that appear in company, industry, or regulatory standards, like design margins, prohibited materials, and so on. Companies normally have well-documented design standards in each design specialty.

5.9   Emitted Noise

This is the limitation on noise that the item is allowed to emit. It is not the noise environment described in Section 5.6. Many engineers get these two items confused. It is important to distinguish between what the item must do (limit its noise output) and what environment it must withstand (noise). We saw in Section 4.7 that the emitted noise limit can be allocated to the various aircraft elements. Of course, noise is highly frequency dependent. The noise limitations in each frequency band should be specified. The two primary noise emitters are the aircraft engines and the fuselage boundary layer. However, many other components are capable of emitting noise, such as air conditioning ducts and hydraulic pumps. Hence, noise emission limits should be laid on all components.

章) 过程。

5. 航线站点数量和运输时间　飞机制造商对于航空公司枢纽机场的数量及其之间的距离几乎没有什么控制力。然而，广义上可以将其视为飞机系统的一部分。航空公司在其航线结构中设立其航线维修站点时，应该考虑这些因素对备件成本的影响。

在飞机中构建可维修性

可维修性领域是集成产品研制 (IPD) 多功能团队最有价值的一个领域。其价值源于一个事实，即一个组织的维修部门从组织上和物理上可能与工程设计部门脱节。IPD 跨职能团队鼓励可维修性工程师成为 12.3 节所述需求及设计过程的一个不可分割的组成部分。通过这种方式，可维修性要求在概念形成之前就纳入设计。即，在系统工程过程中，不是设计形成后，而是将可维修性需求与所有其他需求整合后才进行可维修性评估。

5.8　设计标准

这是另一种类别，它包括出现在公司、工业或监管标准的内容，如设计余量、禁用材料等。各公司通常在每个设计专业都有成文的设计标准。

5.9　发出的噪声

这是允许设备发出的噪声的限制，它不同于噪声环境 (5.6 节)。很多工程师混淆了这两个概念。分辨这两个概念非常重要，即什么是设备必须做到的 (限制其噪声排放)，什么是设备必须承受的 (噪声)。在 4.7 节中，噪声排放的限制指标被分配到飞机各元件。当然，噪声与频率高度相关。因此，应对各频段的噪声限制进行规定。噪声的两个主要来源是飞机发动机及机身边界层。但是，其他部件也会发出

For noise from external sources, there are three possible ways of limiting the noise in the cabin: limiting the source (engine or boundary layer), insulating the cabin, or actively suppressing noise. Trade-offs will determine which of these is most effective.

5.10   Emitted Electromagnetic Interference (EMI)

EMI is a major consideration in aircraft design. All electrical components should be shown not to emit more than a fixed amount of EMI which may interfere with avionics operation.

This is the EMI that an item is allowed to emit and not the EMI that an item must endure. However, it is necessary to consider the emitted EMI constraint jointly with the EMI environment on the vulnerable component, normally an electronics item. Emitted EMI can be constrained, to a certain extent, by insulation. However, the electronics component should be, at the same time, protected by separation from the emitting cables and by shielding.

5.11   Cost

Cost is an important constraint, especially in the aircraft industry. We will see in Section 8.6 how the entire aircraft is designed with cost constraints in mind. In DFMA projects, for example, items are redesigned with a specific cost reduction goal in mind. However, if we design the aircraft *in the beginning* with DFMA in mind, that is, with specific cost constraints, then DFMA will never be needed. To do this, it is necessary to consider as a major cost driver the assembly costs, which are in turn, to a major extent, driven by the number of parts. Hence, parts reduction will be a major consideration in minimizing cost. We saw before how cost can be allocated to the various aircraft elements. Normally, in design-to-cost, the principal cost parameter used is recurring (unit) cost. However, the total direct operating cost (DOC) and non-recurring (development) cost can also be used.

噪声，如空调管道及液压泵。因此，噪声排放的限制应该涵盖飞机所有部件。

对于来自外界的噪声，可能有三种方式限制舱内噪声：限制噪声源（发动机及边界层），隔离座舱，或主动抑制噪声。通过权衡分析可以确定哪种方式是最有效的。

5.10　发射的电磁干扰 (EMI)

电磁干扰 (EMI) 是飞机设计中一个主要的考虑因素。所有电气部件均应表明其电磁干扰 (EMI) 水平低于一个可能会干扰航电设备工作的固定值。

此处关注允许某一设备发射的电磁干扰 (EMI)，而不是某一设备必须承受的电磁干扰 (EMI)。然而，针对易受影响的部件，通常是电子部件，有必要结合电磁干扰 (EMI) 环境考虑对发射的电磁干扰 (EMI) 的约束。通过绝缘，可以在一定程度上限制发射的电磁干扰 (EMI)。但是，同时还应通过与发射电缆隔离，并采用屏蔽保护电子组件。

5.11　成本

成本是一个重要的约束，尤其是在航空工业领域。本书 8.6 节将介绍如何在充分考虑成本的情况下设计整架飞机。例如，在 DFMA 项目中，为达到某个具体的成本削减目标，会对一些项目进行重新设计。然而，如果在设计伊始就考虑 DFMA，即，设定一个具体的成本约束，那么 DFMA 就无存在必要。要达到这个目标，就必须将装配成本视为驱动成本增加的主要原因，反过来说，该成本主要由零部件数量决定。因此，减少零部件数量是削减成本的主要工作方向。通过前文可以看到成本如何分配到飞机各元件。通常情况下，按成本设计时使用的主要成本参数属于重复（单位）成本。然而，也可使用总直接运营成本 (DOC) 和非重复（研制）成本。

5.12 Transportability

Transportability includes all the design limitations on an item from having to be transported. These limitations can be included in weight, dimensions, shock, and vibration. For example, if the fuselage is being assembled by a supplier and transported to the manufacturer's facility by truck, the size of the truck will limit the size of the fuselage segment. Therefore, we should design the fuselage to be transported in segments small enough to be transported by truck. Although transportability is not an issue of major concern in aircraft design. Nevertheless, transportability requirements should be considered whenever they apply.

5.13 Flexibility and Expansion

It is often a design requirement on systems, such as aircraft, to design them for growth. That is, it may be necessary to design them with features, such as a larger wing, which may be needed on *future* versions or derivative models described in Section 2.2. For example, if a future aircraft model is expected to carry a larger payload, then the current aircraft should be designed to meet the future aircraft's performance capability. The wings and propulsion system should be selected, for example, to meet the future range, speed, and cruise altitude requirements.

5.14 Producibility

EIA 632 (1999) states that SE evolves life-cycle solutions for "people, product, and *processes*." It is the process aspect that the producibility requirements address. If two airframe components, for example, the landing gear door and the wing, are expected to meet with a certain tolerance, then requirements should be laid on both the component design and the manufacturing process. First of all, the component mismatch requirements are derived partly from the Generate Aero Forces function. These requirements will establish the limits on the aerodynamic drag caused by the mismatch. A manufacturing analysis will establish the derived requirements for the tolerances in mismatch resulting from the tool design and assembly sequence allocated to both the door and wing assembly processes.

5.12 可运输性

可运输性包括从必须运输的角度出发，对某一项的所有设计限制。这些限制可以包括在重量、尺寸、冲击和振动方面。例如，如果机身由某一供应商组装，并利用卡车运输到制造商的厂房，卡车的尺寸将限制机身段的尺寸。因此，机身段应该设计成足够小，以确保能用卡车运输。尽管可运输性不是飞机设计主要关注的问题，然而，只要涉及运输，就应该予以关注。

5.13 灵活性及扩展性

针对发展的需要进行设计，这通常是系统(比如飞机)的一个设计需求。就是说，可能有必要在设计时保证它们具备某些特征，如一个较大的机翼，未来型号或者2.2节所述衍生型中可能需要它。例如，如果预计将来机型承载更大的商载，则当前飞机的设计应满足未来飞机的性能指标。例如，对机翼和推进系统进行选择，以满足未来的航程、速度及巡航高度方面的要求。

5.14 可生产性

EIA 632(1999) 指出，系统工程针对"人、产品及过程"提供全生命周期解决方案。可生产性需求解决的就是过程方面的问题。如果对于两个机体部件，例如，起落架舱门和机翼，希望它们符合一定的公差要求，那么应该在这两个部件的设计和制造工艺过程都规定需求。首先，从**生成气动力**功能部分地衍生出部件失配需求。这些需求将对由失配导致的气动阻力设定限制。制造分析将针对失配中的容差建立衍生需求，该失配现象源于工装设计及分配给舱门及机翼装配流程的装配工序。

6

Interfaces

The greatest leverage in system architecting is at the interfaces.

Eberhardt Rechtin

One of the goals of SE is *completeness*, that is, to make sure that every aspect of the system has been covered and incorporated into the design. An interface is a boundary between two system elements. Yet it is much more; interfaces are one of those completeness areas which, in traditional engineering, suffer from insufficient treatment, with resultant risk and possible harmful consequences. Alexander (1964) says that a good system is one with the fewest number of misfits. In addition, Rechtin (1991, p. 29) points to the importance of interfaces in the architecting of a system. But, of course, we know that any system should meet its top-level requirements as well, as discussed in Section 4.2.

The purpose of this chapter is to give some basic guidance regarding interfaces, especially pertaining to commercial aircraft, to capture the key aspects of interfaces, and to incorporate them into the design.

There are two types of interfaces: functional and physical. We should, therefore, understand each type of interface and show how interfaces fit into our design.

The importance of interfaces

Interfaces are important from two perspectives, first, a *holistic* perspective of a system, and secondly the system property of *cohesion*. Chapter 4 explains that the requirements for the elements of a system should be determined holistically rather than from a reductionist point of view.

The Vee model explained in Chapter 4 is a top-down, vertical determination of requirements. As pointed out in Chapter 4, this method, used alone, fails to determine the relationships between the elements. Interface requirements are the requirements

第6章 接 口

系统架构中最关键之处在接口。

Eberhardt Rechtin

系统工程的目标之一是完整性，即，确保涵盖系统的各个方面，并在设计中整合这些方面。接口是两种系统元素之间的交界面。但其意义远非仅限如此：接口属于完整性领域，然而在传统工程中未受到充分重视，因而导致风险及可能的不利影响的产生。Alexander(1964) 认为，不协调现象最少的系统才是好的系统。此外，Rechtin(1991，第 29 页) 强调了接口在系统架构中的重要性。当然，我们知道，任何系统必须同时满足其顶层需求，如 4.2 所述。

本章的目的是提供关于接口，尤其是商用飞机领域内接口的、捕获接口的一些关键要素，在设计中整合这个要素的指南。

接口分两种类型：功能接口及物理接口。因此，我们应该了解每种类型的接口，并表明如何将这些接口融入设计之中。

接口的重要性

接口的重要性可以从两方面来看，第一，系统的整体性，第二，系统的凝聚性。第 4 章表明，应该整体地，而不是简化地确定系统元素的需求。

第 4 章介绍的 V 模型是一种自顶向下、垂直化地确定需求的方法。正如第 4

between the elements. If there is a conflict between these two methods, a trade study must be conducted. Chapter 4 also explains requirements trade studies.

In addition, as explained by Hitchins (1993, p. 55) a basic property of a system is *cohesion*. This simply means that for a system to be a system there must be a relationship between the parts. Interfaces define that relationship between the parts.

6.1   Functional Interfaces

Functional interfaces are the most neglected type of interface. And, yet, they are the most important because they characterize the whole purpose, that is, the function of the interface. With the goal of *completeness* in mind, it is well to remember that there are at least two associated functions for every interface.

Since all performance requirements are traceable to functions as explained in Section 4.2, it follows that there will be at least two requirements associated with each interface. Figure 6.1 illustrates this idea:

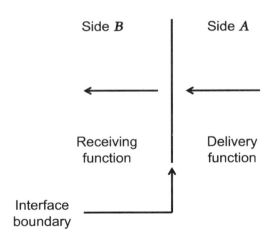

Figure 6.1   Interface functions

We see in this illustration that Side *A* delivers some quantity, say electrical power, to Side *B*. At the same time Side *B* receives that same quantity, electricity, from Side *A*. The two implied requirements are:

1. Side *A* must generate the quantity, and

章所指出的，单独使用该方法无法确定各元素之间的关系。接口需求是各元素之间的需求。当这两种方法发生冲突时，必须进行权衡分析。第 4 章对需求权衡分析也有阐述。

此外，正如 Hitchins(1993，第 55 页) 所强调的，系统的基本属性之一就是凝聚性。这意味着系统之所以成为系统，是因为各个部分之间必定存在关系。而接口则定义了各部分之间的关系。

6.1 功能接口

功能接口是最容易被忽视的接口，然而，它们也最重要的接口，因为它们体现了接口整个用途，即接口的功能。要体现完整论的思路，就应该知道，每个接口都至少有两个相关功能。

因为如 4.2 节所述，所有性能需求均可以追溯到一些功能，因此每个接口，至少有两个相关的需求。图 6.1 表达了这一思路。

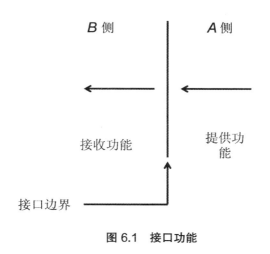

图 6.1　接口功能

如图所示，A 侧提供若干量 (如电源) 到 B 侧。与此同时，B 侧从 A 侧接收同样的量 (即电源)。其中包含两方面的需求：

1. A 侧必须生成这个量，以及

2. Side *B* must perform a function with the quantity that Side *A* delivered.

As we saw in Section 4.1, these requirements should be expressed in verifiable terms. For example, "the [Side *A*] subsystem shall deliver electrical power at 28 volts," as expressed in strict requirements terminology.

The quantity on Side *B* is known as an "antecedent," (Carson, 1995). That is, Side *B* performs a function *using the quantity provided by Side A*. For example, say the element of Side *A* is a fan. Then, the requirement for Side *B* might be: "The fan [Side *B*] shall provide 20 cfm of air, using the 29-volt power supply of the electrical system [Side *A*]."

Hence, we have developed a completeness paradigm. It is a check on the requirements analysis of Chapter 4. If the requirements analysis was performed on this hypothetical fan and failed to identify the two requirements we just identified, then the analysis was not complete or thorough. Now that we have reviewed what functional interfaces are, let's examine a few common ones in commercial aircraft. Remember, each of the parameters listed should be quantified and converted into a *bona fide* requirement as discussed in Section 4.1.

Electrical power interfaces

Electrical power is one of the most common interfaces on the aircraft. However, most electrical interfaces are internal to the electrical subsystem and, therefore, are not subject to the strict interface management that interfaces between elements are. However, the electrical subsystem does not "own" all the electrical elements. For example, the hypothetical fan, discussed above, may belong to the environmental control subsystem (ECS). The typical characteristics (functional interface parameters) of the electrical power are: voltage, current, alternating current (AC) or direct current (DC), power, phase, and high and low values of the power, voltage, and current.

Hydraulic power

Like electrical power, the hydraulic subsystem only has an interface with another element when the other element owns a component utilizing hydraulic power. The primary hydraulic characteristic is hydraulic pressure.

Pneumatic power

2. *B* 侧必须用 *A* 侧提供的这个量来执行某一功能。

如 4.1 节所述，这些需求应以可验证的方式表达。例如，"[*A* 侧] 子系统应提供 28V 的电源"，是一种严格按照需求术语表达的方式。

B 侧的量也被称为"前项"(Carson，1995)。即，*B* 侧利用 *A* 侧提供的量执行某一功能。例如，*A* 侧的元件是风扇。那么，对 *B* 侧的需求可能是："风扇 [*B* 侧] 应利用电源系统 [*A* 侧] 提供的 29V 电源提供 20ft³/min 的空气。"

因此，我们形成了一个完整性范例。这是对第 4 章所述需求分析的检查。如果对该假设的风扇进行需求分析，且未能明确我们定义的两种需求，那么分析过程是不完整或不彻底的。我们已经对功能接口做了回顾，现在将对商用飞机上的一些共同点进行研究。须明确的是，所列的每种参数均需量化，并转化为 4.1 节所述的真实需求。

电源接口

电源接口是飞机上最常见的一种接口。然而，大多数电气接口处于电气系统内部，因此，它们不属于严格意义上的元件间接口管理的范畴。但是，电子子系统并不"拥有"所有电气元件。例如，上面讨论过的假设的风扇，可能属于环控系统 (ECS)。电源接口的典型特征 (功能接口参数) 有：电压、电流、交流电 (AC) 或直流电 (DC)、功率、相位及功率值、电压值和电流值的大小。

液压源

与电源类似，只有在其他元件拥有使用液压源的部件时，液压子系统与此元件之间才存在接口。液压的最重要的特征是液压压力。

气源

气源子系统可以给环控子系统 (ECS) 提供空气，并保证舱内气压。气源也

The pneumatic subsystem may provide air to the ECS and provide cabin pressure. Pneumatic air can also be used to start the engines and to de-ice certain exterior components. Typical pneumatic characteristics include pneumatic pressure and temperature.

Mechanical forces and torques

Mechanical forces and torques constitute the interface functions between many aircraft elements. For example, the control system provides mechanical forces to move the control surfaces, such as ailerons and elevators. The engines provide mechanical torque to run the electrical generators. These are only two among many.

Conditioned air

Conditioned air passes across many boundaries: from the ECS ducts into the cabin, from the cabin into the lavatories, from the cabin into cargo area, from the air ducts into avionics racks, for example. The characteristics of the air are well-known: flow rate, temperature, humidity, quality (absence of particulates and ozone), and pressure.

Heat

Heat is a special case because it comes in two forms: desirable and undesirable. Desirable heat travels across the interfaces between heat exchangers and the fluids, either receiving or delivering the heat, for example. One example of desirable heat exchange is the cooling of engine bleed air.

It is just as important to capture and characterize the undesirable heat transfer across interfaces. This heat transfer is equivalent to a local adverse temperature environment. It is also advisable to cross-reference any adverse environments as part of the local environments in the environmental analysis described in Section 5.6. As an example, say that a component, a duct, for example, is attached to the engine by a brace. Depending on the material it is made of, this brace will transmit heat from the engine to the duct. Considering the high-heat environment of the engine, it is necessary to minimize this heat transfer from the engine to the duct so as not to cause damage to the duct.

Vibration, shock, and loads

可用于起动发动机,对一些外部部件进行除冰。典型的气源特征包括压力及温度。

机械力及扭矩

机械力及扭矩组成了众多飞机部件之间的接口功能。例如,控制系统提供机械力以移动控制面,如副翼及升降舵。发动机提供机械扭矩以驱动发电机。上述例子只是众多例子中的两个。

经调节的空气

经调节的空气流经众多边界:例如,从环控系统 (ECS) 管道进入座舱,从座舱进入盥洗室,从座舱进入货舱区,从空气管道进入航电支架。空气的特性众所周知:流动速度、温度、湿度、质量 (无颗粒及臭氧) 及压力。

热

热是一个特例,因为它有两种形式:有利的和不利的。例如,有利的热会穿过热交换器和各流体之间的接口,吸收或传递热量。有利的热交换的一个例子是发动机引气的冷却。

捕获和定义通过接口传递的不良传热也同样重要。这种热传递相当于一个局部的不利温度环境。如 5.6 节所述,在进行环境分析时,将任何不利环境作为局部环境也是可取的方式。例如,某一个部件,如一根导管,通过一根支柱连接到发动机。这跟支柱根据其组成材料,支架将热从发动机传递到管道。考虑到发动机的高热环境,必须尽量减少从发动机传递到管道的热量,以避免损伤管道。

振动,冲击和载荷

和热类似,振动、冲击及载荷一般会构成不利的接口。结构承受的气动力载荷不属于接口,因为空气不是系统元素。以前文所假设的连接到发动机的管道

Like heat, vibration, shock, and loads, in general, constitute undesirable interfaces. Aerodynamic loads on the structure are not interfaces since the air is not a system element. Take the hypothetical duct connected to the engine in the example above. The brace holding the duct will transmit vibration to the duct in the same way it transmitted heat. Hence, the vibration transmitted is a functional characteristic of the interface which needs to be recorded and incorporated into the design. Furthermore, vibration will be transmitted throughout the entire aircraft through the structure and transmitted to all components with which the structure interfaces. At each interface any transmitted vibration, shock, and loads need to be captured and analyzed. The characteristics of these parameters are well-known. The primary ones are magnitude and frequency.

Signal interfaces

Signal interfaces, also called information interfaces, carry the vast amount of information that is transmitted over the aircraft and between the aircraft and the ground, satellites, and other aircraft. Like electrical power, signal interface parameters include voltage and current, for example. In addition, various signal protocol properties should be specified. Example signals include warnings and fault detection signals from the various subsystems, such as the propulsion subsystem.

To facilitate the integration of aircraft avionics, the Society of Automotive Engineers (SAE) has developed a standard, AS4893 (1996), which defines a set of generic interfaces. The purpose of this standard is to increase the chance that components produced independently will have compatible interfaces. In addition, it provides a basis for commonality for both vendors and users of these components.

6.2 Physical Interfaces

The purpose of the physical interface analysis is to assure that the physical configurations of both system elements are compatible; that is, they fit together. Physical interfaces are one area that is traditionally well handled. However, mistakes do occur and improvement is needed. Two things are needed: First, rigorous SE documentation will assure that all physical aspects are considered. Secondly, new technological advances, such as the electronic developments fixture (EDF) will make it easier to capture the complete physical aspects of the interface. The EDF is discussed below. The interface control drawing (ICD) is the primary tool for recording the physical interfaces.

为例。固定管道的支柱将振动传递到管道，正如其传递热一样。因此，传递的振动是一种功能需求，需要记录并整合到设计之中。另外，振动会通过结构传递到整架飞机，并传递到所有含有结构接口的部件。需要捕获和分析每一接口处传递的振动、冲击及载荷。这些参数的特征是众所周知的，其中最主要的是幅值和频率。

信号接口

信号接口，也称为信息接口，携带大量的信息，这些信息在整架飞机内以及飞机与地面，卫星与其他飞机之间传递。信号接口与电源类似，例如参数也包括电压和电流。此外，还应规定各种信号协议性能。典型的信号包括由各个子系统，如推进子系统发出的警告及故障检测信号。

为便于进行飞机航空电子设备的集成，汽车工程师协会 (SAE) 制定了一项标准，AS4893(1996)，其中定义了一组通用接口。该标准的目的是促进独立生产的部件接口的兼容。此外，它还为这些部件的供应商和用户的通用性打下基础。

6.2　物理接口

物理接口分析的目的是确保两个系统元素的物理构型相互兼容；也就是说，它们能配合在一起。物理接口是传统上处理较好的领域。然而，错误难以避免，改进十分必要。以下两件事情确有必要：首先，严格的系统工程文档管理将确保考虑到所有的物理方面。其次，新的技术进步，诸如电子研制样机 (EDF) 将使捕获接口完整的物理属性更加容易。接口控制文件 (ICD) 是记录物理接口的主要工具。

6.3 External Interfaces

Many aspects of external interfaces are often the most neglected areas of aircraft design. In the context of the entire aircraft external interfaces include communications, service, maintenance, and facilities. In the context of a subsystem or other aircraft element, *external* means external to that element. For example, the external interfaces to a lavatory include the electrical connections needed for lights and fire detection, the plumbing interfaces for water and waste, the air interfaces for environmental control, and the physical interfaces with the surrounding aircraft structure. However, we will focus on interfaces external to the aircraft here.

Service

Aircraft service interfaces are one area for potential neglect since they are often not considered at the beginning until the aircraft design is well under way. Service interfaces include food service, cleaning, cargo loading, fuel loading. In general the service equipment exists and is not subject to redesign for a specific aircraft. However, service equipment may vary widely among airlines and airports. It is advisable to have a complete physical and functional description of all service equipment used by the airline customers and airports. Hence, it is necessary to take all of these variations into consideration. In addition, as part of the design of new types of aircraft, such as the high-speed civil transport (HSCT), the radical changes in configuration may impose new service requirements with resulting impacts on both the aircraft and the service equipment.

Maintenance

The conclusions for maintenance interfaces are generally the same as those for service above. In short, it is necessary to consider the interface of maintenance equipment with the aircraft the same as any other interface. Like service, airlines vary greatly in the maintenance equipment they use. It is desirable to obtain the characteristics of this maintenance equipment because it will affect the design. This list will include all the types of equipment used to test the aircraft, all the heavy-lift equipment, such as fork-lifts, and small tools. All of this equipment constitutes equipment with which the aircraft will interface.

6.3 外部接口

外部接口的许多方面往往是飞机及设计中最容易被忽视的领域。在整架飞机的意义上，外部接口包括通信、服务、维修及设施。而在子系统或飞机其他元件的意义上，外部表示该元件的外部。例如，与盥洗室的外部接口包括灯及火警探测，水及废水的管道接口，用于环境控制的空气接口，以及与周围飞机结构的物理接口。但是，我们将专注于飞机外部的接口。

服务

飞机服务接口是一个潜在的被忽略环节，因为它们往往不是从一开始就考虑的，直到飞机设计开始后才考虑它们。服务接口包括食品服务、清洁、货物配载及加油。一般来说，服务设备业已存在，并且无须针对特定机型重新设计。然而，航空公司及机场的服务设备可能各不相同。明智的做法是，针对航空公司客户及机场使用的所有服务设备，有一份完整的物理及功能的说明。因此，有必要将所有的不同考虑在内。此外，伴随新型飞机,如高速民用运输机 (HSCT) 的设计，构型方面巨大的变化可能会提出一些新的服务需求，最终会对飞机及服务设备双方都产生影响。

维修

维修接口与上述服务情况类似。总之，必须像其他任何接口一样考虑维修设备接口。与服务类似，航空公司使用的维修设备差别很大。获取维修设备的特性十分必要，因为它会影响到设计。这份清单将包括所有用于试验飞机的设备，所有的重物提升设备，如叉式升降机及小工具。所有这些设备均为该飞机接口的设备。

Facilities

Facilities are also a neglected interface area. The passenger access tunnel, commonly called the jetway, interfaces with the aircraft on the ground. It is imperative to consider this fact in both the design of the aircraft and the jetway so that passengers are loaded and loaded optimally and so that no damage is done to the aircraft during that operation by the jetway.

6.4 Internal Interfaces

We have already discussed the types of internal interfaces (to the aircraft) in Section 6.1. The main idea regarding internal interfaces is that all external interfaces at any level of the aircraft hierarchy are internal interfaces at the next higher level. This is a very important and practical idea. However, in order to make sense, it should be combined with the concept that each level of the aircraft hierarchy has an owner. For example, consider the diagram of Figure 6.2.

This figure shows three levels of the aircraft system hierarchy. At the top level the aircraft interfaces with the maintenance system at interface "*A*." This interface is *external* to the aircraft itself. At the second level the ECS interfaces with the electrical subsystem at interface "*B*." Interface "*B*" is, then, *internal* to the aircraft but *external* to both the ECS and the electrical system. This tiering of external and internal interfaces continues to the third level and Interface "*C*" which is internal to the ECS but external to the air supply and the air distribution subsystems.

The importance of this external/internal interface distinction has to do with the level ownership as discussed above, and with the verification of the requirements associated with each interface. Say, for example, that the aircraft program manager owns the top level, the ECS manager owns the ECS, and a supplier owns the air supply. We also learned in Chapter 1 that system verification is bottom-up. All this means that, first, the air supply would be integrated with the air distribution before it is integrated as part of the ECS. The ECS owner would have responsibility for assuring that the level 3 verification is successful. Similarly, the aircraft owner would assure that the level 2 verification is successful. Hence, verification and integration rise to the top with the ownership hierarchy like bubbles in champagne.

设施

设施也是一个容易被忽视的接口。乘客登机通道，俗称登机桥，与地面上的飞机交接。飞机及登机桥的设计必须考虑该因素，以便以最高效的方式使乘客上、下飞机，且在登机桥操作时不会造成对飞机的损伤。

6.4　内部接口

我们已经在 6.1 节讨论了内部接口 (与飞机的) 的类型。与内部接口相关的最重要思路是，飞机架构的任何层级的外部接口在其上一个层级都属于内部接口。这是一个非常重要和实际的概念。然而，为了便于理解，应该与这个概念一同考虑，即飞机架构的每一层级都对应一个所有者。例如，考虑图 6.2 所述情况。

该图显示了飞机系统架构的 3 个层级。在最顶层，飞机与维修系统在接口 "*A*" 处交接。该接口是飞机本身的外部接口。在第二层，环控系统 (ECS) 与电气子系统在接口 "*B*" 处交接。接口 "*B*" 处于飞机内部但处于 ECS 和电气系统外部。外部和内部接口的这种分层继续到第三层级，接口 "*C*" 处于 ECS 内部和气源和空气分配子系统外部。

外部 / 内部接口区别的重要性与前文所述的层级归属有关，也与各接口关联的需求的验证有关。例如，飞机项目经理拥有顶层级，ECS 经理拥有 ECS，供应商则拥有气源。由第 1 章可知，系统的验证是自下而上展开的。所有这些都意味着，首先，在作为 ECS 的一部分进行集成之前，气源必须先与空气分配系统进行集成。ECS 拥有者有责任确保第三层级通过验证。同样，飞机拥有者需确保第二层级通过验证。因此，验证和集成随着归属权层次上升至顶端，就像香槟中的气泡一样。

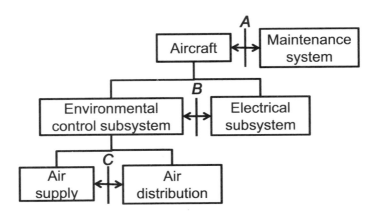

Figure 6.2   Interfaces in the aircraft hierarchy

6.5   Operational Interfaces

Operational interfaces have to do with the interfaces between any two elements of the system, either the aircraft system or external systems, during the operation of the aircraft. These interfaces can be either external or internal.

External interfaces

One of the primary external operational interfaces is communications. Communications are one of the few areas in which no physical interface exists. However, this fact does not diminish the importance of the functional aspects of external communications. The requirements for external communications with airport towers, other aircraft, and other ground nodes are generally well-known and documented.

Other operational interfaces according to the *FAA Systems Engineering Manual* (2014, pp. 57, 67) are the external interfaces, such as navigation and air traffic control (ATC). Factors to be determined through these interfaces are traffic density, flight phases, route configuration, visual flight rules (VFR), and instrument flight rules (IFR).

Human-automation interfaces

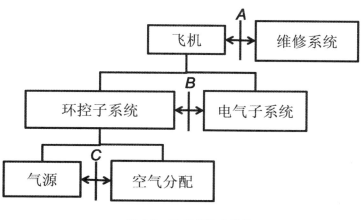

图 6.2 飞机架构中的接口

6.5 运行接口

飞机运行期间，运行接口与系统，无论是飞机系统或外部系统任何两个元件之间的接口有关。这些接口可以是外部的，也可以是内部的。

外部接口

主要的外部运行接口之一是通信。通信是少数几个没有物理接口的领域之一。但是，这个事实并不影响外部通信在功能方面的重要性。与机场塔台、其他飞机及其他地面节点之间的外部通信需求通常为大家熟知，并已形成文件。

根据《FAA 系统工程手册》(2014 年，第 57,67 页)，其他运行接口是外部接口，如导航和空中交通管制 (ATC)。拟通过这些接口确定的因素有交通密度、飞行阶段、航线配置、目视飞行规则 (VFR) 及仪表飞行规则 (IFR)。

人 - 自动化接口

最全面的一套人 - 自动化接口规则由 Billings(1997，第 237 至 246 页) 编制。

One of the most comprehensive sets of rules for human-automation interfaces is the one compiled by Billings (1997, pp. 237–246). Following is a summary of those rules:

- the human operator must be in command;
- to command effectively the human operator must be involved;
- to remain involved the human operator must be appropriately informed;
- the human operator must be informed about automated systems behavior;
- automated systems must be predictable;
- automated systems must monitor the human operator;
- each agent in an intelligent human–machine system must have knowledge of the intent of the other agents;
- functions should be automated only if there is good reason for doing so;
- automation should be simple to train, learn, and operate.

6.6   Interface Management

The very same owners at each level discussed above have another role. That is, one group, department, or company is responsible for assuring that all the features of its side of the interface are correct; that is, each party has designed into its side all the features to make the interface work properly. Typical owners are the manufacturer and the supplier. It is the responsibility of the interface management function to assure that this ownership principle is in place, that both owners recognize their responsibilities, and that they have incorporated the interface characteristics into their design.

It will be a key function of SE management as described in Section 12.1 to schedule and organize the interface meetings between the two owners of the sides. The product of the interface meeting is a simple agreement, sometimes called a *scope sheet*, which does the following:

1. describes the hardware owned by each side;
2. describes the functions received or delivered by each side;
3. obtains the signed concurrence of both sides to the agreement.

The scope sheet can be a simple document of no more than a single page to achieve its purpose.

以下是这些规则的摘要：

- 必须由操作人员发布命令；

- 为有效指挥，必须由人操作；

- 为保持操作人员的参与，必须对其适当告知；

- 操作人员必须了解自动化系统的运行；

- 自动化系统必须是可预测的；

- 自动化系统必须监控操作人员；

- 智能人－机系统里每个代理必须对其他代理的意图有所了解；

- 只有有充分的理由，才允许功能自动化运行；

- 自动化过程应简单、易于培训、学习及操作。

6.6 接口管理

上述每个层级的所有者同样都有另外一项职责。就是说，一个小组，部门或公司都有责任确保其接口端的所有功能都是正确的；也就是说，每一方都已将所有特性融入其一端的设计之中，以确保接口工作正常。典型的所有者是制造商和供应商。接口管理功能的职责是确保有这样一种所有权原则，即，这两个所有者均认识到自己的责任，并将接口特性整合到其设计中。

安排和组织两侧所有者之间的接口会议是系统工程管理关键的一项关键功能（如 12.1 节所述）。接口会议的产物是一份简单的协议，有时被称为一个范围表单，其作用如下：

1. 描述每一端拥有的硬件；

2. 描述每一方接收或提供的功能；

3. 促成双方同意签署该协议。

该范围表单可以是一份不超过一页纸的能达到其目的简单文件。

6.7   The Interface Control Drawing (ICD)

The ICD is the basic product of the interface task. In its final form the ICD is a clear documentation of the interface definition. It shows clearly which items belong to each side. It shows the exact values of the interface functions and their tolerances. However, what may be surprising to some is that the ICD is not used to design any part of the aircraft. That information is on the drawings of each component involved in the interface. The requirements associated with the interface functions are located in the performance sections of the specifications. Hence, the ICD serves as a good coordination document.

6.8   Development Fixtures (DFs)

A development fixture (DF) is a mock-up of the aircraft used during development to assure that the spatial allocation for all components is correct and that they fit correctly. Traditionally, DFs were hard; that is, they were actual physical models of the aircraft, made of wood and other materials. As engineers developed each component or subsystem, they would place a physical mock-up of the part into the DF.

Recent computer technology has made possible the use of the electronic development fixture (EDF). That is, engineers create a three-dimensional model of the entire aircraft on a computer. The EDF makes possible the allocation of space for a component by reserving a zone for it. In spite of large company investments in computers and training required for EDF capability, the benefits are enormous. First, the EDF results in a significant reduction in lead time for development. Secondly, the cost savings resulting from the elimination of the hard DF will go a long way towards offsetting the EDF costs.

EDF has a significant role in the SE interface process. First and foremost, the EDF model will become an integral part of the ICD. Each party to the interface will use the EDF, rather than the traditional two-dimensional drawing, to control the physical interface. Hence, the EDF will add significant value to the interface process.

6.7 接口控制文件 (ICD)

接口控制文件 (ICD) 是接口管理工作的基本产品，ICD 最终形式是对接口定义的纪录。该文件详细说明各个项目分别属于哪一侧，并显示接口功能的准确值及其允差。然而，比较特殊的是，ICD 并不用来设计飞机的任何一个零部件。这些信息包含在该接口涉及的各个部件的图纸之中。与接口功能相关的需求属于规范的性能部分。因此，ICD 是作为一份重要的协调文件而存在的。

6.8 研制样机 (DF)

研制样机 (DF) 是一种用于研制阶段的飞机模型，以确保各部件空间定位及配合准确。传统上，研制样机 (DF) 是实体的，就是说，它们是真实的飞机实体模型，由木材或其他材料制成。当工程师开发部件及子系统时，可将该部件的实体模型放入研制样机 (DF) 内。

现代计算机技术的发展使电子研制样机 (EDF) 变为可能。即，工程师在计算机中建立一个飞机的三维模型。电子研制样机 (EDF) 通过预留空间的方式给各部件分配区域。尽管公司在计算机及电子研制样机 (EDF) 培训方面的投资巨大，但该技术的效益十分可观。首先，电子研制样机 (EDF) 大幅缩短开发过程的交货时间。其次，长远来看，取消实体研制样机 (DF) 节省下来的成本会抵消电子研制样机 (EDF) 的成本。

电子研制样机 (EDF) 在系统工程接口过程中起着非常重要的作用。首先，电子研制样机 (EDF) 模型将成为接口控制文件 (ICD) 的必要部分。接口的每一方均使用电子研制样机 (EDF) 而不是传统的二维图纸控制接口。因此，电子研制样机 (EDF) 对于接口过程意义重大。

6.9   The N^2 Diagram

The N^2 diagram illustrated in Figure 6.3 is a useful tool in SE. It assures that all the functions identified in the functional analyses are reflected in functional interfaces. Each node in the N^2 diagram indicates a possible functional interface. Notice that in the example the Provide Electrical Power and Provide Environmental Control provide power and cooling to the other functions. However, the Navigate Aircraft function does not provide any quantity to the other functions.

In addition to being a useful tool, the N^2 diagram is one of the easiest to implement. Almost any word processing or spreadsheet program can be used to generate it. No special applications are required.

6.10   Interface Requirements

Interface requirements are the requirements associated with the interface functions discussed in Section 6.1 and the physical interfaces discussed in Section 6.2. The certification process requires as discussed in Section 4.9, Item 8 that interface requirements be provided as part of the requirements documentation portion of the certification data package.

	Provide electrical power	Provide environmental control	Provide guidance & navigation
Provide electrical power		Provide power to ECS	Provide power to avionics
Provide environmental control	Cool electrical components		Cool avionics
Provide guidance & navigation			

Figure 6.3   N^2 diagram

6.9 N² 图

图 6.3 所示的 N² 图是系统工程中一种有用的工具。利用该工具可以确保功能分析中定义的所有功能均能在这些功能接口中反映出来。N² 图的每一个节点都表示一个可能的功能接口。值得注意的是，在这个例子中**提供电源**功能和**提供环控**功能给其他功能提供电源和制冷，但是，**飞机导航**功能并不给其他功能提供任何量。

N² 图不仅是一种有用的工具，而且是最容易实施的工具之一。几乎任何文字处理及电子表格程序都能生成它，而无需专门的应用程序。

6.10 接口需求

接口需求是与接口功能 (6.1 节) 及物理接口 (6.2 节) 相关的需求。合格审定过程 (4.9 节、第 8 条) 要求把接口需求作为需求文档的一部分加入合格审定数据包。

	提供电源	提供环控	提供导航
提供电源		给环控系统（ECS）提供电源	给航电系统提供电源
提供环控	冷却电气部件		冷却航电系统
提供导航			

图 6.3 N² 图

6.11   Interface Verification

As shown above in the discussion of internal interfaces, it is not possible or meaningful to verify interfaces, as such. Rather, the correct step is to verify the requirements associated with the interfaces. In addition, during system build-up, it is necessary to verify the functionality of the lower-level elements as a part of the verification of the higher-level elements, as illustrated in Figure 6.2.

6.11 接口验证

如上文内部接口部分所讨论的，对接口进行验证既不可能又无必要。正确的做法是对与这些接口相关的需求进行验证。另外，在系统构建过程中，必须把验证下层级元素的功能性视为上层级元素验证工作的一部分，如图 6.2 所示。

7

Synthesis

Aircraft synthesis is the actual act of designing the aircraft or a segment of it. In the previous chapters we have said nothing about designing the aircraft. That is the basic point of SE: namely, that the functions, performance requirements, and all the constraints will have been so well defined that the design will now be much easier.

We have also seen that the first steps in aircraft system synthesis began when the aircraft architecture described in Section 2.3 and the aircraft system functions in Section 3.2 were defined. Synthesis also occurs when a requirement is allocated (definition 2—Glossary) to an element of the aircraft architecture. The final step occurs when the architecture, functions, and requirements are converted into a design. Hence, synthesis is a collection of steps which occur throughout the SE process. Figure 7.1 illustrates the synthesis process through the path between the hierarchies of functions, requirements, and component architecture.

It must be remembered though that the hierarchical path of Figure 7.1 is only part of the process. The vertical process shown in this chart is the *reductionist* path as explained in Section 4.3. The inclusion of requirements from other sources, such as operations, maintenance, and so forth are required to make the synthesis process *holistic*.

System synthesis has traditionally been considered the domain of the design engineer as distinct from the systems engineer. It is a common misperception that systems engineers overly constrain the design and stifle the creativity of the design. On the contrary, it is the job of the systems engineer to define the system goals and the conditions under which the system must operate so that the designer is free to create the best system possible. Ultimately, though, the ideal state is that SE principles will be so well understood by all engineers that this distinction will no longer be relevant and that SE can be used to enhance creativity rather than stifle it.

It is not the goal of this chapter to present the detailed steps and data to design an aircraft. Such sources as Corning (1977) do an excellent job of that. It is our goal to present the basic aircraft synthesis parameters and to answer the questions: Which parameters are performance requirements? Which parameters are constraints? And

224

第7章 综 合

飞机综合是设计飞机或飞机部段的实际活动。在前面章节中，我们没有提及任何关于飞机设计的内容。这是系统工程的一个基本观点：如果功能、性能需求和所有约束均能够被清晰完整地定义，则设计将极为简单。

同时我们可以看到，飞机系统综合始于飞机架构(2.3节)及飞机系统功能(3.2节)定义之时。当某一需求被分配(定义2—术语表)到飞机架构中某一元件时，也存在综合活动。最终步骤发生在架构、功能及需求转换为设计之时。因此，综合是贯彻整个系统工程过程的一系列步骤。图7.1展示了贯穿于功能架构、需求及部件架构之间的综合过程

必须注意的是，图7.1中的层次化路径只是这个过程的一部分。图中的纵向过程是4.3节所述的简化。其他来源的需求的引入，比如运营、维修及其他需求，将会使综合过程变为整体。

传统上，系统的综合工作被认为属于设计工程师、而不是系统工程师的领域。一种常见的误解是，认为系统工程师经常过度地约束设计，从而扼杀了设计的创造力。恰恰相反，定义系统的目标及系统必须运行的条件正是系统工程师的工作。这样，设计工程师将会有创造可能的最佳系统的自由度。最终，理想状态是，系统工程原则能被所有工程师充分理解，两者之间的区别变得不再重要，且系统工程将用于增强而不是扼杀创造力。

本章的目标不是呈现设计一架飞机的详细步骤及数据，这些在Corning (1977)等一些文献中已经介绍得非常详尽。我们的目标是介绍飞机综合的基本参数，并

225

which parameters are design solutions subject to trade-off? Chapters 8 and 9 will answer these questions. This chapter discusses synthesis in the overall SE process.

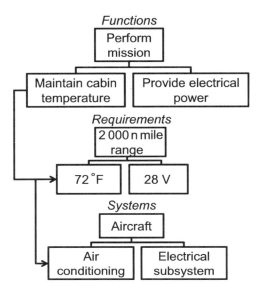

Figure 7.1   The synthesis path

Introducing Holism in the synthesis process

We saw in Section 4.3 that a key principle of SE is *holism*, that is, the creation of a system as a whole and not by looking at the individual parts. The latter is called *reductionism*. If the designer looked at only the requirements flowed down to the individual parts using the Vee model described in Section 4.3, the result would be a reductionist process rather than a holistic process. To obtain a holistic view of the system, the designer needs to look at the requirements from different sources that may be in conflict with the flowed down requirements. These requirements may be from production, operations, maintenance, or any of the other sources discussed in Chapter 4, for example human factors. Section 4.8 also provides several ways the designer can conduct trade-offs among these conflicting requirements. In the end the designer will have requirements for the individual components now holistically determined and not reductionist.

回答如下的问题：哪些参数是性能需求？哪些参数是约束？哪些参数是需要权衡的设计方案？第 8 章和第 9 章将回答这些问题。本章讨论整个系统工程过程中的综合。

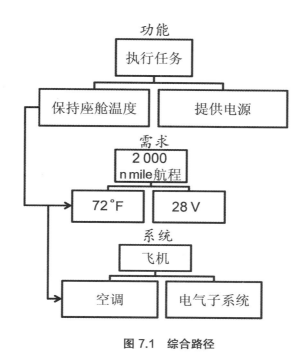

图 7.1　综合路径

介绍综合过程中的整体论

在 4.3 节我们可以看到，系统工程的基本原则基于整体论，即创建作为一个整体的系统，而不是只看个别部分。后者称为简化论。如果设计者只关注使用 V 模型 (4.3 节) 自上而下传递到个别零部件的需求，其结果将是一个简化过程而不是整体的过程。为了获得一个系统的整体视图，设计师需要考虑来自不同来源的需求，而这些可能与自上而下传递而来的需求有所冲突。这些需求可能来自于制造、运行、维修或第 4 章讨论的其他来源，比如人为因素。4.8 节同时也提供一些方法，供设计师在相互冲突的需求间进行权衡研究。最后设计师将获得现在是整体决定的、关于单独部件的需求，而不是简化的需求。

7.1 Aircraft Architecture

We have seen in Section 2.3 that one of the first steps in synthesis is the creation of the aircraft architectural hierarchy shown in Figure 2.1. For a new aircraft in Section 2.2, it is necessary to create and refine the hierarchy throughout the development of the aircraft. For a derivative aircraft also in Section 2.2 or change-based aircraft also in Section 2.2, the architecture will already have been defined. However, elements may be added to or deleted from the original architecture as the new design demands.

7.2 Initial Concept

The first step involving the conceptualization of real hardware and software is the initial concept. For new or derivative aircraft in Section 2.2, this concept would involve the development of a complete aircraft concept, its dimensions, weight, and performance estimate discussed in Section 8.2. This concept is based on the top-level (not derived) requirements. For derivative aircraft, the initial concept would specify major components, for example, an entire wing, fuselage, or tail from a previous design which would be employed. For change-based designs described in Section 2.2, the initial concept would also specify major dimensions, weights, and off-the-shelf items.

The key aspect of the initial concept is that it is *initial*: that is, it is subject to changes based on the trade-offs to follow. The key milestone for the initial concept is the system design review (SDR) described in Section 12.4. The purpose of the SDR is to show that the initial concept meets the top-level requirements and that the trade-offs have been defined which will result in a final baseline configuration at the preliminary design review (PDR) described in Section 12.4.

7.3 Trade-Off Studies

The trade-off process is a key element of SE when applied to aircraft. A trade-off is an analysis conducted to determine the preferred option among two or more options, such as the number of engines, based on a figure of merit, such as cost, weight, or reliability. Trade-offs can be either top level or subsystem level. At the top level, for example, the trade-off between payload and range will be paramount. Chapter 8 will discuss this trade-off and many other top-level trade-offs. Chapter 9 describes subsystem trade-offs, for example, the trade-off between electrical and pneumatic de-icing. There are

7.1 飞机架构

我们在第 2.3 节中看到，综合的第一步，是创建一个如图 2.1 所示的层次化的飞机架构。对于在 2.2 节所述的新飞机，在飞机研制的整个过程中，必须创建并完善这个框架。对于 2.2 节所述的衍生或改装机型，其架构已经被定义，但可以根据新的设计需要，在原有架构上增加和删除一些元件。

7.2 初始概念

第一步是初始概念，把真实硬件和软件概念化。对于 2.2 节定义的新的或者衍生型飞机，这个概念是开发一个完整的飞机概念，包括飞机的尺寸、重量及预估性能 (在 8.2 节讨论)。这个概念基于顶层 (而不是衍生的) 需求。对于衍生型飞机，初始概念将是规定一些可能采用的、来自先前设计的大部件，如整个机翼、机身或者垂尾。对于改装型飞机的设计，初始概念可能也是规定主要尺寸、重量及货架产品。

初始概念的一个重要方面是它是初始的，即，后续的权衡研究将会带来更改。初始概念的关键里程碑是系统设计评审 (SDR)(12.4 节)。SDR 的目的是表明初始概念满足顶层需求，且权衡研究已经完成，其结果将会产生初步设计评审 (PDR)(12.4 节) 时的最终基线构型。

7.3 权衡研究

权衡研究过程是系统工程运用到飞机上的一个关键因素。权衡研究是为在两个或多个选项中确定一个最佳选择而进行的分析。例如，基于诸如成本、重量或可靠性的性能参数确定发动机的数量。权衡研究可以在顶层或者子系统层实施。在顶层，比如在商载和航程间进行权衡研究是非常重要的。第 8 章将主要讨论此

hundreds of other subsystem-level trade-offs.

The importance of the trade-off study is that it is the key step which allows the designer to find the best solution for both the aircraft and its subsystems. Simply identifying a trade-off indicates that the previous design was, perhaps, not the best design for today's environment and the available technology.

Top-level trade-offs will be complete before the SDR so that a top-level design concept can be presented which meets the top-level requirements. Subsystem-level trade-offs are complete before the PDR, at which all requirements are complete. Hence, trade-offs should be conducted top-down. That is, the top-level trade-offs should be conducted before the subsystem-level trade-offs.

An important part of trade-offs is risk as discussed in Chapter 15. When the designer is considering two solutions, the performance, schedule, and cost risks should be considered for each solution. Risk assessment is particularly important when considering new aircraft technologies, discussed in Section 7.6.

7.4 Quality Function Deployment (QFD)

Quality function deployment (QFD) is a process which significantly improves the ability to capture, prioritize, and assess customer requirements using a team approach, and to transform these requirements into concepts. The strength of QFD is its ability to prioritize and weigh customer needs and to differentiate the value among different options. QFD has been shown to achieve satisfaction with both the customer and the manufacturer. Although a major function of QFD is to capture requirements, it is discussed here because it is most useful as a tool for establishing an initial concept.

QFD transforms customer needs into design characteristics and prioritizes the needs based on customer importance. These characteristics and priorities are translated into component characteristics. QFD can examine many factors, including functionality, reliability, cost, technology, and non-quantifiable aspects, such as aesthetic appeal. QFD can be applied at any level of DAC or customer operation, including design, development, manufacturing, training, operation, maintenance, and service.

In summary, some of QFD's strengths are:

1. Its ability to evaluate both quantitative and qualitative needs. It is this ability which provides a bridge between qualitative and quantitative needs. For example, QFD can be used with real airline customers to establish preferences and trade-offs on human factors discussed in Section 5.5 or interior design

项权衡工作和其他一些顶层权衡内容。第 9 章描述子系统的权衡内容，比如使用电除冰还是引气除冰，还有成百上千的其他子系统级的权衡研究。

权衡研究的重要性在于，此工作是使设计师找到对飞机及其子系统两者均为最佳解决方案的关键步骤。只要发现某一项需要进行权衡的内容，就说明之前的设计在当今的环境和可用技术条件下并非最佳设计。

顶层的权衡研究应在系统设计评审 (SDR) 之前完成，从而可提出一个满足顶层需求的顶层设计概念。子系统级的权衡研究应在 PDR 之前完成，在那时所有需求都已完整。因此，权衡研究应按照自上向下的顺序执行，即顶层的权衡研究应在子系统级权衡研究之前进行。

如第 15 章所述，权衡研究的一个重要部分是风险。当设计师考虑两个方案时，应考虑每个方案的性能、进度及成本风险。如 7.6 节所讨论的，当考虑新的飞机技术的时候，风险评估将变得尤为重要。

7.4 质量功能展开 (QFD)

质量功能展开 (以下简称 QFD) 是一个采用团队方法显著改善捕获、评估客户需求、排列这些需求的优先次序，并将其转化为概念的过程。QFD 的优势在于其对客户需要进行优先级排序和加权，并区分不同的选项价值的能力。QFD 已被证明能够满足客户和制造商的要求。尽管 QFD 的一个主要功能是捕获需求，但它也是建立初始概念的一个最常用工具，因此本章将对其进行讨论。

QFD 把客户需要转换成设计特性，并基于客户的重视程度对需要进行优先级排序。这些特性和优先级将会被转化为部件特性。QFD 可以用于检查很多因素，包括功能、可靠性、成本、技术及其他非量化特征，比如外观。QFD 可以用于 DAC 的任何层级或客户运营，包括设计、开发、制造、培训、运营、维修及服务。总之，QFD 的优势包括：

discussed in Section 9.4, fields not normally subject to quantitative evaluation.

2. Its speed and ability to reach important conclusions quickly.
3. Its ability to integrate the opinions and contributions from many design disciplines, customers, suppliers, and decision makers.
4. Its ability to evaluate technical, cost, and schedule risk.
5. Its ability to check decisions against *common sense*.
6. Its ability to identify the drivers which led to any given decision.
7. Its ability to identify key trade-offs.
8. Its ability to identify synergistic solutions: that is, the ability to allocate multiple requirements to a single solution.
9. Its ability to prioritize needs.

It must be remembered, though, that QFD is not a scientific methodology. It is a tool based on reason whose purpose it is to arouse the intuition of the analyst and apply it to the creation of a system. If the results agree with the intuition, then it can be said that the final solution is probably a good one. If not, the analyst should take advantage of the transparency of the methodology and determine how to improve the system.

The primary tool of QFD is the House of Quality. An example of a House of Quality is shown below:

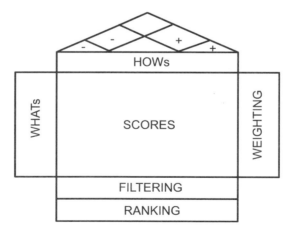

Figure 7.2   The House of Quality

In the House of Quality of Figure 7.2:

1. 其评估定性及定量需要的能力，它能在定性与定量需要之间建立桥梁。例如，能够与真实的航空公司客户一起利用 QFD 方法建立人为因素的基准及权衡内容 (5.5 节)，或内饰设计 (9.4 节)，这些领域通常无法进行定量评估。

2. 快速获得重要结论的能力和速度。

3. 集成不同的设计专业、客户、供应商及决策者意见和建议的能力。

4. 评估技术、成本及进度风险的能力。

5. 通过常识检查决策能力。

6. 辨识导致任何给定决策的动因的能力。

7. 识别关键权衡项的能力。

8. 识别协同综合解决方案的能力，即把多个需求分配给同一个解决方案的能力。

9. 对各种需要进行优先级排序的能力。

需要注意的，尽管 QFD 不是一门科学方法学，但它是一个基于推理的工具，其目的是激发分析者的直觉，并将其用于创建一个系统。如果结果与直觉一致，则认为其最终方案可能是一个好的解决方案。如果不是，分析者应利用此方法清晰明了的优势，确定如何改善该系统。

QFD 的一个主要工具是质量屋 (house of quality)。下面给出一个质量屋的例子：

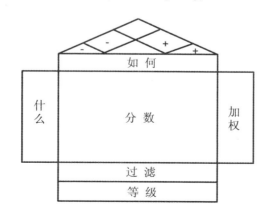

图 7.2　质量屋

1. WHATs are the customer needs or requirements. The WHATs are collected in a group environment from the customer. WHATs can be top-level requirements, such as range, payload, or dispatch reliability; or special customer options, such as passenger entertainment.

2. The establishment of customer needs is an especially critical aspect of the QFD process. This is because customers, or even engineers, may not have a clear understanding of the difference between a need and a solution. It is important at this stage to focus on needs and not solutions.

3. For example, does the customer want an extra door in the cargo area, or do they want improved ventilation? Improved ventilation is a valid need, but an extra door is just one of many possible solutions. The comparison of these solutions is part of the synthesis process. Improved ventilation is the proper need for the QFD process.

4. HOWs are candidate solutions to the WHATs. HOWs are developed by the systems engineer and approved by the customer.

5. A SCORE is a numerical value showing the importance of a particular HOW to contribute to a WHAT. SCOREs are obtained in a joint workshop environment with the customer. Non-linear scoring systems have been found to be useful in identifying the desirable solutions. For example, a scoring system might be as follows: 0 (not important), 1 (slightly important), 3 (moderately important), and 9 (very important).

6. A WEIGHTING is a numerical value reflecting the importance of the WHATs to the customer. WEIGHTINGS are solely determined by the customer.

7. FILTERING is a numerical value designed to account for such factors as cost and schedule risk.

8. The RANKING is a numerical value determined by multiplying the WEIGHTINGS with the SCORES, summing the product for each HOW, and adding the FILTERING value.

The intersections of the roof of the house reflect either synergisms (+) or trade-offs (-) between the HOWs. For example, any customer option which results in an increase in weight will also result in a decrease in the range of the aircraft. When a final concept is selected (a group of HOWs), both the RANKING and the intersections are considered. Figure 7.3 shows a House of Quality with some typical WHATs and HOWs normally encountered in top-level aircraft analysis.

在图 7.2 中的质量屋中：

1. **什么 (WHAT)** 代表客户需要或需求。WHAT 是在来自客户的一组环境中收集起来的，WHAT 可以是顶层的需求，比如航程、商载或签派可靠性，或特定的客户选项，比如旅客娱乐。

2. 建立客户需要是 QFD 过程中一个特别重要的方面。这是因为客户甚至工程师，都可能不大理解需要和方案之间的区别。在这个阶段，最重要的是要关注需要，而不是解决方案。

3. 例如，客户是否需要在货舱段增加一个额外的舱门，或他们是否需要改进通风效果？改进通风效果是一个有效的需要，但是增加一个额外的舱门只是多个可能解决方案中的一个。针对这些方案的比对是综合过程的一部分，改进通风效果才是 QFD 过程中正确的需要。

4. **如何 (HOW)** 是**什么 (WHAT)** 的备选解决方案，**如何 (HOW)** 由系统工程师开发，并由客户批准。

5. **分数 (SCORE)** 是一个数值，这个值体现一个特定的**如何 (HOW)** 对一个**什么 (WHAT)** 贡献的重要程度。**分数 (SCORE)** 是在与客户联合工作的环境中获得的。业已发现，确定优选方案的时候，采用非线性的打分系统是非常有用的。比如：一个打分系统可能是：0(不重要)、1(略微重要)、3(中等重要) 和 9(非常重要)。

6. **加权 (WEIGHTING)** 是一个数值，这个值反映了**什么 (WHAT)** 对客户的重要程度。**加权 (WEIGHTING)** 是由客户单独确定的。

7. **过滤 (FILTERING)** 是一个数值，这个值体现了比如成本和进度风险等因素。

8. **等级 (RANKING)** 是一个数值，由**加权 (WEIGHTING)** 乘以**分数** (SCORE)，求每个**如何 (HOW)** 的乘积之和，最后加上**过滤** (FLTERING) 确定。

质量屋的屋顶的交线反映**如何 (HOW)** 之间是否呈正相关 (+) 或呈负相关 (-)。比如，客户任何可能增加重量的选项都会缩短飞机的航程。当选择最终的概念 (一

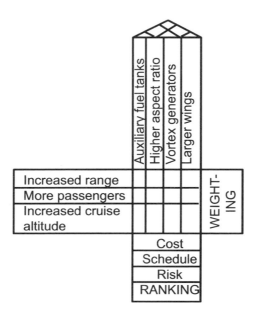

Figure 7.3   The House of Quality with typical WHATs and HOWs

7.5   Safety Features

Through the Preliminary System Safety Analysis (PSSA) discussed in Section 10.2 and the Common Cause Analysis (CCA) also discussed in Section 10.2, the certification process requires that safety become an integral part of the synthesis process. For example, the certification plan requires the description of any unique or novel feature of the design which may be a safety hazard. Other safety-related steps are described in Section 10.2.

7.6   Introduction of New Technologies

New technologies in the SE process

The introduction of new technologies is an integral part of the synthesis process. Section 2.4 presented a list of typical new technologies which may be considered. The question is then how the introduction is accomplished within the SE methodology.

组**如何** (HOW) 时，应考虑**等级** (RANKING) 和这些交线。图 7.3 显示了一个包括一些在飞机顶层飞机分析中遇到的包含典型**什么** (WHAT) 和**如何** (HOW) 的质量屋。

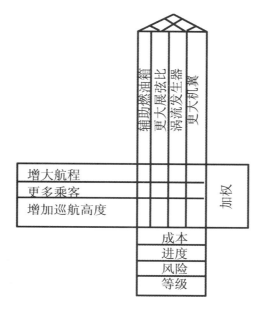

图 7.3　包含典型的什么 (WHAT) 和如何 (HOW) 的质量屋

7.5　安全特性

尽管 10.2 节讨论了初步系统安全性评估 (PSSA) 和共因分析 (CCA)，合格审定过程仍要求将安全性作为综合过程不可或缺的一部分。比如，合格审定计划要求描述任何可能有安全性危害的特有或新颖的设计特征。其他与安全性相关的活动将在 10.2 节中描述。

7.6　新技术的引进

系统工程过程的新技术

引入新技术是综合过程不可或缺的部分。2.4 节表明了可能需要考虑的一系

First, it can be done as part of the initial concept formulation described above. That is, the initial concept will have as an integral feature certain technological features which are prejudged to meet the top-level requirements and have been shown in the initial sizing as discussed in Section 8.2 to be effective both from a weight and DOC standpoint. Examples might be composite structures or fly-by-wire (FBW).

However, the crucial step for the evaluation of new technologies is during the trade-offs to be performed following the SDR. In SE the process of trade-offs leading to a synthesized concept is called *system analysis*. It is here where all aspects of a new technology need to be evaluated. These aspects include the ability to meet performance requirements, cost, weight, and risk.

Evaluation of new technologies

Mackey (1996) has outlined an eight-step strategy for technology management. The first step is to evaluate the requirements for a technology assessment process. These requirements include a thorough evaluation of the mission of the system (the aircraft) and an evaluation of return on investments and benefits of introducing new technologies. The next step consists of evaluating the readiness levels of candidate technologies. Table 7.1 shows the standard readiness levels adapted from NASA.

The next step is to make sure that customers, aircraft manufactures, for example, have direct contact with the universities and other institutions who are developing the new technologies. The fourth step is to gain the commitment of project management. With this commitment comes the necessary funding for the new technologies. The fifth and key step is a process called technology evaluation and adaptation methodology (TEAM) as developed by Loral (1994). TEAM performs a complete mapping of the mission of the system with the needed technologies.

Table 7.1   Technology readiness levels

Level	Description
1	Basic principles observed and reported
2	Technology concept and/or application formulated
3	Analytical and experimental critical function and/or characteristic proof of concept demonstrated
4	Component and/or breadboard validated in laboratory environment

列典型的新技术。问题是如何使用系统工程方法完成新技术引入。首先，引入新技术可以作为上述的初始概念形成的一部分，即这个初始概念应具有作为某一完整特性的一些技术特性，预判这些技术特性满足顶层需求，并在初始参数设计 (8.2 节) 过程表明，无论从重量或者直接运营成本 (DOC) 角度看都是有效的。例子包括复合材料或电传操纵 (FBW)。

然而，评估这些新技术的关键步骤是在系统设计评审 (SDR) 之后进行的权衡研究期间。在系统工程中权衡研究将会产生一个综合化的概念，叫做系统分析。此时需要对一项新技术的所有方面进行评估。这些方面包括满足性能需求、成本、重量和风险的能力。

新技术评估

Mackey (1996) 列出了技术管理的 8 步策略。第 1 步是评估关于技术评估过程的需求。这些需求包括对系统 (飞机) 任务的充分评估和对引入新技术的投资回报 (ROI) 和效益的评估。下一步是评估这些备选技术的成熟度等级。表 7.1 列出了 NASA 使用的标准成熟度等级。

下一步是确保例如客户、飞机制造商能与开发这种新技术的，高校和其他组织直接接触。第 4 步是获得项目管理层的承诺。有了这一承诺，新技术就有了必要的资金支持。第 5 步也是关键的一步，称为技术评估与适用方法 (TEAM)，此方法由 Loral(1994) 开发。TEAM 制订具有所需技术的系统任务的完整路线图。

表 7.1　技术成熟度等级

等级	描　　述
1	发现和报道基本原理
2	形成技术概念和 (或) 应用
3	通过分析和实验对关键功能和 (或) 特性进行概念验证
4	实验室环境下的部件和 (或) 原型样件的确认

(Continued)

Level	Description
5	Component and/or breadboard validated in relevant environment (ground or flight)
6	System or subsystem model or prototype demonstrated in a simulated environment (ground or flight)
7	System demonstrated in flight
8	Actual system completed and flight qualified through test and demonstration (ground or in flight)
9	Actual system flight proven

Source: Adapted from NASA (2012).

It also conducts an assessment of potential payoffs and risks on a technology roadmap. Finally, it develops a plan which documents schedules, funding, and conclusions. The sixth step is the use of a prototype or pilot process. Various types of prototypes can demonstrate requirements, integrate and test functionality, determine system feasibility, create simulated environments, and demonstrate operational concepts. The seventh step is to implement new technology on selected projects. The final step is to summarize and document lessons learned. In this way the steps already taken do not need to be repeated. In summary, in the aircraft industry the introduction of new technology is not just a matter of scientific interest or even of economic profitability. Rather it is a matter of survival.

7.7 Preliminary Design

The last real step in the synthesis process is preliminary design. The name may be somewhat misleading because the design is, in fact, final. The only aspect of the design not done is the detail drawings. At the PDR described in Section 12.4 all the trade-offs identified at the SDR described in Section 12.4 will have been completed. These activities, such as trade-offs, leading to a synthesized design are called *system analysis*. All the functions and requirements will have been completed to the lowest level of the aircraft hierarchy. And finally, the design will be complete to that level with weights and dimensions. These design characteristics are the *design requirements*, which are the product of the synthesis process.

等级	描　　述
5	相关环境下（地面或空中）的部件和/或原理样件的确认
6	在仿真环境下系统或子系统模型或原型样机演示（地面或空中）
7	飞行中的系统演示
8	完成实际系统，并通过试验和演示进行飞行鉴定（地面或空中）
9	实际系统飞行证明

来源：来源于 NASA(2012)。

这一步还对技术路线中的潜在收益和风险进行评估。最终，形成一个包含进度、融资及结论的计划。第 6 步是使用原型样机或先驱过程。不同类型的原型样机可以用于演示需求、集成和试验功能、确定系统可行性、创建仿真环境并演示运营概念。第 7 步是在选定的项目中实施新技术。最后一步是总结并将经验教训归纳成文，这样已实施的一些步骤就不用再重复进行。总而言之，在航空领域引入新技术不仅仅是科学发展或经济利益的问题，而是一个关系企业生存的问题。

7.7　初步设计

综合过程是初步设计真正意义上的最后一步。这个名字可能会产生误解，因为事实上这个设计已经是最终设计了。设计中没有完成的只是详细图纸。在初步设计评审 (PDR)(12.4 节) 时，所有系统设计评审 (SDR)(12.4 节) 中确定的权衡研究工作项都已完成。这些活动，比如权衡研究，将会形成一个综合化的设计，这个活动称为系统分析。所有功能和需求都要完成到飞机架构的最低层级。最终，将在重量、尺寸这一层级完成设计。这些设计特征将成为设计需求，即综合过程的产物。

8

Top-Level Synthesis

For as the aircraft is one system, and hath many subsystems, and all the subsystems of the aircraft, being many, are one aircraft …

Paraphrased from I Corinthians 12: 12–26, *Holy Bible*,
King James (Authorized) Version

This chapter will focus on the creation of an aircraft system at its very highest level. It is not the purpose of this chapter to present the step-by-step procedure for the sizing of an aircraft. For the aircraft, these steps are well-known, for example, in Corning (1977). Rather it is to show how these steps fit into the SE process so that the parameters can be understood in terms of the basic SE categories, namely, performance requirements, constraints, and trade-off parameters. We will also show how these parameters logically flow down to the lower levels of the aircraft hierarchy. In this chapter we will look at how performance requirements and constraints result in both a synthesized aircraft system and a synthesized aircraft at the top level. Of course, the determination of the requirements at all levels will require the implementation of the principle of *holism* discussed in Section 4.3 and Chapter 7.

The creation and building of complex systems is often called *systems architecting*, as defined by Rechtin (1991). Systems architecting goes beyond technical requirements to focus on such concepts as customer satisfaction. This chapter will concentrate on the creation of an aircraft system from verifiable top- level technical and economic requirements.

8.1 The Aircraft System

We have learned in Section 2.3 that the real top level is the *aircraft system*, of which the aircraft is only one of element out of five. These elements are the aircraft, the training equipment, the support equipment, facilities, and personnel. Hence, when performing

第8章 顶层综合

"飞机本身是一个系统，却包含多个子系统，飞机子系统有很多，但飞机只有一个……"

来源于《圣经·哥林多前书》12 章：12-26 节，詹姆斯王（授权）版本

本章主要关注最高层级飞机系统的创建。本章的目的不在于对飞机参数设计的各步流程进行说明。关于飞机的这些设计流程已经有了比较完善的定义，如 Corning(1977) 介绍的。而这里是展示如何将这些步骤融于系统工程过程。从而用基本的系统工程类别来理解各项参数，即性能需求、约束、权衡研究参数等。我们将展示如何将这些参数合理分解到飞机架构中较低的层级。本章我们将讨论性能参数和约束如何形成一个集成化的飞机系统，并在顶层形成一架化成化的飞机。当然，在所有层级上确定需求需要采用整体论 (4.3 节及第 7 章介绍) 的原则。

创造和构建一个复杂系统一般称为系统架构设计，如 Rechtin(1991) 中定义。系统架构设计超出了技术需求的范畴，关注诸如客户满意度之类的问题。本章将重点讨论根据可验证的顶层技术及经济性需求创建飞机系统。

243

8.1 飞机系统

由 2.3 节讨论可知，真正的顶层是飞机系统，其中，飞机只是 5 个元素中的一个。这 5 个元素包括飞机、培训设备、支持设备、设施及人员。因此，当执行

top-level synthesis, it is necessary to synthesize the aircraft system, not just the aircraft. This principle is particularly important for the high-speed civil transport (HSCT) for which radical changes in cargo or passenger loading or maintenance may be required. Following are a few example considerations for specifying the entire aircraft system; although many of these items will affect the design of the aircraft itself, they may also affect the design of the four other aircraft system elements:

1. *Cargo characteristics*   This category includes weight, volume, count, and any special characteristics, such as live animals, explosives, and toxic materials.

2. *Actual origins and destinations*   This item includes particular route capabilities, abilities to operate in severe weather, or at airports with particular characteristics affecting the take-off and landing approaches.

3. *Airport characteristics*   These items include ramp, taxiway, and runway constraints; airport terminal constraints on aircraft dimensions and servicing locations; passenger access equipment; servicing equipment sizes, types, rates, and interfaces; maintenance facilities; lighting; personnel facilities; airline operations facilities; and any special rules.

4. *Configuration change-over times*   This item applies to systems for which the aircraft configuration is required to change between flights.

5. *Utilization rate* This requirement measures how many hours per week is the aircraft required to fly, and how much support this requirement implies.

6. *Reliabilities, both dispatch and economic*   These reliabilities drive both aircraft and support requirements.

7. *Turnaround time*   This requirement drives aircraft cargo and passenger loading characteristics as well as maintenance and servicing features. Adams (1996) shows that on-board data loading can result in intolerably long turnaround times. A shop data loading system is proposed.

8. *Passenger service requirements*   These requirements include food, lavatories, and airport passenger services, such as transportation.

9. *Growth capability*   This requirement influences the need for commonality in all five elements.

10. *Autonomy*   The requirement for aircraft starting and servicing autonomy results from the constraint to utilize less ground equipment.

11. *People-related requirements*   These requirements cover all aspects of the system: for example, number of people, types, quality levels, cross-

顶层综合时，必须针对飞机系统进行综合，而不仅仅是针对飞机。对于高速民用运输机 (HSCT) 这类可能要求在货载、客载或维修工作方面有根本改变的飞机而言，上述原则就显得尤为重要。下面是关于规定一个完整的飞机系统须考虑事项的几个例子，尽管其中许多部分将会影响飞机本身的设计，但它们同时也会影响其他 4 个飞机系统元素的设计。

1. 货载特性　这一类包括重量、体积、数量和其他特定的特征，比如活的动物、爆炸物及有毒材料。

2. 实际出发地和目的地　此项包括特定的航路能力、在严酷天气条件下或在其特定特性可能影响起飞和着陆进近的机场运营的能力。

3. 机场特性　这些包括停机坪、滑行道、跑道约束及机场航站楼关于飞机尺寸和服务点的约束、旅客登机设备、服务设备尺寸、类型、速率和接口、维修设施、灯光、人员设施、航空公司运营设施及其他特殊规则。

4. 构型变更次数　这一项适用于在航班之间需要进行飞机构型更改的系统。

5. 利用率　这个需求测量每周飞机需要飞行的小时数及达到如此飞行小时所需要的支持；

6. 可靠性　包括签派可靠性及经济可靠性。这些可靠性指标将驱动飞机和支持需求。

7. 地面返程准备时间　该需求导致了飞机装卸货和旅客上、下飞机特征及维修和服务特征。Adams (1996) 指出机载数据加载可以导致不可容忍的漫长地面返程准备时间，因此建议采用内场数据加载系统。

8. 乘客服务需求　这些需求包括食物、盥洗室及机场乘客服务，比如运送乘客。

9. 增长能力　这些需求影响所有 5 大元素的共通性需要。

10. 自主化　飞机起动和服务自主化的需求源于使用较少地面设备这一约束。

training requirements, certification needed. Included are dispatch operations, flight crews, cabin crews, servicing, maintenance, overhaul, training, and engineering. People-related requirements should be included in all of the other considerations in this list.

12.*Consumables*   These items are included in all system elements: fuel, lubricants, sealants, coatings, life-limited parts, non-repairable items, interior (for example, seat covers, towels, and carpets), and perishable tools.

13.*Operational requirements*   These requirements include air traffic control (ATC) compatibility, traffic at departures and destinations, communications, navigation, climb rates, initial cruise altitude (ICA), taxi rules, in-flight and ground aircraft and engine wakes, approach speeds, and cruise Mach number.

14.*Exterior noise*   This requirement includes effects on maintenance, ground operations, and flight profiles.

15.*Regulatory environmental requirements*   These include limits on nitrous oxide (NOX), unburned hydrocarbons, heavy metals, paint, coatings and sealants, asbestos, soot and other particulates, fuel and other fluids dropped on ramp, discarded waste, and hazardous waste.

16.*Particular customer requirements*   These requirements include any mandated solutions and all functional and physical interfaces. Also included are customer methods of transferring maintenance and dispatch data. Also included are customer constraints on supplier selection.

17.*Costs*   In addition to the direct operating costs (DOC) discussed below and other indirect costs discussed in Section 8.6, these include all other segment costs. Examples are airport fees driven by exceeding noise and environmental limits, people costs associated with Item 11, spares cost, facilities cost, service costs, and training costs.

An example of an aircraft system-level function requiring synthesis beyond the aircraft level is Provide Category III (all weather) Landing Capability, an operational (Item 13 above) requirement. This function requires synthesis and qualification of the aircraft, the pilot, the maintenance system (support element), and the airport. These elements include the people associated with the elements.

11.**与人相关的需求** 这些需求覆盖系统所有部分，比如人员数量、类型、质量等级、交叉培训需求及需要的资质认证。包括签派员、飞行机组、客舱机组、服务、维修、翻修、培训及工程人员。与人相关的需求应包括表中列出的所有其他考虑。

12.**消耗品** 这些项在所有系统元件中均存在：燃油、润滑油、密封剂、涂料、限寿件、不可维修件、内饰（比如，座椅套、毛巾、地毯）和易损工具

13.**运行需求** 这些需求包括空中交通管制 (ATC) 的兼容性、离场和目的地的交通、通信、导航、爬升率、初始巡航高度 (ICA)、滑行规则、空中和地面飞机和发动机的尾流、进近速度和巡航马赫数。

14.**外部噪声** 这个需求包括对维修、地面运行和飞行剖面的影响。

15.**监管环境需求** 这些包括对氮氧化物 (NO_x)、未燃烧的碳氢化合物、重金属、涂料、涂层和密封剂、石棉、煤烟及其他颗粒物、滴落到停机坪上的燃油或其他液体、丢弃废物及危害性废弃物的限制。

16.**特定的客户需求** 这些需求包括任何强制的解决方案和所有功能和物理接口。同时也包括客户传递维修和签派数据的方法，以及客户对供应商选择的约束。

17.**成本** 除了下面8.6节要讨论的直接运营成本 (DOC) 和其他间接成本之外，还包括所有其他部分的成本。比如，由于超出噪声和环境约束所致的机场罚款、由于第11项带来的人员成本、备件成本、设施成本、服务成本及培训成本。

需要在飞机之上进行综合的一个飞机系统级功能的例子是：提供 CAT III(全天候)3 类着陆能力，这属于运行需求（上述第13条）。这个功能需要针对飞机、驾驶员、维修系统（支持元素）及机场进行集成和鉴定。这些元素还包括与之相关的人员。

8.2 Top-Level Aircraft Sizing

Top-level sizing is the heart of the creative process in aircraft design. It is also the beginning of the synthesis phase in the SE process as we saw in Figure 1.1. The first product of top-level sizing is an *initial concept*. This concept is the result of the top-level functions developed according to the principles of Chapter 3 and the requirements developed in accordance with the principles of Chapter 4. Most of the requirements considered in this step result from the Perform Air Transport Mission function of Figure 3.2. However, as we have seen, all functions should be considered. This initial concept is the subject of the system design review (SDR) discussed in Section 12.4. It is therefore the basis for further trade-offs and optimization.

Wing sizing

During the Perform Transport Mission function discussed in Section 3.2 it is necessary to size the wing by balancing three conditions: take-off, cruise, and landing. The following performance requirements apply:

1. Number of passengers.
2. Weight of cargo.
3. Range.
4. Cruise Mach number.

It is the basic mission of the aircraft to carry a specific number of passengers and cargo a given distance in a given time. Hence, these parameters are performance requirements. The following constraints also are required to size the wing:

1. Field length.
2. ICA.
3. Atmospheric density, pressure, and temperature at each of the three conditions.
4. Approach speed.

The field length is set by the route profile of the customer. Approach speed and ICA can also be traded off against payload weight.

8.2　顶层飞机参数设计

顶层参数设计是飞机设计创造过程的核心，它也是系统工程过程中的综合阶段 (见图 1.1) 的开始。顶层参数设计的产物是*初始概念*，这个概念是根据第 3 章原则形成的顶层功能，以及根据第 4 章原则形成的需求的结果。在这个阶段考虑的大部分需求都是源于**提供空中运输任务**功能 (见图 3.2)。然而，如我们前文所介绍的,应该考虑所有功能。这个初始概念须进行系统设计评审 (SDR)(12.4 节)，因此它是进一步权衡和优化的基础。

机翼参数设计

在 3.2 节中讨论的**执行运输任务**功能，必须通过对起飞、巡航及着陆这 3 个条件的平衡，对飞机机翼进行参数设计。应用下面这些性能需求 :

1. 乘客数量

2. 货物重量

3. 航程

4. 巡航马赫数

在给定的时间内运输一定数量的乘客和货物到一个给定地点是飞机的基本任务，因此，这些参数是性能需求。还需要下面这些约束进行机翼参数设计 :

1. 起飞场长。

2. 初始巡航高度。

3. 3 个情况下每一种情况对应的空气密度、压力及温度。

4. 进近速度。

场长由客户的航线剖面设定，进近速度和初始巡航高度也应对照商载重量进行权衡。

Wing sizing trade-offs

The trade-offs for wing sizing are extensive as listed by Corning, who also provides the detailed equations to conduct these trade-offs.

Supercritical vs. critical wing	Specific fuel consumption (*SFC*)
Aspect ratio, *AR*	Engine type
Sweepback angle, $\Lambda_{C/4}$	Fuel/take-off weight ratio, W_f/W_{to}
Divergence Mach number, Ma_{DIV}	Take-off wing loading, $(W/S)_{TO}$
Average thickness/chord ratio, $(t/c)_{AVE}$	Initial Cruise wing loading, $(W/S)_{IC}$
Maximum lift coefficient, C_{Lmax} (no flap extension)	Initial cruise lift coefficient, C_{LIC}
Maximum lift coefficient, C_{Lmax}(landing)	

In order to find the optimum combination of all these parameters it is necessary to balance the requirements in the three conditions described above, namely, take-off, cruise, and landing. Even then, it will not be certain that the fuel the wings can carry will meet the range requirements. So it will be necessary to repeat these steps as the aircraft is defined.

Engine sizing

The engines are normally sized by the conditions at take-off. For some aircraft the engines are sized by the cruise or climb conditions. Therefore, it will be necessary to repeat the sizing steps for these conditions. The key performance requirement is thrust at take-off. However, it is necessary to derive the engine performance requirement from the results of trade-offs with other aircraft parameters. Other key constraints which are needed to size the engines are:

1. Take-off conditions:
   - Field length.
   - Obstacle height at end of runway.
   - Density, pressure, and temperature at take-off.
2. From wing sizing:

机翼参数设计的权衡研究

Corning 详细列出了机翼参数设计的权衡研究项，并提供了关于进行这些权衡研究的详细公式。

超临界机翼与临界机翼	单位耗油率 (SFC)
展弦比，AR	发动机型号
后掠角，$A_{C/4}$	燃油起飞重量比，W_f/W_{to}
发散马赫数，Ma_{DIV}	起飞机翼载荷 $(W/S)_{To}$
平均厚弦比，$(t/v)_{AVE}$	初始巡航机翼载荷，$(W/S)_{IC}$
最大升力系数，C_{Lmax}（襟翼未放出）	初始巡航升力系数，C_{LIC}
最大升力系数，C_{Lmax}（着陆）	

为了寻求这些参数的优化组合，必须在上述 3 个条件，即起飞、巡航及降落条件下平衡这些需求。尽管如此，还无法肯定机翼中携带的燃油可以满足航程需求。因此，随着飞机的定义，必须重复上述步骤。

发动机参数设计

发动机一般都由起飞条件决定参数指标，有些飞机，其发动机参数指标由巡航或爬升状态决定，因此，必须在上述条件下重复参数设计的步骤。性能需求的关键是起飞推力。但必须根据与其他飞机参数进行权衡研究的结果导出发动机性能需求，发动机参数设计需要的其他关键约束包括：

1. 起飞条件

·场长。

·跑道终端的障碍物高度。

·起飞时的空气密度、压力及温度。

2. 工具参数设计可得到

- Wing loading, W/S.
- Maximum take-off lift coefficient, $CL_{t/o}$.
- Number of engines.

From this information, it will be possible to determine the thrust loading W/T at take-off and the take-off speed, V_{to}. The engine thrust is sized by the thrust required at take-off. The basic parameter required at this point is thrust loading, W/T, at take-off. These basic performance requirements and constraints will provide both thrust loading and the take-off speed, V_{to}.

Take-off weight

The take-off weight is the fully loaded weight as described in Section 5.2. However, in the synthesis process it is not a constraint; it is a derived requirement. However, in the end the airport runway strength will set the maximum value of the take-off weight. So, for that purpose, it is a constraint. It is possible to determine the take- off weight from the following equation:

$$W_{to} = W_{structure} + W_{engines} + W_{fuel} + W_{payload} + W_{fixed\ equipmen}$$

In addition, it is possible to determine the structure weight from:

$$W_{structure} = W_{wing} + W_{fuselage} + W_{landing\ gear} + W_{nacelle\ \&\ pylon} + W_{tail\ surfaces}$$

Each weight component is a function of the total take-off weight and several other parameters. Thus, we can solve for take-off weight. We will treat each one of these separately:

1. Wing weight (W_{wing}) is a function of the take-off weight (W_{to}), the aspect ratio, AR, the taper ratio, l, the thickness ratio (t/c), and the sweepback angle, $L_{C/4}$.
2. The fuselage weight ($W_{fuselage}$) is a function of the number of passengers, the number abreast, and the number of aisles.
3. The landing gear weight ($W_{landing\ gear}$) is normally a fixed fraction of the total take-off weight.
4. The nacelle and pylon weight ($W_{nacelle\ \&\ pylon}$) is proportional to the thrust and

· 机翼载荷，W/S。

· 最大起飞升力系数，$CL_{t/o}$。

· 发动机数量。

根据这些信息，便能够确定起飞时的推重比 W/T 和起飞速度 V_{to}。发动机推力由起飞时所需要的推力决定。在这里所需的基本参数是起飞时的推重比 W/T。这些基本性能需求和约束将提供推重比和起飞速度 V_{to}

起飞重量

起飞重量即满载重量 (5.2 节)。然而，在综合过程中，它不是一个约束，而是一个衍生需求。但最终机场跑道的强度将设定最大起飞重量的值。所以，就此而言，它是一个约束。可以根据下面的公式得出起飞重量：

$$W_{起飞}=W_{结构}+W_{发动机}+W_{燃油}+W_{商载}+W_{固定设备}$$

除此之外，结构重量也能确定如下：

$$W_{结构}=W_{机翼}+W_{机身}+W_{起落架}+W_{短舱及吊挂}+W_{尾翼面}$$

每一个重量分量都是总起飞重量和其他多个参数的函数，因此，我们可以算出起飞重量。我们将单独处理每一个部分：

1. 机翼重量 ($W_{机翼}$) 是起飞重量 W_{to}、展弦比 (AR)、尖梢比 (1)、厚弦比 (t/c) 及后掠角 ($L_{C/4}$) 的函数。

2. 机身重量 ($W_{机身}$) 是乘客数量、座位排数及通道数的一个函数。

3. 起落架重量 ($W_{起落架}$) 通常是起飞总重的一个固定比例。

4. 短舱和吊挂重量 ($W_{短舱及吊挂}$) 与推力成正比，因此是起飞重量和推重比的函数。

hence is a function of the take-off weight and thrust loading.

5. The tail section weight (W_{TS}) is proportional to the wing weight and hence to the take-off weight.

6. Like the nacelles and pylons, the engine weight can be assumed to be proportional to thrust.

7. The fuel weight (W_{fuel}) can be assumed to be a fixed fraction of the take-off weight.

8. The payload weight ($W_{payload}$) is the total assumed weight of the passengers and cargo from the performance requirements as discussed in Section 8.2.

9. The fixed equipment weight ($W_{fixed\ equipmen}$) includes all subsystems including electrical, environmental control, and so forth. Driving factors are the number of passengers, number of engines, and total take-off weight.

When these factors are added, it will be possible to solve for the take-off weight. Yet, even now, we have not finished our trade-offs. We do not know whether the take-off weight exceeds runway limits or whether the aircraft can carry the required fuel weight.

Drag trade-offs

The drag coefficient of the aircraft is determined from the basic equation:

$$C_D = C_{D0} + C_L^2 / \prod ARe + \Delta C_{DC}$$

Each term is affected by a different aspect of the aircraft configuration. C_{D0} is the lift-independent, drag, which is a function of the total aircraft configuration. It contains both form and skin friction drag. The second term, $C_L^2/\prod ARe$, is the induced drag. This factor depends on the total lift coefficient, C_L, which was determined in the wing sizing step above, as that necessary to maintain level flight. The aspect ratio, AR, was also a trade-off parameter in the wing sizing. The lift efficiency, e, and the compressibility drag coefficient, ΔC_{DC}, are also design-dependent factors, and therefore, *derived requirements*, as is the total drag coefficient C_D.

Lift-drag ratio

As we will see later, a key parameter in the determination of aircraft range is the

5. 尾翼重量 (*WTS*) 与机翼重量成正比，因此也与起飞重量成正比。

6. 与短舱和吊挂一样，可以假设发动机重量与推力成正比的。

7. 燃油重量 $W_{燃油}$ 可以假设是起飞总重的一个固定比例。

8. 商载重量 ($W_{商载}$) 是根据第 8.2 节讨论的性能需求假设的乘客和货物的重量之和。

9. 固定设备重量 ($W_{固定设备}$) 包括电气、环控等所有子系统，其动因包括乘客数量、发动机数量和起飞总重。

当加上这些因素后，就能算出起飞重量。然而，由于迄今为止还未完成权衡研究，尚不明确起飞重量是否超出了跑道的极限或者飞机是能否携带所需重量的燃油。

阻力权衡研究

飞机的阻力系数根据以下基本公式确定：

$$C_D = C_{D0} + C_L{}^2 / \prod ARe + \Delta C_{DC}$$

每一项都受飞机构型的不同部分影响，C_{D0} 是与升力无关的阻力，它是飞机总体构型的函数，包括外形阻力和蒙皮摩擦阻力。第 2 项，$C_L{}^2 / \prod ARe$ 是诱导阻力，该项由总升力系数 C_L 决定，由上述的机翼参数指标确定，因其对保持水平飞行十分重要。展弦比 AR 也是一个飞机参数指标的权衡参数。升力效率 e 及压缩阻力系数 ΔC_{DC} 也属于设计依赖项。因此，它与总阻力系数 C_D 一样，也属于衍生需求。

升阻比

稍后我们可以看到，决定飞机航程的一个关键参数是升阻比 L/D，该值目前由如下决定：

lift-drag ratio, L/D, which is now determined by:

$$L/D = C_L/C_D$$

Climb requirements

With the initial engine sized above, the estimated aerodynamics, and the ICA, it is now possible to estimate the amount of fuel required to climb to the cruise altitude and the range to climb, R_{cl}.

Cruise range

The range during cruise (assumed to include descent) can be estimated by the famous Breguet range equation:

$$R_{cr} = (V/SFC)(L/D)\log_{10}(W_1/W_0)$$

where V is the speed, W_0 is the initial cruise weight, and W_1 is $W_0 - W_{fuel}$. Thus, the total range is:

$$R = R_{cl} + R_{cr}'$$

It is possible to vary the range by varying the fuel fraction in the wing sizing step. Hence, an estimate of the total aircraft sizing results.

Top-level derived requirements and allocation management

What we learned in the above steps was not how to size an aircraft, but rather that the sizing process is an initial requirements process in which a set of top-level performance parameters and constraints are established and top-level trade-offs are conducted. These steps as described were only very approximate.

Another product of this process is a large array of *derived requirements*, that is, requirements which are dependent on a solution. Thrust is an example. This requirement is then used to size the propulsion subsystem of the aircraft. All of the other requirements will be used in the flow down of requirements to the aircraft subsystems.

Another aspect of this sizing process is that it gives the SE manager a tool to *manage* the allocated parameters, such as weight. We saw in Section 4.7 how

$$L/D = C_L/C_D$$

爬升需求

利用上述发动机初步参数指标、预估的空气动力学及其初始巡航高度 (ICA)，可以估算爬升到巡航高度所需要的燃油量及爬升距离 R_{cl}。

巡航航程

巡航过程航程 (假设包括下降) 可以利用著名的 Breguet 航程公式估算：

$$R_{cr} = (V/SFC)(L/D)\log_{10}(W_1/W_0)$$

式中：V 是速度，W_0 是初始巡航高度，W_1 等于 $W_0 - W_{燃油}$。因此，总航程为：

$$R = R_{cl} + R_{cr}$$

在机翼参数设计阶段不通过改变燃油重量比来改变航程。因此，它是飞机总体参数设计结果的估算值。

顶层衍生需求和分配管理

从上文可以了解到，上述步骤不是飞机如何进行参数设计过程，而是一个初始的需求过程，在这个过程中建立一系列顶层性能参数和约束，并完成顶层的权衡研究。上述描述的步骤仅仅是非常粗略的。

该过程的另一个产物是大量的衍生需求，即取决于解决方案的需求。推力就是一个例子。该需求用于对飞机的推进子系统进行参数设计。所有其他的需求将会用于将需求向下分解到飞机各个子系统的过程中。

该参数设计过程的另一方面是：赋予系统工程经理一种工具去管理这些分配的参数，比如重量。在 4.7 节中我们看到需求如何在子系统间进行分配，并通过权衡研究调整为更优化的分配。在 12.3 节中我们将看到系统工程经理如何使

requirements could be allocated among subsystems and then adjusted as trade-offs showed more optimal allocations. We will see in Section 12.3 how the SE manager uses these allocations to manage the design. The sizing exercise above provides us with the initial allocation values.

8.3   Other Top-Level Requirements

Remember that we defined many top-level aircraft functions described in Section 3.3. The sizing analysis above will satisfy some of the key functions, for example, Generate Aero Forces also in Section 3.3 and Provide Thrust, also in Section 3.3. However, many of the other functions will result in top-level (not derived) requirements which may have a significant impact on the aircraft. These requirements will depend on the particular needs of the customer.

The Provide Environmental Control function from Section 3.3 is a normal top-level function which may, or may not, reflect any special customer needs. The basic requirements which are *allocated* to this function are the temperature, pressure, humidity, and air quality throughout the aircraft. The importance of listing these as top-level requirements is that many subsystems (for example, ECS, airframe, and propulsion) will contribute to their achievement and that trade-offs may have to be made to determine the optimum allocation of requirements among these subsystems.

The requirements allocated to the Provide Passenger and Crew Accommodations function from Section 3.3 may vary widely among customers. For example, a customer may need special seating arrangements or lavatory or galley configurations. Once again, many subsystems may be involved in satisfying these requirements, and trade-offs may be required.

8.4   System Architecture

Before we can flow down any requirements to the subsystems, we need a *system architecture*, that is, the hierarchy of aircraft elements. We have already formulated a typical aircraft system architecture shown in Figure 2.1. This is not the only possible system architecture. Like the initially sized aircraft above, the system architecture is also subject to trade-off. For example, it is entirely possible to design an aircraft without one or more of the elements shown in Figure 2.1. One could design an aircraft, for example, without hydraulic power, using only electrical power. If such an aircraft could ever be optimum is a matter subject to trade-off. Figure 2.1 shows, of course,

用这些分配管理设计。这些参数设计的经验为我们提供初始分配数值。

8.3 其他顶层需求

值得注意的是，我们定义了 3.3 节所述的很多飞机顶层功能。上述参数设计分析将会满足一些关键功能，比如，第 3.3 节中的**产生气动力**和**提供推力**。然而，很多其他功能将会产生一些可能对飞机有重大影响的顶层（而不是衍生）需求。这些需求将取决于客户的特殊需要。

第 3.3 节中的**提供环控**功能是一个常见的顶层功能，它可能反映了，也可能没有反映特定的客户需要。分配给这个功能的基本需求是整架飞机各处的温度、压力、湿度和空气质量。把这些列为顶层需求是因为很多子系统（比如环控、机体和推进）都会对满足其目标产生影响，所以可能必须进行权衡研究以确定如何把这些需求在子系统之间进行最优化的分配。分配给**为提供乘客及机组起居设施**功能 (3.3 节)，会在不同客户之间存在很大差异。比如，某一个客户可能需要特殊的座位、盥洗室或厨房配置。再强调一遍，满足这些需求可能涉及许多系统。因此，可能需要进行权衡研究。

8.4 系统架构

在将任何需求向下分解到子系统之前，我们需要一个系统架构，即飞机元件的层次架构。我们已经形成了一个典型的飞机系统架构，如图 2.1 所示。该架构并不是系统架构的唯一可能形式。与上述完成初始参数设计的飞机一样，对系统架构也需要进行权衡研究。比如，设计一架没有图 2.1 中一个或几个元件的飞机是完全可能的。例如，可以设计一种完全只用电源，而完全不使用液压源的飞机。这种飞机是否可以进一步优化由权衡研究决定。当然，图 2.1 仅示出了一个顶层系统架构。该架构中每一个元件可能都存在其次级元件，这些元件也需要进

259

only the top-level system architecture. Each element in that architecture would have its own subelements, also subject to trade-off.

A key factor in the development of the system architecture is system safety. As we will see in Section 10.2 the preliminary system safety analysis (PSSA) is used to help develop the architecture.

8.5 Top-Level Constraints

In addition to the performance requirements listed above, and the associated derived requirement we already developed from the initial sizing, we should now address the multitude of top-level constraints.

Airport compatibility will introduce many constraints: For example, the weight limit of the runway and the width of the gates. Support capability at airports will determine the support design philosophy. For example, should the aircraft have autonomous support capability? That is, is it required to land, restart, and take off without any ground support? All of this is determined from the route structure the aircraft is being designed for.

Two key top-level constraints are dispatch reliability and maintenance cost per 1 000 flight hours (MN$/1000FH). As we have seen in Section 4.7, we can allocate these requirements to the subordinate elements. Other top-level constraints are cabin and exterior noise and emissions, which can also be allocated.

Design constraints, environments, and regulatory requirements, as discussed in Chapter 5, can be said to be constraints because they apply to the whole aircraft.

The most important constraint is the cost, normally characterized as DOC. Because of its importance, we will save cost for a fuller discussion in the next section.

8.6 Economic Constraints

In the SE process, cost may be considered a design constraint as valid as any other technical constraint, such as weight or physical limitations. Such a constraint may be incorporated directly into any specification.

In military practice, design-to-cost is normally treated as a trade-off activity to be conducted parallel to the design activity. In contrast, the demands of the competitive market require that cost constraints be treated more directly because market development is based on a given selling price.

Not only can costs be imposed on a system, an aircraft for example, but the cost

行权衡研究。

开发系统架构的一个重要因素是系统安全性。如 10.2 节所示，初步系统安全性评估 (PSSA) 用于帮助开发这种架构。

8.5　顶层约束

除了上面列出的性能需求，以及从初始参数设计过程中开发的相关衍生需求之外，我们也需要处理大量的顶层约束。

机场适应性将会引入多个约束。例如，跑道的重量限制及航站楼登机口的宽度。机场的支持能力将会决定支持设计的理念。比如，飞机是否需要有自主支持能力？即着陆、重新起动和起飞不需要任何地面支持设备？所有这些都与飞机设计所针对的航线结构有关。

签派可靠性及每千飞行小时维修成本 (MN\$/1 000FH) 是两个关键约束。如 4.7 节所示，我们可以将这些需求分配给下一级的元件。其他顶层约束包括舱内和外部噪声和排放，这些都可以被分配。

第 5 章所讨论的设计约束、环境及监管要求可以视为约束，因为它们适用于整架飞机。最重要的约束是成本，通常以直接运营成本 (DOC) 表征。因为其重要性，我们在下面章节中将主要讨论如何节省成本进行。

8.6　经济性约束

在系统工程过程中，成本可能与其他技术约束，如重量或物理限制一样，被视为一种设计约束。这类约束可能会直接纳入一些规范要求中。

在军用领域，按成本设计一般被认为是一个与设计活动并行的权衡活动。与之相反，竞争市场的需要则要求更加直接地对待成本约束，因为市场开发活动基于给定的销售价格。

constraints can be flowed down to the system segments and subsystems just as any other technical parameter can.

To accomplish this flow down, we can divide cost into three categories: non-recurring (development) costs, recurring (unit) costs, and DOC. Within DOC, major categories can be identified and used as design constraints. Although the components of DOC vary slightly from source to source, the major ones cited by Martínez-Val (1994) and Aerospace Engineering (1994) are shown in Table 8.1. All three major categories are, of course, of interest to both the developer and the customer since they affect both system purchase cost and operating cost, and hence profitability of the customer and, therefore, sales by the developer.

Although most aircraft economic analysis is focused on DOC and recurring costs, non-recurring cost is also a major driver in aircraft requirements. Non- recurring cost is the factor which leads to the need for derivative aircraft discussed in Section 2.2. By basing the design on a previous aircraft, the aircraft manufacturer can save a considerable amount of money in development costs. This savings goes beyond the development of the aircraft itself. Derivative aircraft result in savings in tooling and jigs used in the manufacturing process.

Direct Operating Cost (DOC) requirements

Figure 8.1 shows how the SE process can be used to allocate requirements. First, the aircraft DOC goal is allocated to the various aircraft segments in accordance with the items in the following list:

- navigation fees
- landing fees
- ground handling
- crew (cabin, cockpit)
- ownership (depreciation and interest)
- maintenance (fuel and airframe)
- fuel and oil

Figure 8.2 shows how DOC is used to select the design point for a new aircraft. Two types of DOC are important: DOC per seat-mile, and DOC per trip. Design points in the lower left-hand corner are deemed to be economically viable while those in the

成本不仅作用于一个系统(比如一架飞机),与其他技术参数一样,成本约束也可以向下分解到各个系统部段及子系统中。

为了完成分解工作,可以把成本分成3类:非重复(开发)成本(NRC)、重复(单位)成本(RC)及直接运营成本(DOC)。可以辨别 DOC 中的主要类别并将其作为设计约束使用。尽管来源不同的 DOC 的组成部分也稍有不同,由 Martínez-Val (1994) 及《航空航天工程》(1994) 引用的部分的主要内容如表 8-1 所示。当然这三个主要类别与研制方和客户均有关系,因为它们影响系统采购成本和运行成本,因此影响客户的利润和研制方的销量。

尽管飞机的经济性分析大多聚焦于直接运营成本(DOC)及重复成本(RC),非重复成本(NRC)也是飞机需求的一个重要驱动力。非重复性成本(NRC)是衍生性飞机(2.2 节)需要的一个驱动因素。基于已有飞机的设计,飞机制造商可以在研制成本方面节省大量资金。这种节省不仅仅局限于飞机研发本身,衍生机型可以节省制造过程中的工装及夹具费用。

直接运营成本 (DOC) 需求

图 8.1 展示了系统工程过程如何用于分配需求。首先,应依照下列各项,将飞机的 DOC 目标分配到飞机不同的部段:

· 导航费。

· 着陆费。

· 地面运行。

· 机组(客舱、驾驶舱)。

· 所有权(折旧和利息)。

· 维修(燃油和机体)。

· 燃油和滑油。

图 8.2 表明如何用直接运营成本 (DOC) 设计一架新飞机。有两类比较重要

upper right-hand corner are deemed not to be viable. These two regions are the result of an economic analysis which can be conducted either by the airline or the aircraft manufacturer.

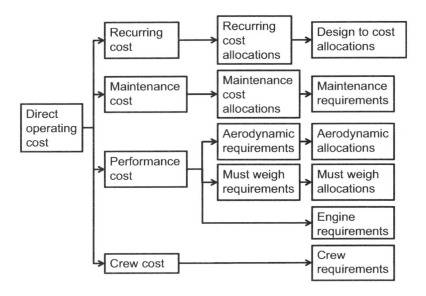

Figure 8.1   Allocation of direction operational cost in design process

Initial Direct Operating Cost (DOC) estimate

Before we can allocate DOC to the various aircraft subsystems, we have to have an initial estimate of the DOC breakdown. If this procedure sounds familiar, it is exactly the same as for the weight breakdown which we saw earlier in this chapter. DOC (in dollars per ton-mile) is comprised of three main factors: flight operations, direct maintenance, and depreciation. This initial DOC estimate is a direct result of the initial sizing discussed earlier in this chapter. The DOC will depend on the top- level requirements and the parameters estimated in the initial sizing, for example, take-off weight and number of engines.

的 DOC：每座英里 DOC 和每航次 DOC。在左下角区域内的设计点是经济可行的，而在右上角区域内的设计点是不可行的。这两个区域是航空公司或飞机制造商进行的经济性分析的结果。

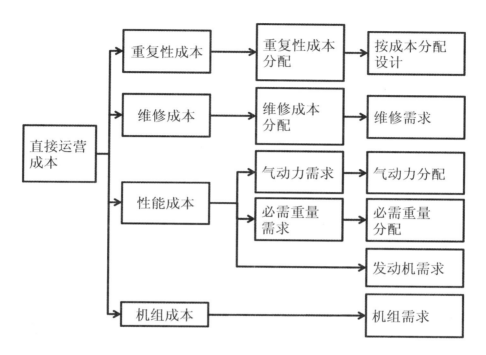

图 8.1　在设计过程中分配直接运营成本

直接运营成本 (DOC) 的初步估算

在将 DOC 分配给飞机不同子系统之前，应有一个 DOC 分解的初步估算。这个流程听上去非常熟悉，它与本章的重量分解是一样，DOC(单位：美元 / 吨 * 英里) 由 3 个主要部分组成：飞机运行、直接维修及折旧。初始 DOC 的估算值是本章前面所述的初步参数设计的直接结果。DOC 由初步参数设计过程中的顶层需求及估算参数决定，如起飞重量及发动机数量。

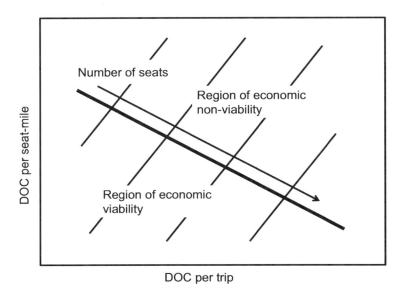

DOC per trip

Figure 8.2 DOC design regimes

Flight operations

Two main factors need to be calculated first to estimate the flight operations component of DOC, namely, block speed, V_b, and block fuel, F_b. Block speed is simply the total range divided by the time from gate to gate. Likewise, block fuel is the total fuel used from gate to gate.

Flight crew

Flight crew costs can be found by available models with cost primarily as a function of aircraft size.

Fuel and oil

Fuel costs were already determined from the block fuel used. Similarly oil costs can be determined from the block time and the number of engines.

266

Hull insurance

The insurance cost is based on the aircraft unit cost and the utilization factor.

Direct maintenance

Direct maintenance cost is comprised of the following three components:

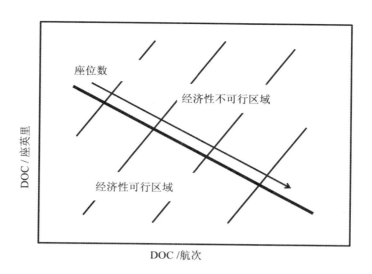

图 8.2　DOC 设计区域

飞行运营

估算 DOC 中飞机运营部分首先需要计算两个主要因素，即轮挡速度 V_b 和轮挡燃油 F_b，V_b 是总航程除以从登机口到登机口的时间。类似，轮挡燃油是从登机口到登机口使用的总燃油量。

飞行机组

飞行机组成本可利用一些已有的模型得出，该模型主要将成本作为飞机大小的函数。

燃油和滑油

燃油承办已根据所用的轮挡燃油量确定。同样，滑油成本也可以根据轮挡时间及发动机数量确定。

机身保险

该保险成本由飞机单位成本和利用率决定。

直接维修

直接维修成本由下面 3 部分组成：

Labor (excluding engines) This is the primary parameter in labor cost models of the aircraft. This factor excludes the labor cost associated with the engines.

Labor (engines) The labor cost associated with the engines can be estimated knowing the size (thrust) and number of engines.

Material Maintenance material cost is most strong correlated with the total cost of the aircraft.

Depreciation cost

Depreciation is the primary DOC component which accounts for the unit (recurring) cost of the aircraft, although the hull insurance component is also dependent on the unit cost. The depreciation is dependent on the unit cost, the depreciation period, and the utilization factor, U, of the aircraft. Normally, the depreciation of the aircraft and depreciation of the engines are calculated separately and added together.

Allocation of cost to system segments

Table 8.1 shows how a specification format can allocate the three cost categories to the major aircraft segments. In the military world, specifications are written to document and allocate requirements at all levels of a system, whether or not one contractor makes that system.

Table 8.1 Allocation of cost constraints

Aircraft segment	Recurring costs	Non-recurring costs	DOC
Airframe	X	X	X
Avionics	X	X	X
Environmental	X	X	X
Mechanical	X	X	X
Electrical	X	X	X
Interiors	X	X	X
Propulsion	X	X	X
Support equipment	X	X	X
Assembly	X	X	N/A
Development	N/A	X	N/A
Production support	X	N/A	N/A
Total cast	X	X	X

人力成本（不包括发动机）这是飞机人力成本模型中的主要参数。该参数不包括与发动机相关的劳动力成本。

人力成本（发动机） 与发动机相关的人力成本可以根据发动机尺寸（推力）数量估算。

材料成本 维修材料成本与飞机总成本的关系最紧密。

折旧成本

折旧成本是 DOC 的主要组成部分，它计入飞机单位（重复）成本中，尽管机身保险部分还取决于飞机单位成本。折旧主要取决于飞机的单位成本、折旧期及利用率 U。飞机和发动机的折旧成本通常是单独计算并进行累加得到。

系统部段的成本分配

表 8.1 中展示了将 3 类不同的成本分配到飞机主要部段的规范格式。在军用领域，这些规范会写成文件，并将这些需求分配到系统的所有层级，无论一家承包商是否制造该系统。

表 8.1　成本约束的分配

飞机部段	重复性成本 RC	非重复性成本 NRC	直接运营成本 DOC
机体	X	X	X
航电	X	X	X
环控	X	X	X
机械	X	X	X
电气	X	X	X
内饰	X	X	X
推进	X	X	X
支持设备	X	X	X
装配	X	X	N/A
研制	N/A	X	N/A
生产支持	X	N/A	N/A
总计	X	X	X

Aircraft cost allocation

We saw above that the recurring cost of the aircraft is inherent in the depreciation and hull insurance components of DOC. With the initial sizing concept we can estimate the total cost of the aircraft using standard industry cost estimating techniques. These cost estimates provide a cost allocation for each subsystem. These cost allocations become the basis for the design-to-cost processes to be used on each subsystem. These cost allocations, also, can become technical performance measures (TPMs) as a subject of program management described in Section 12.7.

Performance allocation

Performance cost allocation must be done in a different manner since it cannot be allocated directly to all segments. There are three primary factors: aerodynamic performance, weight, and engine performance. Each of these factors can, then, be allocated to the appropriate segments. In addition, many components use power and have volume which affects drag. All three must act in concert to maintain the performance cost to a given level. The beginning of this chapter describes how to make an initial estimate of these performance parameters. Each of these parameters becomes a derived requirement which must be managed in order to meet the total estimated DOC.

Aerodynamic parameters

The initial sizing set the aerodynamic parameters (drag and lift coefficients) at the aircraft level to meet the performance and cost goals. The components of the total aerodynamic coefficients can then be established and allocated to the major external airframe components (wing, fuselage, empennage).

Weight allocation

Weight, in contrast, can be allocated directly. Once the total weight requirement of the aircraft is established, the weight of each segment and subsystem can be determined by allocation.

Engine performance

Next, the engine performance requirements established by the initial sizing become the basis for the engine performance contribution to DOC.

飞机成本分配

从上述内容我们可以看到，飞机的重复性成本包括 DOC 中的折旧及机体保险部分。根据初始参数指标概念，我们可以使用工业标准成本估算方法估算飞机总成本。这些成本估算提供了每个子系统的成本分配。这些成本分配构成每个子系统按成本设计过程的基础。这些成本分配还将成为技术性能指标 (TPM)，它是项目管理 (12.7 节) 的主题。

性能分配

性能成本分配必须通过不同方式完成，因为它不能直接分配给所有部段。该过程有 3 个主要因素：气动性能、重量及发动机性能。每个因素都可以分配到合适的部段。除此之外，多个部段使用能源，并拥有体积影响阻力。所有这三者都必须共同作用，以将其性能成本保持在某一给定水平。本章开头描述了如何对这些性能参数进行初步估算。每个性能参数都会变成一个衍生需求，必须予以管理以符合估算出的总 DOC。

气动力参数

初始参数设计过程在飞机级设置气动力参数 (阻力和升力系数)，以满足性能及成本目标。然后可确定总体气动系数的各个分量并将其分配给主要的外部机体部件 (机翼、机身和尾翼)。

重量分配

与之相反，重量可以直接分配。一旦确定了飞机的总重量需求，则每个部段和子系统的重量可以由分配决定。

发动机性能

下一步，由初步参数设计过程确定的发动机性能成为确定发动机性能对 DOC 贡献的基础。

维修成本分配

初始指标量化过程提供了总维修成本的初步估算，单位是美元每 1000 飞行

Maintenance cost allocation

The initial sizing discussed above provided an initial estimate of the total maintenance cost in terms of maintenance dollars per one thousand flight hours (MN\$/1000FH). In practice maintenance cost per flight hour varies significantly among subsystems as shown by historical data. This historical data can be used to establish maintenance cost allocations to the subsystems. Of course, the total maintenance cost cannot exceed the value estimated in the initial sizing.

Flight crew requirements

Finally, the crew requirements can be established from the allocated crew cost. Flight crew requirements are a type of people-related requirements discussed in Section 8.1, Item 11. The requirements include the number and skill types for flight deck crew and flight attendants. These requirements are normally fixed by regulation and do not often figure into aircraft optimization. However, cockpit requirements can be established to minimize the flight crew training.

Indirect cost factors

Like DOCs, indirect costs can also be used to establish design requirements. Major contributors to indirect costs are dispatch and operational reliabilities. For dispatch reliability, for example, the airline will incur direct costs, in terms of lost ticket sales and other factors, for each aircraft which fails to be dispatched on schedule. The impact of dispatch failures may vary greatly among airlines depending on the airline operating characteristics. Other indirect costs include ferry flights for repair and replacement and direct passenger costs for missed connections. Like weight, the dispatch reliability can be allocated to the aircraft segments and to their subsystems. Hence, the standard SE practice of requirements allocation can be used directly to flow down performance requirements to all system segments, subsystems, and components to meet operational direct and indirect cost constraints.

Manufacturer's indirect cost savings

SE is expected to result in other cost savings to the manufacturer. However, these savings cannot easily be converted into design requirements. Many of these savings can be considered as SE *metrics*, that is, a measure of the effectiveness of the process, providing the contribution of SE can be isolated. A discussion of metrics, however, is beyond the scope of this book.

小时 (MN\$/1 000FH)。实际上，通过历史数据可以看出，每飞行小时维修成本在不同子系统之间差别巨大。历史数据可以用于将维修成本分配到子系统。当然，总维修成本不能超过初始参数设计阶段的估算值。

飞行机组需求

最终，飞行机组需求可以通过分配到的机组成本来建立。飞行机组需求是一种与人有关的需求 (8.1 节，第 11 项)。这种需求包括飞行机组和飞行乘务员的数量和技能类型。这些需求通常由规章规定，且通常不反映到飞机优化中。然而，可建立驾驶舱需求用于尽量减少飞行机组培训工作。

间接成本因素

与 DOC 类似，间接成本也可以用于建立设计需求。间接成本主要源于签派及运行可靠性。例如，对于签派可靠性，如果飞机不能按时签派，则航空公司将产生直接成本，包括丧失机票销售和其他因素。签派失效的影响可能在不同航空公司间差别较大，该影响主要取决于航空公司的运营特征。其他间接成本包括飞机为修理和换件进行的空机转场，以及衔接延误导致的直接旅客成本。与重量类似，派遣可靠性可以分配给飞机各部段及子系统。因此，需求分配的标准系统工程做法可以直接用在性能需求分配到所有系统部段、子系统及部件，以满足运行的直接和间接成本约束。

制造商节省的间接成本

人们期望系统工程能使制造商节省其他成本，然而，这些节省不能直接转化为设计需求。很多节省的成本都可以被当作系统工程指标来考虑，即该过程效能的衡量方法，如果系统工程的效果可以独立显现。然而，关于指标的讨论不在本书讨论范围内。

以 Honour 为例，他介绍了项目管理领域使用自动化系统工程工具节省成本的案例 (Honour, 1994)。一个成功的指标是，因图纸发放后重新设计较少在工程

Honour, for example, shows the cost savings in project management from using automated SE tools (Honour, 1994). A successful metric is the cost reduction from fewer redesigns and after drawing release, both in engineering and manufacturing. Others are reduced warranty claims and improved sales. Although SE may have a significant impact on sales, the effect would be virtually impossible to isolate.

Airline customer's indirect cost savings

Similarly SE is expected to result in many other cost savings to the airline customers. These savings also may be impractical to convert into design requirements unless cost models exist which permit this conversion. Indirect airline savings include: savings in reservations and ticket sales, saving in advertising and publicity, reduced maintenance and depreciation of non-flight items, reduced general and administrative (G&A) costs, and reduced passenger services (transportation and hotels). Of these, only those which can be linked to dispatch reliability are usually candidates for design-related costs.

8.7   Top-Level Trade-Offs

In addition to the sizing trade-offs discussed earlier in this chapter, certain other trade-offs count as top-level because they involve trade-offs between (or among) segments. Some typical top-level trade-offs are shown in Table 8.2. Chapter 9 will discuss these on a subsystem-by-subsystem basis.

Table 8.2   Top-level trade-offs

Trade-offs	Subsystems
Mechanical vs. electrical controls	Mechanical, electrical
Electrical vs. hydraulic power	Electrical, hydraulics
Bleed-air vs. self-contained air supply	Pneumatics, propulsion
Noise suppression	Fuselage, interiors, propulsion, electrical (active suppression)
Conventional vs. propulsion control	Propulsion, mechanical, electrical, airframe

和制造两方面降低的成本。其他方面包括担保索赔减少和销量增加。尽管系统工程对销量有较大影响，但其效果难以独立显现。

航空公司客户节省的间接成本

同样，预期系统工程也能为航空公司客户节省许多其他方面的成本。这些节省成本也基本不可能转换成设计需求，除非存在允许这种转换的成本模型。航空公司节省的间接成本包括：在订座和机票销售方面节省的成本、广告和公关方面节省的、减少非飞行件的维修和折旧、减少管理和行政 (G&A) 成本及减少乘客服务 (运送和旅馆)。其中，通常只有与签派可靠性有关的成本才可以作为与设计相关成本的候选项。

8.7 顶层权衡研究

除了本章前面讨论的参数设计的权衡研究之外，还有其他顶层的权衡研究，因为它们涉及两个(或多个部段)之间的权衡。一些典型的权衡研究如表8.2所示。第 9 章将对一些子系统逐个进行讨论。

表 8.2　顶层权衡研究

权衡研究	子系统
机械 vs 电气控制	机械，电气
电源 vs 液压源	电气，液压
引气 vs 自主供气	气源，推进系统
噪声抑制	机体，内饰，推进，电气 (主动抑制)
传统 vs 推进控制	推进，机械，电气，机体

9

Subsystem Synthesis

And the avionics cannot say unto the pilot, I have no need of thee, nor again the empennage to the wings, I have no need of you.

Paraphrased from I Corinthians 12: 12–26,
Holy Bible, King James (Authorized) Version

This chapter will show the primary performance and constraint requirements for each of the major subsystems. It will enumerate and describe the principal subsystem-level requirements parameters and how they are allocated to subsystem hardware and software in the SE methodology.

As we have noted before, subsystems are normally called systems in the aircraft industry. However, in SE terminology *subsystem* better describes where these elements fit within the aircraft hierarchy and methodology.

Virtually all subsystem requirements are derived requirements. That is, they depend on solutions for their values. Therefore, we cannot, in this chapter, say which requirements are allocated to which subsystems. We can only describe typical allocations. Neither can we say, in all cases, what the exact, quantitative requirements are. We can only determine what the functions are that should be converted into quantitative requirements and subsequently allocated to hardware or software.

In addition, it will be remembered that a basic SE philosophy calls for the requirements for the subsystems to be *holistically* determined as described in Section 4.1, and the conflicting requirements must be resolved using the methods of requirements trade-offs described in Section 4.8.

As we saw before in Section 8.4, the flow down of requirements from the top level is dependent on the aircraft architecture selected at that level. We are using the architecture described in Figure 2.1, keeping in mind that other architectures are possible and that top-level trade-offs will be necessary to determine whether the architecture of Figure 2.1 is, indeed, the best architecture.

A key aspect of synthesis is the trade-off. We cannot say exactly what trade-offs

276

第9章 子系统综合

"航电（系统）不能对驾驶员说，我不需要你，尾翼不能对机翼说，我不需要你。"[1]

来源于《圣经·哥林多前书》12 章：12-26 节，詹姆斯王（授权）版本

本章主要介绍每个主要子系统的主要性能及约束需求。本章将列举并描述主要的子系统级需求参数，以及如何使用系统工程方法将其分配到子系统的硬件和软件。

如前面所述，子系统在航空业界中一般称为系统。然而，按系统工程术语，"子系统"这个词能更好地描述这些元素在飞机架构及方法学中的匹配关系。

事实上，所有子系统需求都属于衍生需求。因为这些需求的值取决于解决方案。因此，在本章中我们不能断定哪些需求分配给哪些子系统，只能描述一些典型的分配。同样，我们也不能断定所有情况下准备的定量需求。我们只能确定哪些功能应被转换成定量需求并随之分配给软硬件。

除此之外应注意，系统工程的一个基本理念是要求子系统的需求都以 4.1 节所述形式整体地确定，所有冲突的需求均必须使用需求权衡研究方法 (4.8 节) 来解决。

如我们在 8.4 节介绍的，顶层需求的向下分解取决于此层选定的飞机架构。我们正在使用图 2.1 描述的架构，需要注意的是，其他架构也是可能的，且必须

277

---

1 原文是：And the eye cannot say unto the hand, I have no need of thee: nor again the head to the feet, I have no need of you. 眼不能对手说，我用不着你；头也不能对脚说，我用不着你。说明身体完整重要性，这里 Scott Jackson 进行了修改，目的是说明在子系统综合中的相互关系。——译者注。

need to be made; we can only describe *typical* trade-offs. Subsystems will be involved in trade-offs at two levels: the aircraft and the subsystem level. Although Chapter 8 dealt with aircraft-level synthesis, we will discuss here specific instances in which subsystems may be involved in trade-offs of two or more subsystems. The importance of these inter-subsystem trade-offs is that they should be addressed very early in aircraft development, that is, before solutions are *assumed* based on the experience of a single design discipline.

Implicit in the synthesis of every subsystem discussed below is the principle that quantitative requirements should be developed from each function in accordance with the principles of Chapter 4, and that these requirements should drive the subsystem design. We will continue to capitalize the names of functions in accordance with the conventions of Chapter 3.

Since the segments below are not true subsystems, but rather collections of equipment, the correlation between these segments and the functions of Section 3.3 is only approximate.

9.1   Environmental Segment

Most functions provided by the Environmental Segment emanate from the Provide Environmental Control function in Figure 3.6.

Air conditioning (ATA 21)

The air conditioning subsystem has a number of key functions: Provide Temperature Control of Air Supply, Provide Pressure Control of Air Supply, Provide Distribution of Conditioned Air, Provide Air Filtration, and Provide Ventilation. It is theoretically possible to control the humidity of air in the aircraft. However, conditions rarely warrant humidity control. These functions are subordinate to the Provide Environmental Control function in Chapter 3.

While the traditional method of temperature control is to mix the warm air from the pneumatic subsystem with refrigerated air, this trade-off should be revisited for future designs. Alternatives include an independently controlled autonomous air supply with, perhaps, a ram air supply. A top-level trade-off would then be required to determine whether the ram air drag penalty on the aircraft would be too detrimental.

The design of the air conditioning system is highly dependent on the heat loads. The driving external environment is normally the hot-day ground condition. Internal

进行顶层的权衡研究，以确定图 2.1 所示的架构是否确实为最佳选择。

综合的一个关键方面是权衡研究。我们不能准确说出需要进行哪些权衡研究，只能描述一些典型的权衡研究。在两个层级的权衡研究中都涉及子系统：飞机级和子系统级。尽管第 8 章讨论了飞机级的综合，我们在本章仍将讨论一些具体实例，这些实例中可能涉及两个或者多个子系统的权衡研究。在飞机研制的最早阶段，即在根据某单一设计学科假定一些方案之前，就应认识到这些子系统间权衡研究的重要性。

下面讨论的每个子系统综合中隐含的原则是：应根据第 4 章的原则开发每个功能的定量需求，这些需求应驱动子系统设计。我们将按照第 3 章的术语定义，继续把功能的名字进行首字母大写。[1]

由于下面的部段不是真正的子系统，而是一些设备的组合，这些部段与第 3.3 节中功能的关系只是近似的。

9.1 环境部段

由**环境部段**提供的大部分功能均来源于图 3.6 中的**提供环控**功能。

空调 (ATA21)

空调子系统有许多关键功能：**提供气源的温度控制，提供气源的压力控制，提供调节后空气的分配，提供空气过滤，提供换气通风**。理论上控制飞机内的空气湿度是可能的，然而，调节系统很少确保湿度控制。这些功能都是第 3 章**提供环控**功能的下一级功能。

虽然传统的温控方法把气源子系统暖空气与冷却的空气进行混合，但在未来设计中，应重新对其进行权衡研究。替代方法包括一个独立控制的自主气源，比如冲压空气源。需要通过顶层的权衡研究确定：飞机付出的冲压空气阻力代价

1　中文部分以加粗字体表示。——译者注。

heat loads, such as from electronic equipment, also figure in the sizing.

The Provide Distribution of Conditioned Air function provides the opportunity to minimize the total subsystem weight by optimally placing the air conditioning elements throughout the aircraft. The weight of cables and other components may be decisive factors in the placement trade-off.

In addition, of particular importance are both the *functional* and *physical* interfaces discussed in Chapter 6 the distribution subsystem has with other subsystems. For example, the galley and environmental engineers should be in complete agreement on the flow rate and characteristics of the air to be delivered to the galleys.

While the environmental control engineer has control of defining the air conditioning equipment, the flow of air throughout the aircraft can be considered a top-level issue because the air passes through, under, in, and around components not under the direct control of the environmental control subsystem (ECS). These components include bag racks, equipment racks, lavatories, galleys, tunnels, and cargo areas. This top-level consideration of functions normally considered the purview of subsystem organizations is a key value-added aspect of SE.

Cabin pressure (ATA 21)

Typically, the cabin pressure subsystem does not *provide* the cabin pressure. That is done by the air conditioning subsystem. The main functions of this subsystem are Monitor Cabin Pressure and Control Cabin Pressure Outflow. The sizing of the outflow valves will be dependent on the cabin pressure and the expected rate of out-flowing air to maintain the cabin pressure. The requirements for cabin pressure vary with altitude and are strictly controlled to prevent an excessive pressure change on the pressure shell of the aircraft.

Another function of the cabin pressure subsystem is Control Air Flow in the cabin. It is a constraint of aircraft design that air should flow laterally and not longitudinally in the cabin. Optimal placement of the outflow valves is required to meet this constraint.

Ice and rain protection (ATA 30)

The ice and rain protection subsystem is not a single collection of elements, but rather separate elements each with its own synthesis solution. The functions are Provide Anti-Icing, Provide De-Icing, and Provide Rain Protection.

是否过大。

空调系统的设计很大程度由热负载决定。驱动的外部环境通常是炎热天气下的地面环境。内部热负载，比如来自电子设备的热负载，也应在参数设计中考虑进去。

提供调节后空气分配功能提供一个通过全机优化空调组件的布置，来降低子系统总重。电缆及其他部件的重量可能是布置权衡研究的决定因素。

除此之外，空气分配子系统与其他子系统之间的功能和物理接口也非常重要。比如，负责厨房和环控的工程师应在传递给厨房的空气流速和其他空气特征方面达成完全一致。

虽然环控工程师对定义空调设备有控制权，但整架飞机的空气流量应被当作一个顶层问题来考虑。因为空气要在不受环控系统 (ECS) 直接控制的一些部件中穿过，或在其下方、其内部及周围通过。这些部件包括行李架、设备架、盥洗室、厨房、管道及货舱区域。关于该功能的顶层考虑通常被认为属于子系统组织范围，而这是系统工程关键的增值部分。

座舱压力 (ATA21)

一般而言，座舱压力子系统不提供座舱压力。该工作由空调子系统完成。该子系统的主要功能是**监控座舱压力**并**控制座舱压力放气**。放气阀门的参数指标取决于座舱压力及用于维持座舱压力的预期放气速率。座舱压力需求随高度不同而不同，应严格控制以避免飞机的承压壳体承受剧烈的压力变化。

座舱压力子系统的另一个子功能是**控制座舱空气流量**。它是一个飞机设计约束，空气应能在座舱里横向而不是纵向流动。放气阀门的最佳位置需要满足这个约束。

防冰除雨 (ATA30)

防冰除雨子系统不是多个元件的单一组合，而是各有其综合解决方案的一

De-icing is normally performed, for example, with hot air from the power plant delivered by the pneumatic subsystem. However, as we saw in Section 9.1, air can be provided from other sources. Anti-icing of small structural elements, such as strakes, can be done with a low current electrical heater. All of these solutions involve considerable trade-offs among the ice and rain protection, the electrical subsystem, the pneumatic subsystem, the power plant subsystem, and the fuel subsystem, with the figures of merit being weight, reliability, and fuel consumption.

Rain protection is normally accomplished with wipers and fluids.

Oxygen (ATA 35)

There are three oxygen subsystem functions: Provide Crew Oxygen, Provide Passenger Oxygen, Provide Portable Oxygen. The FARs provide the primary requirements for these functions. The main subsystem trade-off is between chemical and gaseous oxygen supplies. The oxygen duration requirement is driven by the mission profile of the aircraft. That is, the longer it takes to descend to a safe altitude, the larger the oxygen supply should be. Longer durations usually lead to gaseous oxygen supplies.

Pneumatic (ATA 36)

The pneumatic subsystem is similar to the cabin pressure subsystem in that it does not *provide* pressure to anything. The pneumatic subsystem acts as a conduit between the power plant and the air conditioning and ice and rain protection subsystems, for example. The pneumatic subsystem also provides air for engine starting, cargo heating, and water pressurization. Hence, the main function of the pneumatic subsystem is Maintain Pneumatic Pressure. Other functions are Provide Ozone Conversion of the exterior air and Provide Particle Filtration of recirculated air. However, as we saw in Section 9.1, all of these functions can be provided by alternative air supplies and are, hence, subject to top-level trade-offs.

Structural cooling

Conventional subsonic aircraft do not need nor do they have structural cooling subsystems. However, the high-speed civil transport (HSCT) may require cooling of its

些单独元件。功能包括**提供防冰，提供除冰**和**提供除雨**。

例如除冰一般使用气源子系统从发动机引出来的热空气。然而在 9.1 节可以看到，可以由其他来源也可提供空气。小的结构件比如边条翼的防冰，可以用低电流电加热器。所有这些解决方案都涉及在防冰除雨、电气子系统、气源子系统、动力装置子系统和燃油子系统之间，针对重量、可靠性、燃油消耗率等方面大量的权衡研究。

除雨功能一般通过雨刷和除雨液实现。

氧气 (ATA35)

氧气子系统有 3 种功能 ：**提供机组氧气、提供乘客氧气、提供便携式氧气**。FAR 规定了关于这些功能的基本需求。主要的子系统权衡研究是采用化学氧源还是气体氧源。持续供氧时间需求由飞机的任务情况决定。即下降到安全高度时间越长，需要的供氧气量就越大。通常较长的持续时间可能需要气体氧源。

气源 (ATA36)

气源子系统与座舱压力子系统类似，不提供任何压力。气源子系统起到了发动机和空调、防冰、除雨子系统中介的作用。气源子系统也提供空气以供发动机起动、货舱、加温和水的增压。因此，气源子系统的主要功能是**维持空气压力**。其他功能包括对外部空气进行**臭氧转换**并提供再循环空气的**颗粒过滤**。然而，如9.1 节所示，所有功能都能由其他供气系统提供，因此须进行顶层权衡研究。

结构冷却

传统的亚声速飞机既不需要，也不存在结构冷却子系统。然而，高速民用运输机 (HSCT) 可能需要冷却其结构件，尤其是高速飞行状态下的机翼。确定冷却需求及最佳方法需要进行权衡研究。使用管道燃油是进行结构冷却的一种可能

structural elements, primarily the wing, for high-speed flight. Trade-offs will be needed to determine the requirements and optimum technique for this cooling. The use of ducted fuel is one possible method for structural cooling.

9.2   Avionics Segment

Almost all functions provided by the Avionics Segment are subfunctions of the Navigate Aircraft function in Figure 3.6.

Auto flight (ATA 22)

The Navigation Aircraft and Commend and Control Aircraft functions are discussed in Section 3.3; these functions can be allocated either to the auto flight subsystem, to the flight crew, or to both. Factors associated with human flight control are discussed as part of the human factors analysis discussed in Section 5.5. The Provide Auto Flight function is subordinate to the Navigate Aircraft function is the equipment- related part of this function pertaining to the automatic control of the aircraft.

The Provide Auto Flight function uses air data, inertial navigation system (INS), and FMS as inputs and pitch and roll commands to the ailerons, elevators, rudder, and throttle. It provides the macroscopic commands for aircraft take-off, cruise, and landing and also the corrective commands, such as yaw damping, angle of attack correction, and Mach trim. It also performs speed and altitude correction by converting pressure altitude to actual altitude when the aircraft is below 10,000 ft.

Communications (ATA 26-10)

All communications functions are subfunctions of the Communicate Data/ Information function top-level function shown in Figure 3.6. The two principal subfunctions are Provide External Communications and Provide Internal Communications.

External communications

The requirements for external communications are driven by the range of communications, the need to transmit both voice and data, and the need to transmit free of atmospheric interference.

At shorter ranges, for example, nearer airports, very high frequency (VHF) systems have provided voice and data links. This medium is limited to line-of-sight

方法。

9.2 航电部段

航电部段提供的所有功能几乎都是图 3.6 所示**飞机导航**功能的子功能。

自动飞行 (ATA22)

3.3 节讨论了**飞机导航、指挥和控制飞机**功能，这些功能可以分配给自动飞行子系统或飞行机组，或同时分配给两方。与人为飞行控制相关的因素是 5.5 节人为因素分析的一部分。**提供自动飞行**功能是**飞机导航**功能的下一级功能，和飞机自动控制有关，是该功能与设备有关的部分。

提供自动飞行功能使用大气数据、惯性导航系统 (INS) 及飞行管理系统 (FMS)的输入，并提供偏航和横滚指令给副翼、升降舵、方向舵和油门。它提供飞机起飞、巡航和着陆的宏观命令，同时也提供修正指令，包括偏航阻尼、攻角修正及马赫数配平。同时，当飞机低于 10 000 ft 时，通过将压力高度转换为实际高度进行速度和高度修正。

通信 (ATA 26–10)

所有通信功能都是图 3.6 中**数据 / 信息通信**功能顶层功能的子功能，两个主要子功能包括**提供外部通信**和**提供内部通信**。

外部通信

外部通信的需求由通信距离、传输语音及数据的需要及传输无大气干扰的需要决定。

在较短的距离上，比如靠近机场，其高频系统 (VHF) 将提供语音和数据通

communications and is not vulnerable to atmospheric interference.

For worldwide communications, high frequency (HF) links have provided non-line-of-sight communications, however, with the disadvantage of atmospheric interference. A more recent development in worldwide communications is SATCOM. Because of its accuracy and freedom from atmospheric interference and line-of-sight constraints, this medium is capable of replacing both VHF and HF media. Such a system will be essential for the HSCT.

The communications system is only responsible for transmitting voice and data. Process of this data should be accomplished by other systems, such as the ARINC (originally Aeronautical Radio, Incorporated now part of Rockwell Collins) communication addressing reporting system (ACARS) described in Section 9.2.

Another adjunct communications system is selective calling (SELCAL). The purpose of SELCAL is to send a coded signal from the ground through VHF, HF, or SATCOM to ring a chime in the flight deck indicating a desire for communications on the same frequency.

Another external communications system is the airborne telephone. This system operates on L-band on a line-of-sight path to ground stations. Messages are then relayed via ground and satellite to other ground stations throughout the world. SATCOM also has an embedded telephone system.

Because of the importance of external communications, much redundancy is required among external communications systems. Although most communications systems are embedded in the flight deck equipment, some aircraft systems require portable external communications as backup systems.

Static discharge wicks are located on various parts of the aircraft to dissipate static discharge which may cause external communications interference.

A final external communications system is the crash position indicator. An emergency locator transmitter (ELT) emits an emergency locator beacon (ELB) to determine the location of the aircraft.

Internal communications

The purposes of internal communications are for the flight crew and cabin crew to communicate with each other and to the passengers, and for service personnel to communicate with each other or with the pilots.

The passenger address system allows the flight and cabin crew to communicate to the passengers. This can be done either by audio (tape or voice) or by video (tape

信链。传输媒介局限在视距通信范围内，且不易受大气干扰。

在全球范围通信，高频 (HF) 可提供非视距通信，然而，却存在易受大气干扰的缺点。世界范围内通信手段的最新进展是卫星通信 (SATCOM)。由于其准确性及无大气干扰和视距方面约束，这种媒介能够替代甚高频和高频通信。这种系统对高速民用运输机 (HSCT) 而言非常重要。

通信系统只负责语音和数据的传输，处理数据应由其他系统完成，比如 9.2 节中的 ARINC 通信寻址与报告系统 (ACARS)。

另一个辅助通信系统是选择呼叫系统 (SELCAL)。SELCAL 的目的是从地面通过甚高频、高频或者卫星通信发送一个编码信号，使驾驶舱响起钟声，表示在同一频段上有通信请求。

另一外部通信系统是机载电话系统。此系统在 L 波段上与地面站视距范围内工作。消息通过地面或者卫星中继再传递给世界上其他的地面站。卫星通信也是一个嵌入式的电话系统。

由于外部通信的重要性，需要在外部通信系统间有多种冗余度。尽管大部分通信系统都是驾驶舱内设备，一些飞机系统也需要便携式的外部通信作为备用系统。

在飞机不同位置安装 (静电) 放电刷用于释放静电，这些静电可能会造成外部通信干扰。最终外部通信也是坠机位置指示器。紧急定位发射器 (ELT) 可发出紧急定位信标 (ELB)，以确定飞机位置。

内部通信

内部通信用于飞行机组与客舱机组相互通话、与乘客通话，并用于服务人员相互之间通信，或者服务人员与驾驶员通话。

乘客广播系统使飞行或客舱机组能与乘客通话，通话可以通过音频 (磁带或

287

only). Because of the importance of this system in emergency situations, the passenger address system is powered by the backup battery power system described in Section 9.3.

This is another example of incorporating people into the requirements process. People, that is, flight and cabin crew and passengers, are a part of the system to which tasks can be allocated, for example, communicating emergency messages. They are also elements which interface with the system.

The services interphone system allows service personnel to communicate with each other or with the pilots. This function is part of the Provide Internal Communications discussed in Section 3.3 and is implemented by jacks located throughout the aircraft.

The flight interphone system allows the flight deck crew and the cabin crew to communicate with each other.

Other internal communication systems include the call system which allows passengers to signal flight attendants by means of a light. On-board megaphones provide backup communications to the passengers in emergency situations.

Audio mixing

A key requirement is for audio signals from all communication systems to be mixed and provided to the pilots. The system which does the mixing is the digital core avionics system (DCAS). DCAS integrates voice inputs from the headset microphone, the handset microphone, and the oxygen mask microphone, and provides the signals to the headsets and to the loud speakers.

Indicating and recording (ATA 31)

Instrument panel information

The indicating and recording subsystem provides the pilots with the critical information determined by the guidance and navigation subsystem discussed in Section 9.2 and other subsystems. Most requirements of the indicating and recording subsystem derive from the Communicate Data/Information top-level function discussed in Section 3.3. The instruments provide flight information regarding position, attitude, and heading.

The information provided by the instrument panel includes time (GMT) which is used as a time base for maintenance and for the flight recorders.

The layout of the instrument panel is strongly driven by human factors considerations discussed in Section 5.5. These considerations have led to the so-called

语音)或者视频(只有录像带)实现。由于该系统在紧急情况下的重要性,乘客广播系统由备用电池电源系统 (9.3 节) 供电。

这是把人纳入需求过程的又一个例子。这里的人指的是飞行机组、客舱机组和乘客,他们作为系统的一部分,被分配一些任务,比如通知紧急消息。它们也是与系统接口的元素。

服务内话系统允许服务人员相互间通话或与驾驶员通话。这项功能是 3.3 节**提供内部通信**的一部分,由遍布全机的耳机插孔实现。

飞行内话系统能使驾驶舱机组成员与客舱机组相互通话。

其他内部通信系统包括呼叫系统,通过灯光的方式,使乘客招呼飞行乘务员。机上扩音器在紧急情况下为乘客提供备用通信。

音频混合

所有通信系统音频信号的关键需求是应将其混合并提供给驾驶员,用于进行混音的是数字核心航电系统 (DCAS)。DCAS 把从耳塞麦克风、手持麦克风及氧气面罩麦克风来的语音输入进行集成,提供信号给耳机和扩音器。

指示记录 (ATA 31)

仪表板信息

指示记录子系统向驾驶员提供由导引和导航子系统 (9.2 节) 和其他子系统确定的关键信号。指示记录系统的大部分需求由 3.3 节讨论的**数据 / 信息通信**的顶层功能衍生出来。仪表提供关于位置、姿态及航向的飞行信息。

仪表板提供的信息包括时间 (GMT),它用作维修和飞行记录器的时间基准。仪表板的布局在很大程度上由人为因素 (5.5 节) 方面的考虑决定。这些考虑产生了所谓的基本 T 型布局,该布局目前已成为全世界的设计惯例,如图 9.1 所示。

仪表板需要给正副驾驶提供完全相同的基本 T 型布局信息。除此之外,需

basic T layout which, by convention, is used throughout the world, as shown in Figure 9.1.

The instrument panel is required to provide identical basic T information to both the pilot and co-pilot. In addition, a *comparator* determines the differences, if any, between the two readings. If there is a difference, the pilots should rely on the backup systems to determine which is more reliable.

Human factors analysis imposes other requirements to assure readability of the instruments, for example, parallax and glare avoidance, lighting levels, color coding, and font or needle sizes and shapes.

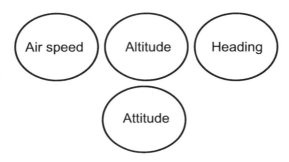

Figure 9.1   The basic T instrument panel layout

A design solution created by human factors is the so-called *dark cockpit* philosophy, which derives from the requirement to avoid distractions to the pilots. In the *dark cockpit* only the information absolutely needed is provided to the pilot.

Annunciation

Another key function of the indicating and recording subsystem is Provide Annunciation. Annunciation is any kind of visual or aural indication of aircraft status. Visual annunciations can take the form of lights or color displays on the instrument panel. Aural annunciations can take the form of many sounds, such as chimes, horns, or human voices. On modern aircraft a central warning system creates synthesized sounds. There are three levels of annunciation:

1. *Warning level*   Immediate action is required.
2. *Caution level*   Action is required but is not immediate.
3. *Advisory level*   An advisory annunciation indicates that an event has occurred.

要有比较器来确定两个读数之间的差异。如果存在差异，驾驶员应依赖备用系统来判断哪一个比较可靠。

人为因素分析产生了其他需求，以确保仪表的可读性，比如避免视差和炫光、灯光等级、颜色编码及字体或指针的尺寸和形状。

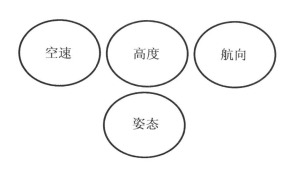

图 9.1　基本 T 型仪表板布局

人为因素产生的一个设计方案是所谓的静暗驾驶舱设计理念，它是由避免使驾驶员分心的需求衍生的。在静暗驾驶舱中，只有绝对需要的信息才会告知驾驶员。

通告

指示记录子系统另一个关键功能是**提供通告**。通告是飞机状态的任何目视或音响的提示。目视通告可以通过仪表板上的灯光或彩色显示器来呈现。音响通告可以使用许多声音，比如，钟鸣、喇叭或真人语音。在现代飞机上有一个中央警告系统用于产生合成的声音。通告有三个等级

1. **警告级**　需要马上行动。

2. **警示级**　需要行动，但不是马上行动。

3. **通告级**　通告级通告表示发生了某一事件。

Satchell (1993) 对飞机的监控与告警系统进行了全面的描述。

Satchell (1993) provides a comprehensive description of monitoring and alerting systems for aircraft.

Data recording

The Record Data function is divided into three subfunctions: Record Cockpit Voice, Record Flight Data, and Record Maintenance Data. The first two are, of course, essential for accident investigations and are imposed by regulatory agencies. The maintenance data are collected to conduct trend analyses for failure prevention and to measure engine degradation, for example. The flight data and maintenance data are often recorded on the same recorder for economy.

Fault collection

A function is required to collect and record root cause information for faults, such as the failure of components. The system which does this is called the central fault collection system. It provides sufficient information so that the cause of the faults can be diagnosed and corrective action can be taken.

Automatic data reporting

Another function of the indicating and reporting subsystem is to provide information to and from ground maintenance personnel during flight. This capability allows ground maintenance personnel to interrogate the aircraft and analyze maintenance data during flight. It also provides automatic reporting of maintenance data to the ground. The system is called the ARINC communication addressing reporting system (ACARS).

Navigation (ATA 34)

The navigation subsystem is an example of a subsystem derived almost entirely from a single top-level function, Navigate Aircraft shown in Figure 3.6. The requirements for the parameter associated with the named subfunctions vary according to (a) the degree of accuracy required, (b) the reference point of the information, and (c) the degree of redundancy required for safety.

The Determine Location of Aircraft subfunction can be divided into several categories. The first major category is dependent position, that is, dependent on

数据记录

数据记录功能被分为三个子功能：**记录驾驶舱语音，记录飞行数据和记录维修数据**。前两个显然是是事故调查的基础，是监管机构要求的。收集的维修数据可用于进行预防失效的趋势分析，例如，用于发动机退化的度量。出于经济性考虑，飞行数据和记录数据往往记载到同一个记录器中。

故障收集

需要有一个功能收集和记录错误，如部件故障的根因信息。这个系统称为中央故障收集系统。该系统提供了充分的信息，用于诊断故障原因并采取修正措施。

自动数据报告

指示记录子系统的另一个功能是在飞行过程中向地面维修人员提供信息，或者接受来自地面维修人员的信息。该能力使地面维修人员可在飞行过程中询问飞机并分析维修数据。该功能还向地面自动报告维修数据。该系统称为 ARINC 飞机通信寻址与报告系统 (ACARS)。

导航 (ATA 34)

导航子系统是一个几乎完全由单一顶层功能衍生的子系统的例子，该顶层功能是图 3.6 中的**飞机导航**。与几个子功能相关的参数需求由于下面三个方面而不同，包括：(a) 所需准确度；(b) 信息基准点，以及 (c) 安全性要求的冗余度。

确定飞机位置子功能可以分成几类。第一大类是相对位置，即相对于地面台的位置。比如**指点信标**位置，给出了相对于跑道阈值的位置。甚高频全向无线电信标 (VOR) 位置给出了相对于正北的方位，测距器 (DME) 提供相对于 VOR 根据台的位置。方位和相对位置的组合可以得出绝对位置。

ground stations. The Marker Beacon position, for example, gives position relative to the runway threshold. The VOR (VHF omni-directional radio) position gives bearing relative to north. The DME (distance measurement equipment) gives position relative to the VOR station. The combination of the two, bearing and relative position, gives absolute position.

The ILS (instrument landing system) gives position relative to airports. The second major category is independent on position, that is, independent of ground stations. This position can be determined by an inertial navigation system (INS). The rate of change of this position provides ground speed. Another, more recent, independent system is the global positioning system (GPS), which determines position from satellite information. Because of the high accuracy of the GPS, this system will probably be the preferred solution of the future.

Other critical pieces of position information are the positions relative to the ground and the prediction of the aircraft position relative to the terrain and to other aircraft. This information is used to provide warning of possible ground or aircraft impact described in Section 9.2. The system which provides prediction of aircraft collision is called the traffic collision avoidance system (TCAS). The system which predicts possible ground impact is called the ground proximity warning system (GPWS).

The basic solution for the Determine Heading of Aircraft function is the horizontal situation indicator (HSI), an inertial instrument. As a backup system, a magnetic whisky (that is, in alcohol) compass is used. The pitch and roll attitudes are determined by the inertial attitude direction indicator (ADI).

The navigation subsystem uses air data to satisfy a number of navigation functions. These include Determine Air Speed, Determine Angle of Attack and Slip, Determine Air Temperature, and Determine Impending Stall.

A major navigation function is Provide Flight Management. The purpose of the flight management function is to provide the correct aircraft commands to both the pilot and the auto pilot discussed in Section 9.2. The flight management system (FMS) is a computer which integrates the navigation data described above and provides this information to the Flight Director, another computer, which generates the commands. These commands are provided to the pilot through the instrument panel described in Section 9.2 and directly to the auto pilot. On some aircraft, pilot commands are provided to the aircraft directly from the Flight Director. The considerations involved in the trade-off between these two approaches are load feel characteristics, weight,

仪表着陆系统 (ILS) 给出了相对于机场的位置。第二大类与位置无关，即不依赖于地面站。该位置信息可由惯性导航系统 (INS) 确定。通过该位置的变化率生成了地速。近年来，出现了另一种独立系统，即全球定位系统 (GPS)，它通过卫星信息确定位置。由于 GPS 的高准确度，该系统在未来或许是优选方案。

位置信息的其他关键部分是相对于地面的位置，以及飞机相对于地形和其他飞机的预测位置。这个信息用于提供可能撞地或撞机警告 (9.2 节)。提供飞机相撞预测的系统称为空中防撞系统 (TCAS)，而预测可能撞地的系统称为近地警告系统 (GPWS)。

确定飞机航向功能的基本方案是水平状态指示器 (HIS)，它是一个惯性仪表。同时还使用磁罗盘作为备用系统。俯仰和横滚姿态则由姿态方向指示器 (ADI) 确定。

导航子系统使用大气数据以实现一系列导航功能。这里包括**确定空速、攻角和侧滑角、确定空气温度及确定即将失速**。

提供飞行管理功能是一种主要的导航功能。飞行管理功能的用途是为驾驶员和自动驾驶仪 (9.2 节) 提供正确的飞机指令。飞行管理系统 (FMS) 是一台计算机，它集成上述导航数据并提供给另一台计算机，即飞行指引仪，并由飞行指引仪生成指令。如 9.2 节所示，这些指令通过仪表板提供给驾驶员，或直接提供给自动驾驶仪。在一些飞机上，飞行指引仪直接向飞机提供相关的驾驶指令。这个两种方式的权衡考虑包括载荷和感觉特性、重量、可靠性、维修成本及培训成本。

reliability, cost maintenance, and training costs.

9.3 Electrical Segment

The primary top-level function in Figure 3.6 that the Electrical Segment supports is the Provide Power function.

Electrical power (ATA 24)

The primary electrical power functions are Provide AC Power and Provide DC Power. Necessary adjunct functions include Provide Backup Power and Provide Load Distribution. These functions are all subordinate to the Provide Power function discussed in Section 3.3.

The normal way of providing electrical power is by a generator powered by mechanical torque from each engine. This engine produces alternating current (AC) power, which must be converted to direct current (DC) power for the DC electrical components. The DC power is also used for recharging batteries for backup power. Similarly, the DC power from the batteries is converted to AC power to provide power to the AC components which are needed during emergency operation: that is, they are part of the MMEL discussed in Section 5.4.

Load distribution is achieved by analyses during the synthesis phase. The objective of load distribution is to assure a balance of electrical loads so that no elements of the electrical system are overloaded when some components fail.

The electrical power subsystem synthesis is replete with constraints. Typical are the limitations on emitted EMI discussed in Section 5.10. EMI can be limited by sufficient cable shielding. Shielding also protects other components against surges in electrical power. All electrical power elements must be grounded, namely, to the airframe.

Other electrical power sources include the auxiliary power unit (APU) discussed in Section 9.7 and the ground power unit. Both of these may be used to provide electrical power when the aircraft is on the ground and to assist in engine starting.

A major top-level trade-off is the method of power generation. This trade- off is top-level because it trades the mechanical compensation for engine speed variation with electrical compensation. There are three major ways: First, the traditional method is called a constant-speed drive (CSD). With CSD the variation in engine speed is reduced to a constant generator speed entirely mechanically with a resulting mechanical complexity. In the second method, integrated drive generator (IDG), the mechanical

9.3　电气部段

电气部段支持的主要顶层功能是图 3.6 中的**提供能源**功能。

电源 (ATA 24)

电源的主要功能是**提供直流电**和**提供交流电**。必要的附属功能包括**提供备用电源**和**提供负载分配**。这些功能都属于 3.3 节中的**提供能源**功能的下一级功能。

提供电源的正常途径是由每台发动机通过机械扭矩提供动力的发电机。发电机产生交流电，它必须转换成直流电给直流电气部件供电。直流电同时还作为备用电源的蓄电池充电。与此类似，应急运行状态下，来自蓄电池的直流电也可以转换成交流电提供给所需交流部件，即这些都属于 MMEL(5.4 节) 的组成部分。

在综合阶段，通过分析完成电气负载分配，其目的是确保电气负载的均衡，使得当一些部件失效时，电气系统中没有任何一个部件过载。电源子系统的综合存在许多约束。一个典型的例子是发射 EMI 限制 (5.10 节)。EMI 可以通过有效的线路屏蔽来限制，同时屏蔽也可以保护其他部件免受电涌影响。所有电源元件都必须接地，即接到飞机机体上。

其他的电源来源包括辅助动力装置 (APU)(9.7 节) 及地面电源装置。当飞机停放在地面上时，它们能为飞机提供电源，并帮助发动机起动。

另一个主要顶层权衡研究是关于电源产生方法的。将这个权衡研究归于顶层权衡是因为它要对发动机转速变化的机械补偿与电气补偿进行权衡。该过程有三种主要方法：首先，传统的称为恒速传动装置的传统方法。使用恒速传动装置，变化的发动机转速将会完全通过机械的方式，降低到一个恒定的发电机转速，这会导致机械的复杂性。另一种方法是采用综合传动发电机 (IDG)，将降低机械转速和发电的功能合并到一个单元中。第三种方法是变速恒频发电机 (VSCF)，该

speed reduction and generation functions are combined into a single unit. In the third method, variable-speed constant frequency (VSCF), the mechanical drive is allowed to vary while the generator produces a constant frequency output through electrical conversion. Factors affecting the final selection include efficiency, cost, reliability, and weight.

Shipside lighting (ATA 33-30, -40, -50)

Lighting functions are simple: Provide Exterior Lighting, Provide Lighting for Cargo and Bays, and Provide Exterior Emergency Lighting. All lighting functions can be said to derive mainly from the Communicate Data/Information function discussed in Section 3.3. Included are flight deck instrument panel lights, warning and caution lights, flood lighting, and other flight deck lighting. Requirements for flight deck lighting should consider the lighting level over the entire surface of the illuminated objects.

The lighting subsystem must provide lighting to cargo and service areas. As for the flight deck, the human factors aspects should be considered in determining the lighted areas and lighting intensities, that is, the actual lighting required for flight and service crews to perform their tasks.

Exterior lighting includes landing and taxi lights, anti-collision and position lights, ground floodlights, logo lights, wing and engine nacelle scan lights, and emergency lighting.

9.4   Interiors Segment

Most of the requirements associated with the interiors segment flow from the top-level function, Provide Passenger and Crew Accommodations discussed in Section 3.3. However, we will see below that this segment satisfies other emergency, information, and lighting functions as well.

Crew accommodations (ATA 25-10)

The key crew accommodations functions are to provide crew seating, storage, and equipment and furnishings. Other functions are to provide life support and evacuation.

The flight deck is a prime example of the integration and synthesis of several segments at a higher level than any individual segment. The flight deck should be laid out with many considerations in mind. Other segments, besides the interiors segment,

方式允许这种机械传动装置的转速变化，但发电机通过电气转换产生恒频输出。影响上述最终选择的因素包括效率、成本、可靠性及重量。

照明 (ATA 33–30, –40, –50)

照明功能非常简单：**提供外部照明**，**提供货物和货舱照明和提供外部应急照明**。可以说所有照明功能主要都是由**数据 / 信息通信功能 (3.3 节)** 衍生而来。包括驾驶舱仪表板照明、警告和警示灯、泛光照明及驾驶舱其他照明。驾驶舱灯光的需求应考虑到被照明物体整个表面的光亮等级。

照明子系统必须提供货舱和服务区域的照明。与驾驶舱一样，在确定照明区域和照明亮度时应考虑人为因素，即飞行和服务机组执行其任务实际需要的灯光。

外部照明包括着陆和滑行灯、防撞和航行灯、地面泛光灯、标识灯、机翼和发动机短舱灯及应急照明。

9.4　内饰部段

与内饰部段相关的大部分需求来自 3.3 节所述的顶层功能，**即提供乘客和机组起居设施**。然而，下面我们将会看到，该部段同时也满足其他紧急、信息及照明功能。

机组起居设施 (ATA 25–10)

机组起居设施的关键功能是提供机组座位、存放、设备及装备，其他功能包括提供生命支持和撤离。

驾驶舱是多个部段在高于任何一个单独部段的层级上集成和综合的极好范例。驾驶舱需要考虑多方面的因素。除了内饰部段，需要驾驶舱评估的其他主要

with major assets in the flight deck include the avionics, mechanical, environmental, propulsion, airframe, and electrical segments. All of these elements should be integrated from a top-level point of view with a single purpose in mind: to allow the flight crew to fly the aircraft safely. The interiors segment plays a major part in this synthesis. The interiors segment is responsible for assuring the comfort of the pilots and the convenience of all flight deck equipment.

We saw in Section 7.4 that quality function deployment (QFD) provides an excellent methodology for transforming the qualitative requirements for interiors, such as passenger comfort, into an actual concept which meets customers' expectations.

Passenger accommodations (ATA 25-20)

The primary passenger accommodations functions are to provide passenger seating, provide passenger entertainment, provide storage for passenger items, and provide fixed interior items. In addition, the interiors segment provides acoustic and thermal protection with the sidewalls.

The most important considerations in the passenger cabin synthesis pertain to human factors and safety. We have seen in Section 5.5 that human factors determine the requirements for comfort, posture, reach, controls, and convenience. Safety considerations described in Section 10.2 drive many of the cabin requirements, for example, for aisle spacing, flame resistance, and many other factors. These factors are all clearly laid out in the FARs. As for the flight compartment, the interiors segment must provide life support, fire protection, and evacuation capabilities.

A key feature needed in passenger accommodation synthesis is *flexibility*. That is, the interior design should be flexible enough to satisfy the individual needs of various airline customers. Flexibility can be achieved through component modularity. Through modularity various components, for example, seats and overhead storage racks, can be designed to be rearranged in different customer configurations.

Among all segments, the interiors segment is very largely driven by direct customer requirements. Many interior items and materials are directly required by the airline customer. Of course, these direct requirements do not relieve the interiors engineer from a rigorous investigation of the interfaces with other elements as described in Chapter 6.

Water, waste, lavatories, galleys, and plumbing (ATA 25-30, -40, -38)

部段包括航电、机械、环控、推进、机体及电气部段。所有这些部段都应以顶层视角进行集成，目的只有一个：使飞行机组能安全驾驶飞机。内饰部段在综合中起主要作用。内饰部段负责确保驾驶员的舒适，以及所有驾驶舱设备的操作便利性。

在7.4节中我们看到，质量功能展开(QFD)提供了一个非常有效的方法，将内饰的定性需求，如乘客舒适度，转换为满足客户期望的一个实际概念。

乘客起居设施(ATA 25-20)

乘客舱位安置的主要功能是提供乘客座位、娱乐及行李存放，并提供固定的内饰件。除此之外，内饰部段还向乘客提供侧壁声及热保护。

人为因素及安全性是客舱综合中最重要的方面，在5.7节可以看到，人为因素决定了舒适性、姿势、可达性、控制和便利性等需求。很多座舱需求由安全性(10.2节)方面的考虑产生，如过道间距、防火及许多其他方面的因素，这些因素在FAR中已经详细列出。和驾驶舱一样，内饰部段还必须提供生命支持、防火及应急撤离能力。

乘客舱位安置综合的一个关键因素是灵活性。即，内饰设计应足够灵活，以适应不同航空公司客户的个性化需求。灵活性可以通过部件的模块化来实现。通过模块化，不同部件(比如座椅和顶部行李架)可以被设计成按不同航空公司客户需要重新布置的构型。

在所有部段中，内饰部段很大程度上由客户需求直接驱动。很多内饰件和材料都是直接由航空公司客户决定的。当然，这些直接需求并不能替代内饰工程师分析与其他元件接口时的细致工作。

水、废水、盥洗室、厨房和管路(ATA 25-30, -40, -38)

该子系统的主要功能是**提供盥洗室**功能和**提供厨房**功能，次级功能包括**提**

The principal functions of this subsystem are Provide Lavatory Capability and Provide Galley Capability. Subordinate functions include Provide Water and Provide Waste Disposal.

We have combined these five interiors subelements into a single subsystem because of the integral nature of their operation. Both the lavatories and the galleys utilize water and dispose of waste, and they are interconnected by the same plumbing system. Hence, in order to minimize the weight of the whole subsystem, it should be designed as a unit and not as individual pieces.

The modularity concept discussed in Section 9.4 is of particular importance to this subsystem because the movement of lavatories and galleys is difficult and involves many interfaces, primarily with the airframe and electrical subsystems and also many internal plumbing interfaces.

Emergency provisions (ATA 26-60)

The functions of this subsystem are to provide evacuation, flotation, interior fire extinguishing, and miscellaneous emergency capabilities. For evacuation, slides, life rafts, and assist lines for evacuation over the wing are provided. Miscellaneous emergency equipment includes fire axes and flash lights.

Signs and lights (ATA 33-10, -20)

This subsystem provides lights for passengers and crew and also provides the escape lights for evacuation. It also provides the placards for instructions and warnings. Considerations such as placards in appropriate languages, readability, and visibility are important.

Interior design

Interior design is one of the major factors in aircraft sales. Yet it is one of the most difficult to implement from an SE point of view. Such functions as Provide Passenger Comfort and Provide Passenger Entertainment are difficult to specify in verifiable terms. QFD described in Section 7.4 is a valuable tool in such circumstances. QFD allows the manufacturer, in cooperation with airline customers, to identify specific

供水和**提供废水处理**功能。

因为这五类舱内子部件工作的相互关联性，此处将它们归为一个子系统。由于盥洗室和厨房都会使用水和处理废水，它们都用同一个管道系统相连。因此，为了降低整个子系统的重量，应将其设计为一个单元而不是各自独立的部分。

模块化概念 (9.4 节) 对于该子系统而言尤为重要，因为盥洗室和厨房的移动十分非常困难，且涉及很多接口，主要是与机体及电气子系统的接口，同时还有很多内部的管道接口。

应急装置 (ATA 26-60)

该子系统的功能是提供撤离、漂浮、舱内灭火及其他应急能力。针对撤离，提供了滑梯、救生筏及机翼上的辅助撤离绳索，其他紧急设备包括消防斧头和闪光灯。

告示和灯 (ATA 33-10, -20)

该子系统为乘客和机组提供灯，同时也提供应急撤离灯。同时提供警示牌，此类警示牌应重点考虑合适的语言、可读性及可见性。

内饰设计

内饰设计是飞机销售的一个主要影响因素，然而从系统工程视角来看，这是难实现的部分。比如**提供乘客舒适性**和**提供乘客娱乐**便难以用可验证的词语规定。在这种情况下质量功能展开 (QFD)(7.1 节) 是一个非常有用的工具。QFD 使得制造商可与航空公司客户一起制定满足客户需要的内饰设计的具体可验证属性。这些属性可能包括座椅宽度和间距、电视视角、乘客与顶部行李架的可达性需求等。QFD 还有助于提供一种更受驾驶员欢迎的驾驶舱布局。

verifiable attributes of the interior design which will meet the needs of the customer. These attributes might include seat width and spacing, television viewing angles, and passenger reach requirements for overhead storage. QFD will also assist in providing a cockpit layout pleasing to pilots.

9.5   Mechanical Segment

Flight controls (ATA 27)

The major function of the flight control subsystem is Transfer Pilot and Autopilot Commands to Control Surface Deflection. An adjunct function is Provide Load Feel, which is a subfunction of the Command and Control Aircraft function shown in Figure 3.6.

Typical subsystem-level trade-offs include: hydraulic vs. electrical (Fly- by-Wire) control. Another advanced option is Fly-by-Light (FBL) described in Section 2.4. With this option, control signals are transmitted by fiber-optic lines. Hydraulic power is generally more reliable and lighter than electrical. However, electrical systems can eliminate complex cable systems. Since the FARs require redundancy in control systems, both techniques can be used. In addition to these methods, propulsion controlled aircraft (PCA) present another control technique discussed in Section 9.6 for redundancy.

For control surface actuation subsystems, electrical, hydraulic, and pneumatic actuators are employed. Trade-offs to be considered include redundancy vs. dispatch reliability.

Hydraulic power (ATA 29)

The principal hydraulic function is Provide Hydraulic Power. Of course, hydraulic power is itself a solution. Therefore, the top-level trade-off is electrical vs. mechanical vs. hydraulic power. Auxiliary hydraulic power is a redundancy requirement for safety. Since hydraulic pumps are driven by mechanical torque supplied by the power plant, another top-level trade-off is autonomous vs. power plant driven hydraulic power. Some aircraft employ self-contained electrical motors and hydraulic pumps near the control surfaces they are moving. This is also subordinate to the Provide Power function discussed in Section 3.3.

A subsystem-level trade-off is hydraulic pressure vs. actuator size. Higher pressures can result in smaller actuators with a resultant greater risk.

9.5　机械部段

飞控 (ATA27)

飞控子系统的主要功能是**传递驾驶员及自动驾驶仪的指令以控制操纵而偏转功能。**另一个辅助功能是**提供载荷感觉，**它是图 3.6 中的**指令及控制飞机**功能的子功能。

典型的子系统级权衡研究包括：液压和电气 (电传) 控制。另一个先进的选择方案是 2.4 节所述的光传操纵 (FBL)。该方案采用光纤来传输控制信号。液压源通常比电源更为可靠和轻便，但是采用电气系统可以消除复杂的电缆系统。由于 FAR 要求保证控制系统的冗余度，因此这两种技术都可以使用。除了这些方法之外，推进控制飞机 (PCA)(9.6 节) 也可以作为另外一项保证冗余度的控制技术。

针对控制面作动子系统，可采用电气、液压及气源作动器。拟考虑的权衡包括冗余度与签派可靠性之间的权衡。

液压源 (ATA29)

液压的主要功能是**提供液压源，**当然，液压源自身就是一种解决方案。因此，顶层权衡研究在电气、机械及液压源之间进行。辅助液压源是安全性方面的冗余度需求。由于液压泵由动力装置提供的机械扭矩驱动，另一项顶层的权衡是关于自主液压源与动力装置驱动液压源的选择。一些飞机在操控面附近部署独立的电动机和液压泵。它也是 3.3 节**提供能源**功能的次级功能。

在子系统级，需要在液压压力大小和作动器尺寸间进行权衡。较高的液压压力意味着更小的作动器，从而伴随更大的风险。

辅助液压源可以由一个液压泵提供，该泵由 APU 供电的电动机驱动，也可

Auxiliary hydraulic power may be provided by a hydraulic pump run by an electric motor powered by the APU. This same pump can be powered by the ground electrical power system.

Landing gears and brakes (ATA 32)

The landing gears and brakes functions emanate from the Provide Ground Movement function described in Section 3.3. In addition, this subsystem may receive requirements derived from the Generate Aero Forces function described in Section 3.3. Primary subfunctions are: Provide Braking, Provide Carriage, and Provide Steering. Other subfunctions include Provide Retraction and Extension, and Provide Access and Cover for Gear Retraction and Extension.

Gear sizing is driven by the requirement for aircraft rotation. Stroke length is driven by the required energy absorption on landing. Backup gear retraction can either employ the free-fall technique or the use of an auxiliary hydraulic actuator. As usual the primary figures of merit for all trade-offs are cost and weight.

Top-level trade-offs involving the landing gears and brakes are, first, main vs. nose gear steering and, secondly, the determination of the number of wheels. The decision to employ main gear steering is determined by the required turn radius of the aircraft and by wheel loading. The number of wheels is determined primarily by the aircraft gross weight. The number of wheels is traded off against the tire size, which is determined by aircraft space limitations. These factors are also traded off against the scrub angle of the wheels (the angle between the plane of the wheel and the direction of motion). Large scrub angles cause increased tire and strut seal wear and limit the ability to tow the aircraft.

Another top-level trade-off is retraction time vs. aerodynamic performance. Quick retraction times require greater hydraulic power, while slower retraction times result in a reduced climb gradient. Another trade-off involving aerodynamic performance is the decision to use a wheel well cover or not. All major aircraft use wheel well covers to reduce drag. However, some smaller aircraft have open wheel wells.

The three most common landing gear arrangements are:

1. *Conventional tail wheel*   This configuration is not used on any modern jets. Its main advantage is that it is the lightest option. However, its disadvantages are that it has bad ground handling qualities and requires a fuselage inclination.

以由地面电源来驱动。

起落架和刹车 (ATA32)

起落架和刹车功能源于**提供地面运动**功能 (3.3 节)。除此之外，该子系统还接受**产生气动力**功能 (3.3 节) 衍生的需求。主要子功能包括 :**提供刹车、提供承载和提供转向**。其他子功能包括**提供收放和提供起落架收放的检修口盖**。

起落架参数设计由飞机起飞滑跑抬前轮的需求驱动。减震支柱长度由着陆时需要吸收的能量决定。备用起落架收起可以使用自由坠落技术或辅助液压作动器。权衡的主要目标通常是成本和重量。

涉及起落架及刹车系统的顶层权衡首先是主轮转弯和前轮转弯之间的选择，其次是确定机轮的数量。

采用主起落架转向的决定由所需的飞机转弯半径和轮载确定。机轮数量主要由飞机总重决定。机轮数量与轮胎尺寸之间将进行权衡，它由飞机空间限制决定。这些因素与机轮刮擦角度 (机轮平面和运动方向之间的夹角) 之间也需进行权衡。大的刮擦角将增加轮胎和支柱的密封件磨损，并限制拖曳飞机的能力。

另一个顶层权衡是关于收起时间和空气动力性能。快速的收起需要更大的液压动力，较慢的收起将会减小爬升梯度。涉及空气动力学性能的另一项权衡是关于否采用轮舱口盖的决策。所有较大的飞机均采用轮舱口盖的方式减小阻力，但有一些小型飞机也采用开放式的起落架轮舱。

三种最常用的起落架布局包括：

1. 传统尾轮式 现代喷气式飞机不采用这种构型，其主要优点是重量最轻。但缺点是地面运行品质差，且需要机身倾斜。

2. 自行车式 这种构型最容易收起起落架。然而，这种方式最难着陆，并很难维持正确的机翼角度。

3. 三点式 该布局是现代亚声速喷气机上最常见的起落架布局。三点式布

2. *Bicycle gear*   This configuration is the easiest for gear stowage. However, it is the hardest to land and maintain the proper wing angle of incidence.

3. *Tricycle gear*   This is the most common gear arrangement on modern subsonic jets. It has good handling characteristics, primarily because the main gear is near the center of gravity. It provides the largest moment arm for fuselage rotation.

9.6   Propulsion Segment

The primary top-level function supported by the Propulsion Segment is the Provide Total Impulse function in Figure 3.6.

Fuel subsystem (ATA 28)

The two primary functions of the fuel subsystem are Provide Fuel to the Engines and Provide Fuel to the APU. Functions which follow from these are Provide Fuel Storage, Provide Fire Protection, Provide Fueling and Defueling, Provide Dumping, and Provide Fuel Control. We saw in Section 9.1 that the fuel system has also inherited another function, namely, Provide Wing Anti-Icing. Other aspects of the power plant subsystem described in Section 9.6 also put demands on the fuel consumption, namely, electrical and hydraulic power and bleed air. Like the wing, the fuel subsystem also provides wing load alleviation because of its weight. The fuel system can also be used to cool the engine oil.

Another aspect of the fuel subsystem is fuel management which emanates from the Manage Fuel function discussed in Section 3.3. The primary function of fuel management is to maintain fuel circulation to prevent the formation of ice in the fuel. Other key functions include the control of the center of gravity (c.g.) and the cross-feeding of fuel between the engines to maintain a balance in the availability of fuel and to prevent a lateral imbalance.

The fuel system synthesis is inextricably linked to the top-level aircraft synthesis. We have seen in Section 8.2, Item 7 that the fuel weight is one of the earliest top-level *derived* requirements. This requirement immediately flows to the fuel subsystem for sizing that subsystem. Fuel storage is normally limited to the wing because of the inherent efficiency of integrating the wing structure and the fuel storage. Hence, major trade-offs are required between fuel storage requirements and aerodynamic performance. Longer aircraft ranges demand more fuel. This demand is compatible with thicker wings and/or wings with increased chord length. If the wing span is

局具有良好的操作特性，因为其主起落架靠近重心，因此为机身抬前轮提供了最大的力臂。

9.6 推进部段

推进部段支持的主要顶层功能是图 3.6 所示的**提供总推力**功能。

燃油子系统 (ATA 28)

燃油子系统的两个主要功能是**提供燃油给发动机**及**提供燃油给 APU**。由这些产生的功能包括**提供燃油存储、提供防火、提供加油和放油，提供燃油倾倒**和**提供燃油控制**。在第 9.1 节中我们看到燃油系统还隐含其他功能，即**提供机翼防冰**。9.6 节所述动力装置子系统的其他方面也需要消耗燃油，即电源、液压源及引气。与机翼类似，因为其重量的原因，燃油子系统也提供机翼载荷减缓功能。燃油子系统还能用于发动机滑油的冷却。

燃油子系统的另一个部分是燃油管理功能，它源于 3.3 节所述的**管理燃油**功能。燃油管理的主要功能是保持燃油循环从而防止燃油结冰。另一项主要功能包括控制重心 (c.g.) 并在发动机之间交输供油，以保持可用燃油的平衡，并防止横向不稳定。

燃油系统综合与飞机顶层综合联系紧密。根据 8.2 节介绍，燃油重量是最早的顶层衍生需求之一。该需求立即被分解到燃油子系统，以用于子系统的参数设计。燃油存储主要局限于机翼，因为将机翼结构和燃油存储集成的方式效率较高。因此，需要在燃油存储需求和气动性能之间进行主要的权衡研究。航程较长的飞机需要更多的燃油。该需求应与更厚的机翼或增大弦长的机翼相匹配。如果翼展受限于机场约束，则应进行展弦比与更大弦长之间的权衡研究 (8.2 节)。这些权衡研究都源自于燃油需求。

limited by airport constraints, then the aspect ratio discussed in Section 8.2 needs to be traded against the larger chord lengths. These are the trade-offs which result from fuel requirements.

Pylon (ATA 54 through 54-80)

The pylon has two primary functions: Provide Support for Engines and Provide Conduit for Subsystem Functions to and From Engine. The pylon can also be thought of as part of the airframe segment; however, because of its importance to the engine, we have included it as part of the propulsion segment. The pylon is an important element whether the engines are wing-mounted or body-mounted, a top-level trade-off. The pylon is also an important element in the Provide Fire Protection function. Another function of the pylon is Provide Aero-Elastic Damping to prevent aerodynamic flutter.

Power plant (ATA 71)

The primary power plant function is Provide Thrust discussed in Section 3.3. However, a number of other functions have been imposed on the power plant, namely, Provide Bleed Air to Pneumatic Subsystem, Provide Mechanical Torque for Hydraulic Power, and Provide Mechanical Torque for Electrical Power. Provide Fire Protection is a necessary adjunct function. All of these functions are subordinate to the Provide Total Impulse function discussed in Section 3.3.

Other functions which have emerged through practice are Alleviate Wing Load and Provide Lift Augmentation. The propulsion system alleviates wing loads by off-setting the lift load with its weight. It provides lift augmentation through strakes which cause the turbulent boundary layer to stay attached and thus improves the maximum lift coefficient, C_{Lmax}, during take-off. Another secondary, but important, function is Provide Noise Attenuation. The power plant achieves this function by integrated noise treatment involving many components. Berry (1993) cites the inlet, fan, compressor, fan nozzle, burner, turbine, and primary nozzle as being the principal noise sources.

An important subfunction is Start Engine. To accomplish this function, the engine needs a source of compressed air. On the ground this air can be provided either by the ground equipment or by the APU. In flight the engine can be restarted by free air or by the cross-feed of air from another engine.

Of the four primary functions, the most important is Provide Thrust. All of

吊挂 (ATA 54 至 54–80)

发动机吊挂有两项主要功能：**提供发动机支撑和提供子系统与发动机之间的通路**。发动机吊挂可以认为是机体部段的一部分，然而，由于发动机的重要性，我们可以将其视为推进部段的一部分。当进行机翼吊挂发动机的权衡研究时，吊挂是非常重要的元件。吊挂同时也是**提供防火**功能的一个重要元件。吊挂的另一个功能是**提供气动弹性阻尼**，以防止气动颤振。

动力装置 (ATA 71)

动力装置的主要功能是 3.3 节所述的**提供推力**功能，然而，动力装置还有很多其他功能，包括**提供引气给气源子系统**，**提供机械扭矩给液压源**，**提供机械扭矩给电源**。**提供防火**也是一个必要的辅助功能。所有这些功能都是 3.3 节中**提供总推力**功能的次级功能。

通过实践产生的其他功能包括**机翼载荷减缓和提供增升**。推进系统通过自重偏移升力载荷减缓机翼载荷。推进系统通过边条导致湍流边界层保持吸附，从而在起飞过程中改善最大升力系数 C_{Lmax}，提供增升。另一个重要的辅助功能是**提供噪声衰减**。动力装置通过集成涉及多个部件的噪声处理方法来实现该功能。Berry(1993) 指出进气口，风扇、压气机、风扇喷嘴、燃烧室、涡轮及主喷嘴是主要噪声源。

动力装置的一个重要的子功能是**起动发动机**，为完成该功能，发动机需要一个压缩空气源。在地面上时，空气可以由地面设备或者 APU 提供，飞行中发动机可以通过自由空气或从另一台发动机交叉引气重新起动。

这四项主要功能中，最重要的是**提供推力**。所有其他功能都基于多年经验的权衡研究结果产生。尽管气源、液压源及电源可以通过其他方式获得，但经验表明，动力装置仍是最高效的来源。通过较好地实施系统工程，这些决策将随着

the others have been arrived at by trade-offs over decades of experience. Although pneumatic air, hydraulic power, and electrical power could be obtained by alternative means, experience has shown that the power plant is the most efficient source. With good SE, these decisions will be revisited as technology and economic factors shift.

We have seen in Section 8.2 that the number of engines is a top-level trade-off. For every mission (payload, range, speed) there will be an optimum number of engines which can be found through the trade-offs described in Chapter 8. In addition, FAR constraints will establish the minimum number of engines, thrust, and predicted reliability required for emergency conditions, such as engine-out over water.

Another top-level trade-off is engine placement. Wing-mounted engines result in a lighter wing weight. In addition, their location farther away from the fuselage makes it easier to maintain lower interior noise levels. However, the advantage of rear-mounted engines is that the aircraft is easier to control in an engine-out situation.

For wing-mounted engines, a key trade-off is the placement of the engines on the wing. Although engines nearer the tip are preferred for wing loading, engines near the fuselage are superior for controllability in a single engine-out condition.

Berry (1993) provides a comprehensive set of design considerations and trade-offs for the propulsion segment.

Thrust management (ATA 76)

The Manage Engine Thrust function is the process of applying forward or reverse forces as required for each phase of flight and landing roll-out. The key subsystem-level trade-off is manual vs. automatic thrust management. In automatic thrust management the thrust is managed by a computer. The pilot's role is to set the throttle at various appropriate aircraft phases, such as take-off, climb, cruise, descent, or landing. A secondary possible thrust management function is Provide Aircraft Control. By automatically managing the thrust of the engines, the thrust management subsystem can control the aircraft in roll, pitch, and yaw.

Propulsion monitoring

The propulsion monitoring subsystem maintains a constant vigilance of the propulsion segment's health for maintenance. In addition, it assures that all elements of the segment adhere to their red-line limits for safety.

技术及经济因素的转变得以重新审视。

根据 8.2 节介绍, 发动机数量是一个顶层的权衡研究项。针对每一项任务 (商载、航程及速度), 可以通过第 8 章所述权衡研究得出发动机的最优数量。除此之外, FAR 约束还规定了发动机最少数量、推力及紧急情况, 如在水面上发动机停车时所要求的预计可靠性。

另一项顶层权衡研究是发动机的布置。翼吊发动机使机翼重量较轻, 除此之外, 由于发动机距离离机身更远, 使得更容易维持比较低的舱内噪声。然而, 尾吊发动机的优点是在发动机失效的情况下飞机更容易操控。

对于翼吊发动机, 一个关键权衡研究是发动机在机翼上的位置。尽管发动机靠近翼尖对机翼载荷更有利, 但在单发失效的情况下, 发动机靠近机身的飞机的可操纵性更优越。

Berry (1993) 提供了关于推进部段的一系列设计考虑和权衡研究内容。

推力管理 (ATA 76)

管理发动机推力功能是按照飞行和着陆滑跑每个阶段的需要施加正向和反向作用力的过程。子系统级关键权衡研究是关于手动或自动推力管理的选择。自动推力管理是指推力由计算机进行管理。驾驶员的作用是在不同的特定飞行阶段, 比如起飞、爬升、巡航、下降和着陆时设定油门。第二个可能的推力管理功能是**提供飞机控制**。通过自动管理发动机推力, 推力管理子系统可以控制飞机的横滚、俯仰及偏航。

推进监控

推进监控子系统对推进部段的健康状况始终保持警觉, 以用于维修目的。除此之外, 它还保证推进部段的所有元件都在其安全红线之内。

9.7 Auxiliary Segment (ATA 49)

An APU is not an absolute necessity on an aircraft. However, most commercial aircraft have an APU to provide functions which may not be available at some airports. In addition, it provides a redundant source of power for aircraft functions in the event the engines are inoperable. Its primary functions are Provide Auxiliary Electrical Power, Provide Pneumatic Power for Engine Starting discussed in Section 9.1, and Provide Auxiliary Pneumatic Power for Air Conditioning. The electrical power function is provided as a backup to the primary electrical power source in flight. The pneumatic functions are primarily ground functions. The auxiliary power functions are subordinate to the Provide Power function discussed in Section 3.3.

As for the power plant, the Provide Fire Protection for the APU is a necessary adjunct function. The primary reason for the APU's existence is to provide the autonomous starting capability at airports where no external ground power sources are available. This capability emanates from a very top-level aircraft requirement for autonomous starting capability, if required for the route structure of the aircraft. The other functions, electrical and pneumatic power, are backups to the primary subsystems. These functions emanate from safety and operational considerations. Other auxiliary functions could be performed by the APU, such as Provide Hydraulic Power, if necessary.

9.8 Airframe Segment

All of the airframe components listed below have many aspects in common. They are subject to similar performance requirements and constraints, and many of the same trade-offs are the same. Functions in common include Sustain Loads and Maintain Aerodynamic Profile. These functions support the top-level functions, Provide Aerodynamic Performance discussed in and Maintain Structural Integrity both discussed in Section 3.3.

A key trade-off for the airframe is between conventional materials, e.g. aluminum, vs. newer composite materials. Composite materials provide the potential of lighter weight and greater strength. These benefits should be traded against higher recurring costs and possibly higher maintenance costs. All of these factors fit into the DOC discussed in Section 8.6. That is, if the reduced weight and fuel used with composite materials more than compensate for the increased manufacturing and maintenance

9.7　辅助动力部段 (ATA 49)

APU 并非飞机上绝对必需的部分。然而，大部分商用飞机都有 APU，以提供在一些机场无法使用的功能。除此之外，APU 可在发动机不能工作的时为飞机功能提供了冗余动力源。其主要功能包括**提供辅助电源**、**为发动机起动提供气源** (9.1 节) 并**提供辅助气源给空调系统**。提供电源功能作为飞行中作为主电源的备份源。气源功能主要是地面功能，辅助动力功能是 3.3 节所述**提供能源**功能的次级功能。

与发动机类似，为 APU **提供防火**是非常必要的辅助功能。APU 存在的主要理由是在无外部地面电源的机场提供发动机自主起动的能力。该能力源于一个非常顶层的飞机需求：飞机的航线结构是否需要有自主起动能力。其他功能，包括电源及气源，是作为主子系统的备份。这些功能都是基于安全性和运行方面的考虑。必要时可由 APU 执行其他辅助功能，如提供液压源。

9.8　机体部段

下列所有机体部件有很多共同之处。它们都要满足类似的性能需求及约束，很多需要同样的权衡研究。共通的功能包括**承受载荷**和**保持气动外形**。这些功能支持 3.3 节所述的**提供气动性能**和**保持结构完整性**功能。

机体的一个关键权衡研究是关于传统材料 (如铝) 与新型复合材料之间的权衡。复合材料具有重量轻和强度大的优势。这些都应和更高的重复性成本和可能的高维修成本进行权衡。所有因素都需要考虑归入 8.6 节所述直接运营成本 (DOC)。即，如果使用复合材料减轻的重量和减少使用的燃油比增加的制造和维修成本高，则直接运营成本 (DOC) 将会降低，因此可以认为飞机更加经济且市场化更好。

costs, the DOC will decrease, thus making the aircraft more economical to operate and therefore more marketable.

Constraints discussed in Chapter 5 are particularly important in airframe design. Environmental constraints pertaining to corrosion and shock loads discussed in Section 5.6, for example, apply to all airframe elements. Especially important are durability requirements. Each airframe member should be shown to last a given number of years without failure. Within the category of qualitative safety requirements discussed in Section 10.2, the principle of fail-safe design applies. That is, if one structural member fails, due to cracking, for example, another should not exceed its limit load, that is, two-thirds of its ultimate load.

Fuselage (ATA 57)

In addition to the two primary functions discussed above, the fuselage must perform the following functions: Maintain Pressure, Provide Access for Ingress and Egress; Provide Cargo Loading; Provide Space for Passengers, Crew, Cargo, and Subsystems; Provide Cargo Fire Protection; and Provide Pilot Visibility. Provide Pilot Visibility is a subfunction of the Communicate Data/Information top-level function discussed in Section 3.3.

Of these, the Maintain Pressure function is perhaps the most demanding. For the HSCT, this function will be one of the most critical from a safety point of view since loss of pressure will most likely be a catastrophic event, unlike with subsonic aircraft.

Empennage (ATA 55)

In addition to the primary functions, the key function of the empennage, also called the tail, is Provide Aerodynamic Control. This function includes both the aerodynamic stability which the vertical and horizontal stabilizers provide to the whole aircraft as well as the vertical and lateral control provided by the vertical (rudder) and horizontal (elevator) moving surfaces. Of course, with the implementation of propulsion control discussed in Section 2.4 there would be no need for the moving surfaces. This trade-off is anticipated to result in a safer aircraft because of the reduced number of control systems.

A key tail top-level trade-off is the horizontal tail placement. The location of the horizontal tail is dependent on the location of the engines. The T-tail configuration

第 5 章的讨论的约束在机体设计中非常重要。关于 6.5 节所述的腐蚀和冲击载荷的环境约束适用于所有机体部件，尤其是对耐久性需求而言尤为重要。每个机体部件均应表明部件可使用一给定年限，而不会中途失效。在 10.2 节讨论的量化安全性需求的类型内，适用失效安全设计准则。即，如果一个结构件失效，如由于裂纹产生导致，则另一个结构件不应超出其限制载荷，即其极限载荷的三分之二。

机身 (ATA57)

除了上述两个主要功能之外，机身还必须执行如下功能：**保持压力，提供登机和离机通道，提供货物装卸，为乘客、机组、货物和子系统提供空间，提供货舱防火，提供驾驶员视景**。提供驾驶员视景是 3.3 节**数据 / 信息通信**顶层功能的子功能。

其中，**保持压力**功能可能是最重要的。对于 HSCT，从安全角度看，该功能是最关键的，因为失压非常有可能是一种灾难性事件，这一点与亚声速飞机不同。

尾翼 (ATA55)

除了主要功能之外，尾翼的关键功能是提供**气动控制**。该功能既包括垂直安定面及水平安定面对整架飞机提供的气动稳定性，也包括垂直活动面（方向舵）和水平活动面（升降舵）提供的垂直和横向控制。当然，通过实行 2.4 节所述的推进控制，也可能不需要活动面。可以预期的是，这项权衡预期会由于减少了控制系统的数量而使飞机更为安全。

尾翼关键的顶层权衡之一是水平尾翼布置。水平尾翼的位置取决于发动机的位置。T 型平尾配置主要针对机身安装的发动机，翼吊发动机飞机最好采用安装在机身上的水平尾翼。

is almost certainly required for fuselage-mounted engines, while fuselage-mounted horizontal tails are preferred for wing-mounted aircraft.

Wing (ATA 57)

The primary function of the wing, it is assumed, is to provide the lift that keeps the aircraft aloft. However, this function is just part of the Maintain Aerodynamic Profile discussed above. The wing also serves to maintain a minimum aerodynamic drag. The use of winglets has increased in recent years to accomplish both of these functions.

In addition to the two principal functions above, the wing must perform the following functions: Transfer Loads to Fuselage, Provide Storage for Fuel, Provide Aerodynamic Control, and Provide Space for Subsystems. We saw in Section 8.2 how key wing requirements (wing area, aspect ratio, sweepback angle, and taper ratio) are the subject of aircraft-level trade-offs. The primary wing design trade-offs are driven by the Provide Storage for Fuel function as discussed in Section 9.6. Ailerons, strakes, slats, and spoilers have been used for many years to facilitate the Provide Aerodynamic Control function.

9.9   Allocation to Software

From an SE point of view, software is exactly the same as any other part of the aircraft. It has functions, and it has requirements. Every subsystem listed above will have software. So, we have not listed software here to imply that it is a separate subsystem. We only want to show that it should be included it in the whole SE scheme during the formulation phase of the aircraft or any part of it. Section 10.3 presents a more extensive discussion of the development and certification requirements and constraints that are normally imposed on software.

9.10   Subsystem Constraints

For subsystem constraints, the general principles of Chapter 5 apply. However, it is worth emphasizing the following points: First, the general industry and regulatory design constraints which were laid on the aircraft as a whole apply to the individual subsystems as well. Hence, their inclusion in the design is just as subject to scrutiny in design reviews and verifications. Secondly, the allocation procedures (4.7) apply to all subsystems. The most notable ones are weight and dispatch reliability, among others.

机翼 (ATA57)

机翼的主要功能是提供飞机空中飞行所需的升力。然而，该功能只是上述讨论的**保持气动外形**功能的一部分。机翼同时也用于保持最小的气动阻力。近年来翼梢小翼使用的增多就是为了满足上述功能。

除了上述两个主要功能之外，机翼还必须执行其他功能，包括：**传递载荷到机身，提供燃油存储，提供气动力控制及提供子系统空间**。在 8.2 节我们介绍了如何对关键机翼需求（机翼面积、展弦比、后掠角及尖梢比）进行飞机级权衡研究。机翼设计的主要权衡研究由 9.6 节所述的**提供燃油存储**功能驱动。利用副翼、翼板、缝翼及扰流板**提供气动力控制**功能的方法已经应用多年。

9.9　分配给软件

从系统工程的视角来看，软件和飞机其他部件完全一样，既有功能也有需求。前文所述的所有子系统都有软件，因此，我们这里不专门列出软件表明它是一个单独的子系统。我们只想表明在飞机及其任一部分形成阶段，应将其包括在整个系统工程过程之中。10.3 节对通常与软件相关的开发、合格审定要求及约束进行了更详细的讨论。

9.10　子系统约束

对于子系统约束，第 5 章的一般原则都是适用的。然而，应着重注意下面几点：首先，适用于整架飞机的业界及监管的通用设计约束，同时也适用于单独的子系统。因此，该设计中是否包含这些内容也要接受设计评审及验证活动的详细检验。其次，分配程序 (4.7 节) 也适用于所有子系统。其中，最重要的是重量及签派可靠性。

10

Certification, Safety, and Software

The objective of the certification process is to substantiate that the aircraft and its systems comply with applicable requirements.

ARP 4754A (2010, p. 76)

The certification process, as indicated by the quotation above, is focused on the safety aspects of the aircraft development as determined by such agencies as the Federal Aviation Administration (FAA) and the Joint Aviation Authorities (JAA). These requirements are spelled out in documents, such as Federal Aviation Regulations (FARs) and Joint Airworthiness Regulations (JARs). The FAA has taken a major step in incorporating SE principles into the certification process by the publication of ARP 4754 (1996) and later in ARP 4754A (2010). This document is a set of guidelines compiled and published by the Society of Automotive Engineers (SAE) in cooperation with the FAA. Hence, it represents a look into the future of certification rather than present day practices. However, this look is extremely important from an SE point of view since it demonstrates the FAA's and SAE's commitment to the SE process.

A second aspect of certification which supports SE principles is verification. While the certification process focuses on those verification techniques which support airworthiness, it nonetheless forms part of the total SE verification process dedicated to the goal of 100 percent verification.

This chapter does not intend to present a complete description of the certification requirements and process, but rather to outline basic features of the process and to show how this process is integrated with and is mutually compatible with the SE process. Thus, the reader should not rely on the contents of this chapter as a source of certification requirements but rather should obtain original source material or contact the appropriate regulatory agencies. As stated in the Preface, this book does not intend to replace the official standards and guidelines, or to be a definitive interpretation of them, but rather to be a *pointer* to them and to show how they can be seen in the SE context, and how these processes can be adapted to the commercial aircraft domain.

第 10 章 合格审定、安全性及软件

合格审定过程的目标是证实飞机及其系统满足适用的要求。

ARP 4754A (2010, 第 76 页)

如上述引用所示,合格审定过程重点关注 (美国) 联邦航空局 (FAA) 或 (欧洲) 联合航空局 (JAA) 这类机构确定的飞机研制安全性方面的内容。这些要求形成了一些文件，如 (美国) 联邦航空条例 (FAR) 及 (欧洲) 联合航空条例 (JAR)，这些文件对相关的合格审定要求进行了详细阐明。FAA 在 1996 年发布了 ARP 4754，后续于 2010 年发布了 ARP 4754A，在将系统工程原则纳入合格审定过程方面取得了长足进展。ARP 4754A 是一份包含众多指南的文件，由 (美国) 汽车工程师协会 (SAE) 与 FAA 合作编制并发布。因此，该文件代表了合格审定的未来视角，而不是目前的实践。然而，从系统工程角度来看，这种视角极其重要，因为它表明 FAA 和 SAE 开始推行系统工程过程。

合格审定支持系统工程原则的第二个方面是验证。尽管合格审定过程重点关注用于支持适航性的验证手段，但它无疑是整个系统工程旨在完成 100% 验证目标的验证过程的一部分。

本章无意于详尽地描述合格审定要求及过程，而是对此过程的基本特征进行概述，介绍该过程如何与系统工程过程集成,以及它们之间如何相互兼容。因此，读者不应将本章的内容作为合格审定要求的来源，而应该获取原始资料或者以相应监管机构意见为准。正如 " 前言 " 所述，本书无意取代官方标准及指南，或者

321

Although safety is the primary focus of certification, the certification requirements affect the design, design processes, and management processes beyond the realm of safety. For example, aircraft manufacturers must submit a description of the features of their design processes and configuration management processes, which are used in both safety related and non-safety related requirements, design, and verification processes.

The terminology used in ARP 4754A sometimes differs slightly from the terminology used in SE. ARP 4754A uses the term *system* in the traditional aircraft sense, that is, to mean a subsystem in the SE context. We have used SE terminology here, that is, *subsystem* where appropriate. ARP 4754A refers to passenger safety as a *function*. Many systems engineers, the author included, would refer to safety as a constraint or specialty requirement. Since SE has not totally standardized its own terminology, these differences cannot be regarded as significant. More important than the terminology is the assurance that the SE process is complete and that all of these considerations are included and analyzed properly.

This chapter also discusses the development and certification of systems with software. The reason for the grouping of software with certification and safety is that software development is a critical safety element and contains its own certification process requirements within the overall aircraft certification process.

Finally, it should not be inferred that the discussion of the certification processes in this chapter forms a definitive interpretation of the regulatory documents discussed in it. The reader is advised to refer to those documents when engaged in actual certification processes. The purpose of this chapter is only to show that the certification processes are compatible with SE principles.

10.1   Certification

This section will discuss how the certification requirements outlined in ARP 4754A are compatible with the SE process. ARP 4754A was written primarily, but not exclusively, with electronic subsystems in mind; however, the principles apply to any type of system or subsystem. Table 10.1 shows how the required elements of certification data are in agreement with SE principles.

10.2   Safety

Safety is the primary concern of the certification process. The term *safety* covers both safety related design constraints as well as the quantitative safety requirements. For

作为这些标准及指南的权威解释，而是作为它们的指引，并展示它们如何融入系统工程环境，以及这些过程如何应用于商用飞机领域。

尽管安全性是合格审定的主要关注点，但合格审定要求还影响安全性领域之外的设计、设计过程及管理过程。比如，飞机制造商必须提交其设计过程及构型管理过程的特点描述，在与安全性相关或与安全性无关的需求、设计及验证过程中都要应用这些过程。

有时候，ARP 4754A 中使用的术语可能与系统工程中使用的术语略有不同。ARP 4754A 使用传统飞机"系统"一词，在系统工程语境下指的是"子系统"。我们这里使用系统工程术语"子系统"更为合适。ARP 4754A 将乘客安全性称为功能，很多系统工程师，包括作者自己，会将安全性称为约束，或者专业需求。由于系统工程术语尚未完全标准化，所以这些差异并非重点。比术语更重要的是要保证系统工程过程的完整性，并且保证所有这些事项均已包括在内，并进行了适当的分析。

本章也讨论了包含软件的系统研制及合格审定。之所以将软件的合格审定以及安全性归为一组，是因为软件开发是一个关键的安全性元素，并且在飞机整个合格审定过程中包含其自身的合格审定过程需求。

最后，本章关于合格审定过程的考虑并不是规章文件的权威解释。建议读者在合格审定过程中，参考规章文件。本章的目的仅在于表明合格审定过程与系统工程原则之间相互兼容。

10.1 合格审定

本节将讨论 ARP 4754A 中概述的合格审定要求如何与系统工程过程兼容。ARP 4754A 主要针对电气子系统，但其内容却不仅限如此。然而，系统工程原则适用于任何类型的系统和子系统。表10.1示出了合格审定数据所需元素如何与系统工程原则相统一。

certification, safety constraints consider both the availability (continuity) as well as the integrity (correctness of behavior) of the function. In other words, safety analysis is concerned both with *whether* a function is performed as well as with the risks inherent in not achieving the desired performance of the function.

Quantitative safety requirements

All quantitative safety requirements result from the FAR 25.1309 requirement, which states:

> The airplane equipment and associated components, considered separately and in relation to other [subsystems] must be designed so that ① any catastrophic failure condition is (a) extremely improbable, and (b) does not result from a single failure; and ② any hazardous failure condition is extremely remote, and ③ any major failure condition is remote.

Table 10.1   Compatibility of the SE and certification processes

SE Elements	Certification Aspects
Functional analysis (Chapter 3)	Aircraft-level functional requirements, allocation of aircraft functions to systems, functional hazard analysis (FHA) discussed in Section 10.2
Requirements development and allocation (Chapter 4)	Requirements categories discussed in Section 4.9, allocation of item requirements to hardware and software, preliminary safety assessment (PSSA) discussed in Section 10.2, validation plan and data
Synthesis (Chapters 7–9)	Development of system architecture, common cause analysis (CCA) discussed in Section 10.2, system implementation
Verification and validation (Chapter 11)	Verification data, system safety analysis (SSA) discussed in Section 10.2, inspection and review process
SE management (Chapter 12)	Certification plan, development plan, configuration management plan and data, process assurance plan and evidence, certification summary

This requirement can be paraphrased to say that the hazard probability must be inversely proportional to the hazard severity. This FAR sets the total maximum allowable probability (MAP) of a catastrophic event at 10^{-7}. This requirement means

10.2 安全性

安全性是合格审定过程的主要关注点。安全性一词既包括与安全性相关的设计约束，也包括定量安全性需求。对于合格审定，安全性约束考虑功能的可用性（连续性）及功能完整性（行为的正确性）。换句话说，安全性分析不仅仅关注某一功能是否执行，还包括未达到该功能的预期性能可能导致的风险。

定量安全性需求

所有的定量安全性需求都来自 FAR 25.1309 条的要求，该条款指出：

飞机设备与有关部件的设计，在单独考虑以及与其他子系统一同考虑的情况下，必须设计成：①造成任何灾难性失效状态的概率 (a) 为极不可能；并且 (b) 不会导致单点失效；②发生任何危害性失效状态的概率为不可能，并且③发生任何重大的失效状态的概率是微小的。

表 10.1　系统工程与合格审定过程的兼容性

系统工程元素	合格审定方面
功能分析（第 3 章）	飞机级功能需求，飞机级功能分配至系统，10.2 节中讨论的功能危害性分析 (FHA)
需求开发和分配（第 4 章）	4.9 节中讨论的需求分类，项目需求分配至硬件及软件，10.2 节中讨论的初步系统安全性评估 (PSSA)，确认计划和数据
综合（第 7 至 9 章）	系统架构开发，10.2 节讨论的共因分析 (CCA)，系统实现
验证和确认（第 11 章）	验证数据，10.2 节讨论的系统安全性评估 (SSA)，检查和评审过程
系统工程管理（第 12 章）	合格审定计划，研制计划，构型管理计划和数据，过程保证计划和证据，合格审定总结

本条需求可以解释为，危害事件的发生概率必须与危害程度呈反比关系。FAR 25.1309 将灾难性事件最大允许发生的总概率 (MAP) 设定为 1×10^{-7}。该要求

that for each of 100 possible events, for example, the probability cannot exceed 10^{-9}.

Functional hazard assessment (FHA)

The functional hazard assessment (formerly called functional hazard analysis) is an integral part of the safety process. The FHA is tied to the SE concept of functional analysis discussed in Chapter 3. It is a systematic and comprehensive examination of a subsystem's functions. Its purpose is to determine potential hazards a subsystem can cause or contribute to, not only if it malfunctions, but also in its normal operation. The FHA provides the results of this examination as an assessment at the overall system level. The FHA requires a hazard assessment for each aircraft function and for each *combination* of aircraft functions. Hence, the FHA has introduced the philosophy that hazards may be caused by combinations of functions.

The development of the functions themselves may seem like a formidable task, and it is. The saving grace is that most of the task only has to be done once since most of the higher-level functions are the same from aircraft to aircraft. New functions will only have to be developed as new features are introduced onto aircraft. These functions will normally appear at lower levels of the functional hierarchy. This factor will tend to simplify the FHA.

Preliminary System Safety Assessment (PSSA)

The PSSA evaluates the proposed architecture and compares it to the failure conditions in the FHA described in Section 10.2 in order to determine the safety requirements of the subsystems or items. For example, the PSSA could determine the degree of redundancy required to mitigate a hazardous condition.

System Safety Assessment (SSA)

An SSA is a systematic, comprehensive analysis of the system functions to show that the safety requirements have been implemented in the design. The SSA evaluates all aspects of the system concept from a safety point of view. It reviews the functions, interfaces, event probabilities, failure conditions, combinations of functions, maintenance factors, and the results of all other analyses. Hence, the SSA is an integral

意味着对于每 100 个可能的事件，灾难性失效出现的概率不得超过 $1×10^{-9}$。

功能危害性评估 (FHA)

功能危害性评估 (之前被称为功能危害性分析) 是安全性过程的组成部分。FHA 与系统工程功能分析概念 (如第 3 章中所讨论) 相结合，对子系统功能进行系统和全面的检查。其目的在于确定某一系统不仅仅在功能异常情况下，也包括正常工作情况下，能够导致或者促成的潜在危害。FHA 提供此检验的结果作为整个系统级的评估，并且需要对各个飞机功能和各个飞机功能的组合进行危害性评估。因此，FHA 引入了危害可能由功能组合引起的理念。

功能自身的开发似乎是，也确实是一项很有挑战的工作。有利的一面在于大部分的工作只需要做一次，因为飞机之间的大部分较高层级功能是一致的。当飞机引入新的特征时，才需要开发新的功能。这些功能通常出现在功能架构的一些较低的层级上，这种情况下，FHA 往往进行简化。

初步系统安全性评估 (PSSA)

初步系统安全性评估 (PSSA) 对建议的架构进行评估，并将其与 10.2 所述 FHA 中的失效状态进行对比，以确定子系统及项目的安全性需求。例如，PSSA 能够确定减轻危害条件所需的冗余程度。

系统安全性评估 (SSA)

系统安全性评估 (SSA) 对系统功能进行系统性的综合分析，以表明设计已经满足了安全性需求。SSA 从安全性视角全方位评估系统概念方案，评审系统的功能、接口、事件发生概率、失效状态、功能组合、维修因素及所有其他分析结果。因此，SSA 是系统工程验证过程中 11.2 节所述分析验证过程不可或缺的组成部分。

part of the SE verification process as verification by analysis described in Section 11.2.

Common Cause Analysis (CCA)

The purpose of the CCA is to ensure the independence of subsystems which have faults in common. The potential for common faults is most common in subsystems which rely on redundancy or on the same software which is used by more than one subsystem. Hence, CCA falls within the synthesis phase of development described in Chapter 7.

The CCA performs a zonal safety analysis to ensure multiple subsystems in the same aircraft zone do not interfere with each other. The CCA performs a particular risk assessment to examine events and influences which may affect more than one subsystem. The CCA performs a common mode analysis to determine whether or not subsystems are truly independent.

Safety and the SE processes

Table 10.1 shows that the safety assessment required by the certification process is compatible with SE process. FHAs described in Section 10.2 are conducted at both the aircraft and subsystem levels. The PSSA described in Section 10.2 inserts safety squarely in the middle of the requirements analysis early in the SE process. The SSA described in Section 10.2 contributes to the verification of the implemented system.

Organizational safety

The concept of organizational safety is a product of SE in its broadest sense because it considers not only the operational aspects of the aircraft but also the development as well. Studies have shown that the common factor in major catastrophes, such as the Space Shuttle *Challenger* and the North Sea disaster, was not technological but rather organizational as described by Paté-Cornell (1990). Organizational factors identified were: time pressures, failure to observe warnings of deterioration and signals of malfunction, lack of an incentive system to handle properly the trade- offs between productivity and safety, failure to learn from mistakes and motivate reporting of problems, and lack of communication and processing of uncertainties. It is the responsibility of the program management in cooperation with the Chief Systems Engineer (CSE) to assure that these factors are minimized as discussed in Section 13.2. Although organizational safety is not specifically mentioned in either standard SE

共因分析 (CCA)

共因分析 (CCA) 的目的是确保可能产生共性故障的子系统的独立性。依赖冗余或者依赖不止一个子系统使用的相同软件的子系统内，最有可能发生共因故障。因此，共因分析应该包含在第 7 章所述的研制综合阶段之中。

共因分析 (CCA) 进行区域安全性分析，以避免飞机同一区域内的多个子系统相互干扰。共因分析 (CCA) 进行特定风险评估，以检查可能影响多个子系统的事件或影响因素。共因分析 (CCA) 进行共模分析，以确定子系统是否真正独立。

安全性和系统工程过程

如表 10.1 所示，合格审定过程要求的安全性评估与系统工程过程是相互兼容的。飞机及系统层级均进行 10.2 节所述的 FHA。在系统工程过程早期，PS-SA(如 10.2 节所述) 均需将安全性全面纳入需求分析过程。系统安全性评估 (SSA)(如 10.2 节所述) 有助于实施系统的验证。

组织安全性

广义上，组织安全性的概念是系统工程的产物，因为组织安全性不仅仅考虑飞机的运行，也考虑飞机的研制。1990 年，Paté-Cornell 的研究表明，重大灾难 (如挑战者号航天飞机和北海灾难事件) 中的共性因素不是技术上的，而更多是组织上的。已认定的组织因素包括：时间压力、未能观测到性能退化的警告及功能异常的信号、缺乏合理控制生产率和安全性之间权衡的激励体系、未能从错误和问题的动机报告中汲取经验教训及缺乏不确定事件的沟通和处理。确保这些因素最小化 (如 13.2 节所述) 是项目经理和与其协作的系统工程总师 (CSE) 的职责。尽管标准系统工程或合格审定指南都没有具体提及组织安全性，但组织安全性可以作为过程保证 (见表 10.1) 的隐含部分考虑。此外，上述讨论的证据表明，

or certification guidelines, it could be considered to be an implied aspect of process assurance shown in Table 10.1. In addition, the evidence discussed above shows that it warrants increased attention.

Qualitative safety

Qualitative safety requirements are all those requirements necessary to meet the quantitative safety requirements of Section 10.2. Many of these requirements result from the PSSA described in Section 10.2, the CCA also described in Section 10.2, or the FHA also in Section 10.2. In each case the requirement must be verifiable as shown in Section 4.1.

The purposes of the qualitative safety requirements are typically: to restrict the severity of failures, to assure that one failure does not cause another failure, or to assure redundancy of critical functions. A typical example is the requirement for double retention of bolts. That is, if a bolt fails, the structural load will be sustained by another bolt.

10.3 Software Development and Certification

Software constitutes a major portion of the development costs of an aircraft. In addition, it is a major focus area in certification. Hence, software development should be approached as methodically as the rest of the aircraft; that is, the software should be considered part of the total aircraft, not a separate system to be developed separately. The basic principles discussed in this book (for example, performance requirements, constraints, synthesis, and verification) apply equally to software and firmware as to hardware.

Software has two characteristics which make it unique and require special treatment: The first is its importance with respect to safety. Virtually every subsystem on the aircraft contains software, avionics more than the others. The second is the difficulty in developing and verifying it. These two aspects make it a critical factor in safety studies and in certification.

Specific methodologies and standards have been produced for the development and certification of systems with software. Notable among these is RCTA/DO- 178B (1994), which is the primary standard for software for commercial aviation. It is not the purpose of this section to describe the software development and certification process, but rather to show how that process already employs SE principles and therefore is compatible with the SE process. Following are a few of the principal considerations.

Software relation to system

330

组织安全性应当获得更多的关注。

定性安全性

定性安全性需求是为了满足 10.2 节所述定量安全性需求的所有必要需求。这些需求很多都产生于 PSSA 及 CCA 或者 FHA(如 10.2 节所述)。需求在所有情形下都必须是可验证的 (如 4.1 节所述)。

定性安全性需求的目的是：限制失效的严重程度，保证一个失效不会造成另一个失效，或者保证关键功能的冗余。典型的例子是对双螺栓紧固的需求，即，如果一个螺栓失效，另一个螺栓将能够承受结构载荷。

10.3 软件开发及合格审定

软件是一架飞机研制成本的重要组成部分。此外，也是合格审定重点关注的领域。因此，软件开发应该像飞机其余部分一样尽可能有序地进行。就是说，应将软件视为整架飞机的一部分，而不是一个独立的系统。本书所讨论的基本原则 (比如，性能需求、约束、综合及验证)，与硬件一样，同样适用于软件和固件。

软件的独特性源于两个特点，这两个特点都必须特别处理。第一是其安全性方面的重要性。实际上，飞机上的每个子系统都包含软件，只是航电部段包含的软件比其他部段更多。第二是软件开发和验证的难度。这两个方面使得软件成为安全性研究及合格审定方面的重要因素。

对于带有软件的系统，其研制和合格审定有特定的方法和标准。其中最重要的是 1994 年发布的 RCTA/DO-178B，它是商用航空软件的首要标准。本节不着重描述软件开发及合格审定的过程，而是要表明这些过程如何应用系统工程原则，因此，它们与系统工程过程是兼容的。下文是一些主要考虑的方面。

331

The development of the system and the software is a two-way street. The following aspects of the system must be developed and provided to the software: First, the system requirements are allocated to the software in accordance with the SE allocation process described in Section 4.5. Secondly, the established software levels are assigned described in Section 10.3, Item 1) in accordance with the criticality and risk of the particular software application. Other software design constraints described in Section 10.3 are assigned, and hardware is defined. Similarly, the software requirements and architecture are introduced as an integral part of the entire system. Error sources have either been identified or eliminated for safety assessment, and fault containment boundaries have been established.

The software life-cycle is analogous to the aircraft and SE life-cycle functions described in Section 3.1. First, there is the planning process which lays out the entire software development process. Secondly, there is the development process itself. Finally, there is a set of activities which should be conducted concurrently. These include verification, configuration management, quality assurance, and the certification liaison process. All of these processes occur for the aircraft as a whole. Table 10.2 shows how the elements of software development and certification fit into the SE process.

Most importantly from an SE point of view, the certification authorities do not consider software to be a separate entity from the rest of the aircraft. Therefore, the aircraft manufacturer is obliged to submit certification compliance evidence in a total aircraft context. Hence, the certification basis as shown in Table 10.1 for the entire aircraft must include the software aspects of certification. The certification authorities will consider the plan for software aspects of certification along with other aircraft data required for certification. The software accomplishment survey will be an essential ingredient in certification compliance.

Software constraints

Like other subsystems in the aircraft architecture, software must adhere to prescribed constraints. The following are the principal software constraint categories. These constraints are normally documented in the system, that is, the aircraft specification because they affect the quality of the entire aircraft. These constraints also apply to firmware:

1. *Safety level*   In SE terms, all software requirements are *derived* requirements.

软件与系统的关系

系统和软件的研制是一条双向道。系统研制必须为软件提供如下输入：第一，根据系统工程分配过程，将系统需求分配至软件（如 4.5 节所述）。第二，根据具体软件应用的关键程度及风险，指定确定的软件层级（如 10.3 节、第 1 条所述）。指定其他软件设计约束（如 10.3 节所述），并且对硬件进行定义。同样，软件需求和架构被视作整个系统不可或缺的部分。安全性评估识别或者消除差错源，并建立了故障包容边界。

软件生命周期与飞机及系统工程生命周期功能（如 3.1 节所示）类似。第一，软件生命周期规划过程展示了软件的整个研制过程。第二，包含研制过程。最后，包含一系列需要同时进行的活动，包括验证、构型管理、质量保证及合格审定联络过程。应用于飞机时，所有这些过程须协同进行。表 10.2 展示了软件开发及合格审定元素如何融入系统工程过程。

从系统工程视角来看，最重要的是合格审定当局认为软件并不是与飞机其他部分无关的独立实体。因此，飞机制造商有义务提交考虑整架飞机环境的合格审定符合性证据。因此，整架飞机的合格审定基础（见表 10.1）必须包括软件合格审定方面的内容。合格审定当局将考虑合格审定软件方面的计划，同时考虑合格审定所需的其他飞机数据。软件完成综述 (SAS) 将作为合格审定符合性的重要组成部分。

软件约束

与飞机架构中的其他子系统一样，软件必须符合规定的约束。以下是主要的软件约束类别。这些约束通常以文件形式记录于系统中，即飞机规范，因为它们影响整架飞机的质量。这些约束同样适用于固件：

That is, functional requirements are allocated to both hardware and software after a design concept has been selected and it has been decided where the software will fit into the system. When it has been decided where the software is going to fit in, what functions the software will perform, and what the hazard categories of the functions are from the system safety analysis (SSA) described in Section 10.2, a software level can be assigned to the software item. These levels are labeled A, B, C, D, and E and establish the level of quality and testing necessary for the software. Software levels must be approved by the certification authorities shown Table 10.2.

Table 10.2   Compatibility of the SE and software development and certification processes

SE Elements	Software Development and Certification Aspects
Functional analysis (Chapter 3)	Software functional analysis, system functions allocated to software, criticality of software functions through FHA discussed in Section 10.2.
Requirements development (Chapter 4)	Software requirements development, allocation of system requirements to software, software levels discussed in Section 10.3, Item 1), design constraints, hardware definition to software, software requirements, software redundancy requirements through PSSA discussed in Section 10.2.
Synthesis (Chapters 7–9)	Software synthesis, software architecture, error sources identified and eliminated, fault containment boundaries established, software independence through CCA discussed in Section 10.2.
Verification and validation (Chapter 11)	Software testing and analysis, evaluation of software implementation SSA discussed in Section 10.2.
SE management (Chapter 12)	Software planning, software configuration management (SCM), software configuration index (SCI), software quality assurance (SQA), plan for the software aspects of certification (PSAC), liaison with certification authorities.

2. *Protection requirements*   It is necessary to protect software functions from each other and from other functions. Software protection is an essential requirement in software development shown in Table 10.2 and usually takes the form of partitioning by time and space.

3. *Software standards*   Software standards determine the requirements for programming language, control structures, and other design constraints. These standards are also another part of software development shown in Table 10.2.

1. **安全性等级** 按照系统工程术语，所有的软件需求都属于衍生需求。就是说，当选择了设计概念，并决定了软件嵌入系统的位置后，功能需求既分配给硬件也分配给软件。当已决定软件将在何处嵌入，软件将执行什么功能，以及根据系统安全性分析 (10.2 节)，决定了这些功能的危害性属于什么类别时，便指定了该软件的等级。这些等级被标记为 A、B、C、D 和 E，这些等级决定了软件的质量等级和必要的测试。如表 10.2 所示，软件等级必须由合格审定当局批准。

表 10.2　系统工程和软件开发及合格审定过程之间的兼容性

系统工程元素	软件开发及合格审定方面
功能分析 (第 3 章)	软件功能分析，分配至软件的系统功能，通过 FHA(10.2 节) 得出的软件功能重要性
需求开发 (第 4 章)	软件需求开发，系统需求分配至软件，软件等级 (如 10.3 节、条目 1 所述)，设计约束，软件的硬件定义，软件需求，通过 PASA 确定软件冗余需求 (如 10.2 节所述)
综合 (第 7 至 9 章)	软件综合，系统架构，识别和消除的差错源，故障包容边界建立，CAA 中确定的软件独立性 (如 10.2 节所讨论)
验证和确认 (第 11 章)	软件测试和分析，软件实施 SSA 评估 (如 10.2 节所讨论)
系统工程管理 (第 12 章)	软件规划，软件构型管理 (SCM)，软件构型索引 (SCI)，软件质量保证 (SQA)，软件合格审定计划 (PSAC)，与合格审定当局联络

2. **保护需求** 必须防止软件功能相互影响，以及受其他功能的影响。软件保护，在软件开发 (见表 10.2) 中是基本的要求，并且通常采取时间和空间分隔的形式。

3. **软件标准** 软件标准决定了程序语言、控制架构及其他设计约束的需求。这些标准也是软件开发 (见表 10.2) 的另一个部分。

4. **内存和时间消耗** 内存需求规定对于任何给定应用的所需额外内存。

5. **存储保护** 开发存储保护需求以避免计算机存储器受到一些因素的影响，

4. *Memory con timing consumption*   The memory requirement specifies the required excess memory for any given application.

5. *Memory protection*   Memory protection requirements are developed to protect the computer memory from influences, such as electrical power source transients or EMI, HIRF, or lightning generated transients.

6. *Program storage con integrity*   It is the purpose of program storage and integrity requirements to specify the reliability, availability, and other features of data integrity when it is stored in memory.

7. *Computer anomalies*   The computer anomaly requirements specify the timing and detection of potentially unsafe conditions in a computer and the change to a safe condition. Littlewood and Stringini (1992) emphasize the difficulty in predicting these anomalies and argue that the use of computers for performing complex decisions should be avoided.

8. *Fail-safe design*   The requirements for a fail-safe design specify the principles to be employed to ensure a safe design. The objective of this requirement is to make the design less sensitive to failure modes.

9. *Robustness*   The robustness requirements specify the ability of the software or firmware to operate in the face of external failures or invalid inputs.

10.4   Commercial Aviation Safety Team (CAST)

According to the overview, the Commercial Aviation Safety Team (CAST 2011) is an international partnership of aircraft manufacturers, employee groups, regulatory authorities, and aircraft operators working together to enhance safety in the commercial aviation domain. CAST makes recommendations to all elements of the aviation domain for ways to enhance safety. Their record is impressive. According to the FAA (2013), just one of the CAST members, CAST recommendations have resulted in a decrease in the fatality rate of 83 percent in the United States.

Recommended safety enhancements for manufacturers

CAST periodically publishes safety enhancement summaries detailing recommended enhancements to aircraft intended to improve safety. Enhancements include actions by manufacturers, regulators, and air carrier associations. Following is a summary of such a safety enhancement sheet published by CAST (2012) and (2014); the summaries below are highly condensed and paraphrased; refer to the original

例如电源瞬变现象、电磁干扰、高能辐射场或雷电产生瞬变现象。

6. *程序储存和完整性* 程序储存和完整性需求的目的是当程序储存在内存时，规定其可靠性、可用性及数据完整性的其他特性。

7. *计算机异常* 计算机异常需求规定了对计算机中潜在不安全状态的计时、探测及安全状态的变更。1992 年，Littlewood 和 Stringini 强调了预测这些异常的困难性，并强调应避免使用计算机执行复杂决策。

8. *失效 - 安全设计* 失效 - 安全设计需求规定了为保证安全设计所采用的原则。失效 - 安全设计需求的目标是降低设计对失效模式的敏感性。

9. *鲁棒性* 鲁棒性需求规定了软件或固件在面对外部失效或无效输入时工作的能力。

10.4　商用航空安全小组 (CAST)

根据概述介绍，**商用航空安全小组 (CAST 2011)** 是飞机制造商、员工团队、监管当局以及飞机运营商共同参与的国际合作机构，旨在提高商用飞机领域的安全性。商用航空安全小组 (CAST) 为提高航空领域所有元素的安全性给出建议，他们取得的成绩令人印象深刻。FAA 只是 CAST 会员之一，据其 2013 年描述，商用航空安全小组 (CAST) 的建议使美国航空致人死亡事件的发生率降低了83%。

为制造商建议的安全性改进

商用航空安全小组 (CAST) 定期发布安全性改进总结，详细介绍针对飞机的推荐改进措施，以提高安全性。包括改进制造商、监管当局及航空承运人协会的活动。以下是商用航空安全小组 (CAST) 于 2012 年和 2014 年发布的一份安全性改进单的摘要 ；下述摘要经过了高度精简和改述 ；关于准确的措辞和相关责任，

for exact wording and for responsibilities. Enhancements underway at the time of publication of the fact sheet are not listed here. In addition, the following list contains items that are in progress and not completed at this time.

Category: Controlled Flight into Terrain (CFIT)

Enhancement: Terrain Avoidance Warning System (TAWS)

- Install TAWS on all newly manufactured aircraft.
- Retrofit TAWS on existing aircraft.
- Institute system to support TAWS including installation, maintenance, and training.
- Develop standard operating procedures (SOP) for flight deck crew members.
- Include vertical angles in instrument approach procedures.
- Incorporate a digital elevation model to determine minimum vectoring altitudes (MVA) to reduce TAWS alerts.
- Develop procedures to provide better separation from terrain at selected sites.

Enhancement: Runway lighting

- Install visual glide slope indicators (VGSI) on all runways used by carriers.

Enhancement: Distance measuring equipment (DME)

- Install DME on older aircraft.

Enhancement: Improved area navigation (RNAV) procedures

- Include vertical guidance in RNAV procedures.
- Incorporate required navigation procedures (RNP) technology to allow for more precision landings.
- Incorporate advanced precision approach procedures into the Next Generation Air Traffic System design.
- Institute plan for periodic check of the minimum safe altitude warning (MSAW) system.

Enhancement: Proactive safety plans—Flight Operational Quality Assurance (FOQA) and Aviation Safety Action Plan (ASAP)

- Develop voluntary procedures and protocols for trends and corrective actions.

请参见原文。此情况说明书发布时，正在进行的改进没有列出。此外，如下清单包括了现在正在进行，且尚未完成的条目。

类别：可控撞地飞行 (CFIT)

改进：地形提示和告警系统 (TAWS)

- 在所有新制造的飞机上安装 TAWS。

- 在现有飞机上加装 TAWS。

- 建立体系以支持 TAWS，包括安装、维修及培训。

- 制订驾驶舱机组成员标准操作程序 (SOP)。

- 在仪表进近程序中包括垂直角度。

- 采用数字标高模型确定最小引导高度 MVA 以减少 TAWS 告警次数。

- 研制程序为选定地点的地形提供更好的间隔。

改进：跑道照明

- 在承运人使用的所有跑道上安装目视下滑道指示器 (VGSI)。

改进：测距器 (DME)

- 在较旧的飞机上安装 DME。

改进：改进的区域导航程序 (RNAV)

- 在 RNAV 程序中包括垂直导引。

- 采用必需的导航程序 (RNP) 技术以考虑到更精准的着陆。

- 在下一代空中交通管制系统设计中采用先进的精密进近程序。

- 为最低安全高度警告 (MSAW) 系统的定期检查制定计划。

改进：主动的安全性计划 — 飞行操作品质保证 (FOQA) 和航空安全行动计划 (ASAP)

- 为安全性趋势和修正措施制订自愿的程序和方案。

改进：改进的机组资源管理 (CRM) 程序

Enhancement: Improved crew resource management (CRM) procedures

• Promote SOP for CRM.

Enhancement: TAWS Improved Functionality.

• Install GPS sensors on all aircraft.
• To achieve the capability to update terrain databases for operators.
• Update TAWS underlying algorithms.
• Evaluate TAWS features currently not used.

Enhancement: Improved CFIT training

• Add CFIT training to all required air carrier curricula.
• Add CFIT training for all controllers.

Category: Approach and Landing Accident Reduction (ALAR)

Enhancement: Airplane flight manuals (AFMs)

• Provide inspectors with latest AFM database.

Enhancement: Flight deck equipment upgrade

• Develop advisory material for checklists and alerts for new type designs.
• Incorporate FAA Human Performance Considerations in checklists.
• Provide automatic aural altitude call outs.

Enhancement: Implementation plan for aircraft design

• Ensure continuing airworthiness processes and incorporate risk management techniques; monitor fleet performance and prioritize safety critical threats and corrective actions. (See Chapter 15 for an elaboration of risk management considerations and Section 4.3 of the *FAA Systems Engineering Manual* (2014).)

Enhancement: Promote safety culture

• Promote safety culture for all chief executive officers (CEOs) and directors of safety (DOSs).
• Incorporate safety culture information in manuals.

- 为 CRM 促进 SOP。

改进：TAWS 改进功能

- 在所有飞机上安装 GPS 传感器。

- 获得为运营商更新地形数据库的能力。

- 更新 TAWS 潜在算法。

- 评估目前尚未使用的 TAWS 特征。

改进：改进 CFIT 培训

- 为有需要的航空承运人增加 CFIT 培训。

- 为所有空中交通管制员增加 CFIT 培训。

类别：减少进近及着陆事故 (ALAR)

改进：飞机飞行手册 (AFM)

- 为检查员提供最新的 AFM 数据库。

改进：飞机驾驶舱设备升级

- 为新型号设计制订检查单和告警制订咨询材料。

- 在检查单中纳入 FAA 人类表现考虑。

- 提供自动音响高度呼叫。

改进：飞机设计实施计划

- 保证持续适航过程，并纳入风险管理手段；监控机队绩效并优先处理安全性关键威胁及修正措施。（参见第 15 章关于风险管理考虑的详述，以及 2014 年出版的《FAA 系统工程手册》第 4.3 节）

改进：提升安全性文化

- 提升所有首席执行官 (CEO) 和安全性主管 (DOS) 安全性文化。

- 在各种手册中纳入安全性文化内容。

改进：维修规则

Enhancement: Maintenance rules

- Re-emphasize maintenance rules for landing struts in maintenance programs.
- Re-emphasizemaintenancerulesforlandingstrutsformaintenancesuppliers.
- Increase oversight of maintenance procedures for suppliers.
- Include minimum equipment list (MEL) in maintenance procedures.
- Directs DOS to determine maintenance deficiencies.

Enhancement: Installation and improvement of flight deck equipment

- Implement electronic checklist and smart-alerting systems.

Enhancement: Flight crew training

- Implement flight crew training volutarily over ALAR topics.

Enhancement: Aircraft design

- Incorporate fault tolerant design principles.

This recommendation is in agreement with the resilience principles described in Chapter 16.

Category: Loss of control

Enhancement: Inform personnel and flight crew of policies and procedures

- Review processes for distributing essential information.
- Distribute information to flight crews and maintenance personnel.
- Include essential information in training programs and flight manuals.

Enhancement: Standard operating procedures (SOP)

- Publish and enforce and provide training for SOP for all phases of flight.

Enhancement: Risk assessment and management

- Implement methods to prioritize safety-related decisions.

This recommendation is in agreement with the methodology of Chapter 15 on Risk Management. However, this recommendation focuses on safety-related risks,

- 再次强调维修大纲中关于起落架支柱的维修规则。

- 再次针对维修供应商强调关于起落架支柱的维修规则

- 加强对供应商维修程序的监督。

- 在维修程序中纳入最低设备清单 (MEL)。

- 指派 DOS 以确定维修缺陷。

改进：飞机驾驶舱设备的安装和改进

- 采用电子检查单和智能化告警系统。

改进：飞行机组培训

- 自愿进行飞行机组关于减少进近及着陆事故 (ALAR) 主题的培训。

改进：飞机设计

- 采用损伤容限设计原则。

此条建议与第 16 章所述恢复力原理一致。

类别：失控

改进：将政策和程序通知飞行机组和相关人员

- 评审分发重要信息的过程。

- 将信息分发给飞行机组及维修人员。

- 将重要信息纳入培训大纲和飞行手册中。

改进：标准操作程序 (SOP)

- 发布，推行所有飞行阶段的 SOP，并提供相应的培训。

改进：风险评估和管理

- 实施优先处理安全性相关决策的方法。

此条建议与 15 章 "风险管理" 的方法相一致。然而，此建议更关注与安全性相关的风险，而 15 章讨论的风险范围更大。

改进：人为因素及自动化

343

while Chapter 15 discusses a broader range of risks.

Enhancement: Human factors and automation

- Compile a list of policies and procedures dealing with mode awareness and energy state awareness.
- Disseminate improved automation policies and procedures to operators and manufacturers. (The Billings (1997) list of rules for the interaction of humans and automation is a potential source of policies and procedures for this recommendation.)

Enhancement: New aircraft designs

- In the design of new airplanes incorporate angle of attack and low-speed protection, thrust-asymmetry compensation, and bank-angle protection.
- Airbus (2013) describes this technology called flight envelope protection and its implementation. This technology is among the new technologies discussed in Section 2.4.
- In the design of new airplanes incorporate features that will minimize thrust asymmetries, yield to manual force control when necessary, incorporate annunciations when necessary, and include low-speed protection.
- For all aircraft incorporate vertical situation displays in aircraft designs.
- For new aircraft incorporate displays and annunciations that will reduce accidents due to loss of control.
- For new aircraft designs which do not incorporate evaporative anti-icing systems, incorporate advanced anti-icing features.

Enhancement: Flight crew proficiency

- Develop voluntary program to improve flight crew proficiency.

Enhancement: Advanced maneuvers training

- Develop training program to prevent and recover from flight conditions outside the normal operating envelope.

Enhancement: Runway incursion

- Provide enhanced air traffic control (ATC) training and flight crew training for runway incursion (RI).

- 编制一系列处理模式感知和能量状态感知的政策及程序。

- 向运营商及制造商分发改进的自动化政策及程序。(1997 年，Billings 列出的针对人和自动化之间相互作用的规则是本建议中政策和程序的一个潜在来源。)

改进：新飞机设计

- 在新飞机设计中，纳入攻角和低速保护、推力不对称补偿及倾角保护。

- 2013 年，空客介绍了称为飞行包线保护的技术及其实施。该技术是 2.4 节讨论的新技术之一。

- 在新飞机设计中，采用将最大限度地减小推力不对称，必要时使用力控制，必要时有告示信号及包括低速保护等设计特征。

- 对于所有飞机，在飞机设计中采用垂直状态显示。

- 对于新飞机，纳入可以减少因失控导致事故的显示器和通告牌。

- 对于不采用蒸汽防冰系统的新飞机设计，采用先进的防冰设计特征。

改进：飞行机组熟练程度

- 开发自愿程序，提高飞行机组熟练程度。

改进：先进的机动飞行培训

- 开发培训项目，以防止超出正常运行包线的飞行条件并从这种情况中改出。

改进：跑道侵入 (RI)

- 提供强化的关于跑道侵入 (RI) 的空中交通管制 (ATC) 培训和飞行机组培训。

- 为空中交通管制员提供改进的 CRM 培训。

- 为商用航空地勤人员、通用航空地勤人员、牵引车操作人员、车辆操作人员建立、编制和培训 SOP。

- Provide enhanced CRM training for air traffic controllers.
- Establish, document, and train SOP for commercial aviation ground operators, general aviation ground operators, tow tug operators, vehicle operators.
- Incorporate new technologies and new procedures that will improve situational awareness for air traffic controllers.
- Clarify procedures for air traffic control instructions and mandatory readbacks.

Enhancement: Turbulence

- Standardize methodologies for improved situational awareness and procedures during flight turbulence.

Enhancement: Uncontained engine failures (UEF)

- Implement advanced methods and technologies to detect potential engine defects. This item is in agreement with the concept of *latent* flaws discussed in Chapter 16.

Category: Maintenance

Other sources, for example Reason (1997, pp. 85–105), have concluded that faulty maintenance is a root cause of many aircraft and other accidents. The enhancements listed below address some of the deficiencies inherent in the maintenance process. Reason mentions American Flight 191 and Japan Airlines Flight JL 123 as examples.

Enhancement: Advanced circuit design

- Implement technologies to determine fault sources for maintenance.
- Implement advanced circuit breaker technology.
- Implement processes to certify advanced circuit breaker protection.
- Implement and certify advanced circuit breakers on in-service airplanes.

346

Enhancement: Maintenance policies and procedures

- Assure that work cards, procedures, and manuals are complete, accurate, available, and appropriately used.
- Assures that maintenance data are collected and reported to the original equipment manufacturer (OEM). This item is in progress.

- 采用能够提升空中交通管制员情境感知的新技术和新程序。

- 澄清空中交通管制指令和强制复诵的程序。

改进：湍流

- 将飞机遭遇湍流时，改进情境感知方法和程序标准化。

改进：非包容性发动机失效 (UEF)

- 实施先进的方法和技术，探测潜在的发动机缺陷。本条与 16 章讨论的潜在缺陷概念一致。

类别：维修

例如，Reason(1997，第 85 至 105 页) 等其他一些来源总结认为，错误的防护是很多飞机事故和其他事故的根源。下面列出的改善意见阐述了维修过程中固有的缺陷。Reason 以泛美航空 191 航班和日本航空公司 JL 123 航班作为例子。

改进：先进的电路设计

- 采用能够为维修确定故障源的技术。

- 采用先进断路器技术。

- 采用过程对先进断路器保护的取证。

- 在在役飞机上应用先进断路器，并取证。

改进：维修政策及程序

- 保证工卡、程序及手册是完整、准确和可用的，并保证被恰当使用。

- 保证收集维修数据并提交给原始设备制造商 (OEM)。该项在进行中。

- 找出并排除维修过程中可能影响飞机安全性的差异项。该项在进行中。

- 在维修时提供空速管的可视化标记，以加强飞行前巡检。

改进：错误的跑道起飞

- 评审高风险机场的现有标牌和标记计划，识别潜在危害，制订减缓计划，

- Identify and correct gaps in the maintenance process that may affect aircraft safety. This item is also in progress.
- Provide visible tagging of pitot tubes during maintenance when covered to enhance preflight walk-arounds.

Enhancement: Wrong runway departure

- Review existing signs and marking plans at high-threat airports, identify potential hazards, develop mitigation plans, and incorporate necessary changes.
- Incorporate wrong-runway operations in pilot and controller training programs and in Airport Layout Reviews, in policies for early take-off clearances, and in the installation of moving map displays and runway awareness systems.

Category: Cargo

Enhancement: Fire containment

- Develop standard fire suppression and/or containment systems.

Enhancement: Load training and SOP
- Plan and enforce contractor training for cargo loading.

Enhancement: Hazardous materials
- Implement improved regulations, methods, technology, and training for detecting and preventing hazardous materials.
- Develop systems to contain and suppress fires resulting from hazardous materials.

Enhancement: Processes and oversight of cargo-related issues

- Enhance legal processes for dealing with fires resulting from hazardous materials.
- Enhance oversight of cargo-related issues.
- Develop safety culture for dealing with cargo issues.

Category: Icing

Enhancement: Avionics

- Design and install smart pitch guidance systems.

并采用必要的更改。

- 在驾驶员和空中交通管制员培训项目中，在机场布局评审，在提早起飞放行许可政策中，以及移动地图显示器和跑道感知系统装置中，纳入错误跑道操作内容。

类别：货物

改进：着火包容
- 开发标准的着火遏制和（或）包容系统。

改进：装载培训和 SOP
- 制定并加强货物装载承包商的培训。

改进：危害性材料
- 制定针对探测和防止使用危险材料的改进的规章、方法、技术和培训。
- 开发能够包容和遏制危险材料引起着火的系统。

改进：货物相关问题的过程和监督
- 加强处置危害性材料引起着火的法律过程。
- 加强对货物相关问题的监督。
- 开发处理货物问题的安全性文化。

类别：结冰

改进：航电
- 设计和安装智能化俯仰导引系统。

改进：培训
- 编制处置发动机与结冰相关事件的培训材料。

Enhancement: Training
- Develop training material for dealing with engine events related to icing.

Category: Midair collisions

Enhancement: Terrain Collision Avoidance System (TCAS) policies and procedures
- Design B/C/D airspaces to make visual flight rules (VFR) more easily usable. (Note: B/C/D refers to defined airspaces in which instrument flight rules (IFR), special VFR, and VFR apply.)
- Conduct study to determine if TCAS alerts can be reduced or eliminated at high elevation airports. This item is in progress.
- Conduct study to determine if an adjustment of lateral or vertical air traffic separation minima would reduce or eliminate the number of TCAS alerts between IFR and VFR traffic. This item is in progress.
- Conduct study to determine whether current TCAS can support NextGen traffic levels. Create new strategy for NextGen. This item is in progress.

Category: Airplane state Awareness (ASA)

Enhancement: Low speed alerting

- Implement manufacturer service bulletins to provide low speed alerting on existing transport type designs.

Enhancement: Crew state awareness

- Specify non-standard, non-revenue flights for functional checks.
- Conduct risk assessment and develop guidelines for non-standard, non-revenue flights.
- Develop SOP to reduce flight crew loss of state awareness and develop training programs.
- Verify and validate crew capability with third party training providers in executing SOP for state awareness after receiving training.

This third party verification and validation is in agreement with the independent review principle described in Chapter 15, Risk Management.

- Implement standard practices with respect to upset (loss of control) or stall resulting

类别：空中相撞

改进：空中防撞系统 (TCAS) 政策及程序

- 设计 B/C/D 空域，使目视飞机规则 (VFR) 更加容易使用。(注 : B/C/D 系指应用仪表飞行规则 (IFR)、特许 VFR 和 VFR 的规定空域)

- 进行研究以确定在高标高机场能否减少或者消除 TCAS 告警。

- 进行研究以确定调整空中交通最小横向和垂直间隔能否减少 IFR 和 VFR 飞机之间的 TCAS 告警数量。此项正在进行中。

- 进行研究以确定当前的 TCAS 是否可以支持 NestGen 交通等级。创造 NestGen 的新战略。此项正在进行中。

类别：飞机状态感知 (ASA)

改进：低速告警

- 实施制造商服务通告 (SB) 以提供现有运输机型号设计上的低速告警。

改进：机组状态感知

- 规定非标准、非营业飞行的功能性检查。

- 对非标准和非营业飞行实施风险评估并制定指南。

- 开发 SOP 以减少飞行机组状态感知丧失，并制订培训项目。

- 在进行培训后，与第三方培训提供商一起验证和确认飞行机组实施状态 感知 SOP 的能力。

这里的第三方验证及确认与 15 章 " 风险管理 " 中描述的独立评审原则一致。

- 实施因状态感知缺失导致飞机翻转 (失控) 或失速时的标准做法 ；采用模拟器和实际的场景。

- 注重非标准情况下，飞行的控制和稳定。

- 采用实际的复飞情景下的培训。

from lack of state awareness; incorporate in simulators and in realistic scenarios.

- Emphasize control and stabilized flight in non-standard situations.
- Incorporate realistic training in go-around scenarios.
- Incorporate CRM in training.
- Incorporate virtual meteorological conditions (VMC) in displays.
- Incorporate bank angle alerting and recovery guidance displays on new aircraft and FBW programs.
- Study incorporation of bank angle protection and energy state cues in FBW, existing non-FBW systems, and out of production aircraft. This item is in progress.
- Study new technologies to improve state awareness. This item is in progress.
- Conduct research to improve state awareness; display in an intuitive manner; prioritize and escalate alerts; improve technology readiness levels (TRL) as shown in Table 7.1; this item is in progress.
- Conduct research to determine the benefits of using various levels of prototype advanced aerodynamic modeling of full stall characteristics to perform full stall recovery training; this item is in progress.
- Conduct research to assess crew performance with respect to state awareness; this item also includes attention-related performance; this item is in progress.

Conclusions

The following conclusions can be drawn from the above CAST recommendations: First, recommendations are just recommendations; they are not mandatory requirements. They will not be mandatory until the regulatory agencies make them mandatory. In some cases the manufacturers will implement them even if they are not mandatory. Secondly, most of the recommendations are procedural or administrative. Therefore, the cost of implementing them will be minimal. Hence, the benefit to cost ratio can be expected to be large.

10.5 Fatality Rate History

Figure 10.1 shows the fatal accident rate for Part 121 that is scheduled carrier, operations over more than 30 years. It is apparent that safety recommendations, such as those made by CAST, have had a significant impact on safety.

- 培训中纳入 CRM。

- 显示中纳入虚拟气象条件 (VML)。

- 在新飞机和电传操纵 (FBW) 项目中，采用倾角告警和恢复导引显示。

- 研究 FBW、现有的非 FBW 系统和停产飞机中倾角保护和能量状态信号的体现；此项正在进行中。

- 研究新的技术以提高状态感知。此项正在进行中。

- 进行研究以提高状态感知；直观显示；按优先级顺序排列告警；提高技术成熟度 (TRL)(如表 7.1 所示)；此项正在进行中。

- 进行研究，以确定采用不同等级全失速特性先进气动模型样机进行全失速改出培训的益处；此项正在进行中。

- 进行研究，以评估机组在状态感知方面的表现；此条目也包括与注意力相关的表现；此项正在进行中。

结论

根据上述 CAST 建议，可以得出如下结论：

首先，在监管机构要求强制执行之前，建议仅仅是建议，不是强制性要求。在某些情况下，即使它们不是强制性的，制造商也会执行。第二，大部分的建议是程序性和管理性的。因此，实施它们的成本将会微乎其微，而预期会带来很大的收益成本比例。

10.5 致死率历史

图 10.1 展示了按 121 部运营的定期航班承运人在超过 30 年的运营中发生的致死事故率。可以明显看出，安全性建议，例如 CAST 所制定的，对安全性产生了显著影响。

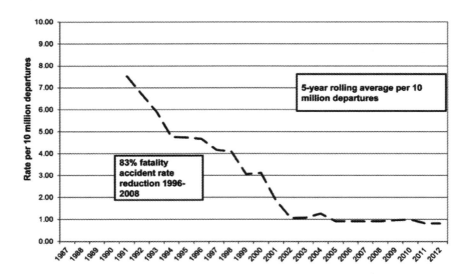

Figure 10.1   Fatal accident rate for Part 121 operations (scheduled carriers)

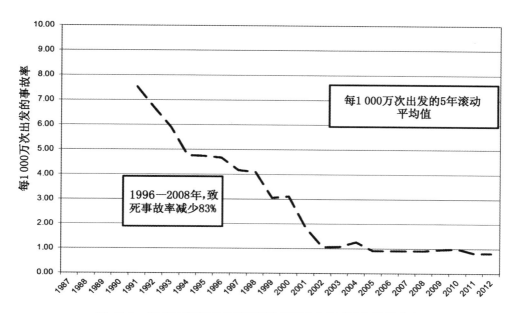

每1 000万次出发的5年滚动
平均值

1996—2008年,致
死事故率减少83%

图 10.1　按 121 部运营（定期航班承运人）的致死事故发生率

11

Verification and Validation

[Verification is] the evaluation of an implementation of requirements to determine that they have been met.

ARP 4754A (2010, p. 13)

[Requirements validation is] the determination that the requirements for a product are correct and complete.

ARP 4754A (2010, p. 13)

[Product validation is] actions to confirm that the behavior of a developed system meets user needs.

Stevens et al. (1998, p. 364)

Verification of requirements in the commercial aircraft industry is an extremely complex process. Verification of certification-related requirements is controlled by Federal Aviation Regulations (FARs) and thoroughly conducted and monitored as shown in Table 10.1. These requirements and the balance of the requirements are verified by a variety of methods, including ground tests, simulations, analyses, examinations, and flight test. Similarity analyses are based on flight histories of specific components and subsystems. In any event, the SE principle of complete verification is paramount. This chapter will show how 100 percent verification is accomplished for aircraft systems. It also shows how traditional aircraft processes, such as certification, are incorporated into the SE verification process.

This chapter also explains the difference between verification and validation and how requirements validation differs from product validation.

356

11.1 The Verification Matrix

All verification starts with the verification matrix. This matrix is a standard part of the specification format as shown below. Basically, the verification matrix assigns a

第 11 章　验证及确认

[验证是] 评估需求的实现，以确认已经满足需求。

ARP 4754A (2010, 第 13 页)

[需求确认是] 确定产品需求是正确和完整的。

ARP 4754A (2010, 第 13 页)

[产品确认是] 确认所开发系统的行为满足用户需要的活动。

Stevens et al. (1998, 第 364 页)

商用飞机工业领域的需求验证是一个极其复杂的过程。合格审定相关的需求验证受 (美国) 联邦航空条例 (FAR) 控制, 其实施和监控的全过程如表 10.1 所示。这些需求和需求之间的平衡是通过大量的方法进行验证的, 这些方法包括地面试验、模拟、分析、检查及飞行试验。相似性分析基于具体的设备和子系统的服役历史。在任何情况下, 验证完整性都是系统工程最重要的原则。本章将介绍飞机系统的 100% 验证是如何完成的。本章还介绍了合格审定等传统飞机过程如何与系统工程验证过程相结合。

本章也讨论了验证与确认之间的区别, 以及如何区分需求确认和产品确认。

11.1　验证矩阵

所有的验证活动都始于验证矩阵。如下所述, 矩阵是规范格式的标准部分。

verification method to each requirement, thus fulfilling the SE goal of 100 percent verification.

For the certification data package, the verification matrix has specific required contents, as follows:

1	Requirement.	4	Verification method(s) applied.
2	Associated function.	5	Verification conclusion (pass or fail).
3	Development assurance level.	6	Verification coverage summary.

11.2   Traditional SE Verification

Traditionally, SE has recognized four categories of verification: test, demonstration, analysis, and examination. Examination is sometimes called inspection. Some sources include simulation and similarity as separate categories. We have included these as types of analysis.

Test

A test is a type of verification which requires instrumentation. These tests can be, for example, pressure tests, wind tunnel tests, or flight tests. There is sometimes confusion between development tests and verification tests. A development test is a test conducted to reach a desired design and to satisfy requirements. Trade-offs can be conducted during development tests. Wind tunnel tests can be development tests. They can also be verification tests. In fact, the same test can be both a development test and a verification test. The difference is that only the test which represents the final solution will be the verification test. It is the test that will verify (or be part of the verifications), for example, the requirement to provide a given lift capability. All the other test points qualify as development tests.

Testing is an essential element in the certification process. Testing assures that every aircraft element performs the function it was intended to perform, performs it to the expected level of performance, and does not perform functions it was *not* intended to perform.

Analysis

总的来说，验证矩阵将验证方法分配到每个需求，以完成系统工程 100% 验证的目标。

对于合格审定数据包，验证矩阵要求的具体内容如下：

1	需求	4	采用的验证方法
2	相关的功能	5	验证结论（通过与否）
3	研制保证等级	6	验证覆盖率总结

11.2 传统系统工程验证

传统上，系统工程识别出四种验证类别：试验、演示、分析及检查。检查 (examination) 有时被称为检验 (inspection)。将模拟和相似性分析作为独立类别也是有出处的，我们将模拟和相似性分析纳入分析的类别。

试验

试验是验证的一种类别，该过程需要试验设施。这些试验可能包括压力试验、风洞试验或者飞行试验等。研制试验与验证试验之间有时会发生混淆。进行研制试验的目的是为了得到预期的设计，并满足相关需求。在研制试验中可以进行权衡分析。风洞试验可以是研制试验，也可以是验证试验。实际上，同一个试验既可以是研制试验，也可以是验证试验。区别是，只有代表最终设计方案的试验才是验证试验。比如，验证提供给定升力能力需求（或作为验证的一部分）的试验就是这种试验。所有的其他试验点都可以视为研制试验。

试验是合格审定过程的必要组成元素。试验确保每一个飞机元件都执行其预定执行的功能，达到期望的性能等级，而且不执行非预定的功能。

分析

分析是为验证某一无法通过其他方式验证的需求而进行的任何一种数学、计

Analysis is any kind of mathematical, computational, or logical task performed to verify a requirement which cannot be verified in any other manner. In addition, analysis is used as an initial type verification to assure that the aircraft meets the requirements early in the development phase. The results of the analysis will be confirmed by actual flight tests later in the program. Analysis can also be used in conjunction with testing to extend the envelope of the test results. For example, if the aircraft is tested at certain points in the flight envelope, computer simulations can be used to predict the aircraft behavior in regimes beyond and in between the actual flight data points.

Analysis, from the certification point of view, does not include similarity, which is discussed in Section 11.2.

Simulation

Verification by simulation is a type of analysis using computers. Simulation is used extensively in the commercial aircraft industry. The flight handling characteristics and aircraft performance characteristics can be simulated and used to verify performance requirements long before they are tested in flight. Simulation can also be used in ergonomics to simulate the movement of the human body when interacting with aircraft components. Analysis by computer simulation is also a valid method of verification for certification. Typical simulation types include finite element analysis (FEA) to analyze structures. Another type of simulation is computational fluid dynamics (CFD) which analyzes air flow around the aircraft.

In-service data and similarity

Analysis by similarity is a frequently used method of verification. It is based on the assumption that another component which has met *the same performance requirements* and operated *in the same environment* will meet its own performance requirements. In the aircraft industry this similarity is based largely on *in-service data*, that is, on past experience of another component in another aircraft. Great care should be taken, however, to assure that the operating conditions of the comparison aircraft are equivalent to the aircraft being designed. In fact, no two aircraft are identical and no two operating conditions are identical. One aircraft may have operated in extremely cold climates, while the other may have operated in very hot climates. When the supplier wishes to use an existing component as a solution, a Letter of Similarity should be submitted to the manufacturer justifying the past use as a verification of the requirements. In-service data is also a valid method of verification for certification.

算或者逻辑工作。此外，分析被用作一种初始型号验证手段，确保飞机在研制阶段早期满足相关需求。分析的结果通过项目后期飞行试验来确认。可以结合试验来进行分析，以扩展试验结果的包线。比如，飞机在飞行包线内的某些点上进行试验，可以用计算机模拟预测飞机在超出飞行包线的区域及这些实际飞行数据点之间的行为。

从合格审定的观点看，分析不包括相似性分析 (如 11.2 节所述)。

模拟

模拟验证是使用计算机进行的一种分析方法。模拟分析广泛用于商用飞机领域。该方法可以模拟飞行操控特性和飞机性能特性，并用于在试飞前验证性能需求。模拟分析也可以用于人体工程学，以模拟与飞机部件交互时人体的运动。计算机模拟分析也是一种有效的合格审定验证方法。典型的模拟类型包括有限元分析 (FEA)，用来分析结构，另一种类型的模拟是可用于分析飞机周边气流的计算流体力学 (CFD)。

在役数据及相似性

相似性分析是一种常用的验证方法。相似性分析基于一种假设，即只要满足相同性能需求，并且在相同环境下工作，另一个部件也将满足其自身性能需求。在飞机工业领域，相似性分析在很大程度上基于在役数据，即基于另外一架飞机上部件过去的经历。然而，确保对比机型的运行条件和所设计飞机之间运行条件的等效性是重点考虑内容。实际上，没有两架飞机和两个运行条件是完全一致的。一架飞机可能在极端寒冷的天气下运行，然而另外的飞机可能在非常炎热的天气下运行。若供应商希望使用现有部件作为解决方案，应向制造商提交相似性信函，论证以既往的使用经验作为需求验证的合理性。在役数据通常也是有效的合格审

System safety assessment (SSA)

Analysis also includes the SSA, which is one of the elements required by the certification process described in Section 10.2. SSA is used to verify that the system meets the safety requirements.

Demonstration

Demonstrations are similar to tests but do not require any instrumentation of any sophistication. Typical demonstrations might include the functioning of an emergency alarm, for example. The human actions simulated in Section 11.2 during concept development can now be demonstrated with real components and real humans.

Examination

Examinations are the easiest type of verification. They are simply a visual confirmation that a requirement has been met. There can be both drawing examinations, to determine that the required equipment has been included in the drawing, and hardware examinations, to confirm that a piece of equipment has been installed on the aircraft. This category is also called inspection.

In the certification process, inspection and review is an official verification category. In addition, certification calls for reviews to assure that the product complies with the requirements.

11.3 Verification of Regulatory Requirements

Certification calls for a complete verification program and for the manufacturer to document the program of verification in the certification plan shown in Table10.1. Of particular interest are the regulatory requirements and the specified test conditions found in the FARs. But, from an SE point of view, verification of the regulatory requirements is only part of the total picture.

定验证方法。

系统安全性评估 (SSA)

分析也包括系统安全性评估 (SSA)，它是合格审定过程所需元素之一（如 10.2 节所述）。SSA 用于验证系统是否满足安全性需求。

演示

演示与试验类似，但是不需要任何精密仪器。典型的演示可能包括应急告警功能。在概念开发阶段，11.2 节模拟的人的活动现在可以通过真实的部件和真实的人来进行演示。

检查

检查是最简单的验证类别。检查仅仅是目视确认需求是否得到满足。检查的方式可能是图纸检查，即确定所需设备已经体现在图纸上，检查还有可能是硬件检查，即核实设备已经安装在飞机上。此类验证也称为检验。

在合格审定过程中，检查和评审是正式的验证类别。此外，合格审定要求进行评审以确保产品符合相关需求。

11.3 验证规章要求

合格审定要求有完整的验证程序，要求制造商在合格审定计划 (CP)（如 10.1 节所示）中记录这些验证计划。规章要求的验证特别关注 FAR 中的规章要求和规定的试验条件。但是，从系统工程观点来看，规章要求的验证仅仅是整个验证工作的一部分。

11.4   Verification of Customer Requirements

In the traditional world, verification of customer requirements is a most often neglected task. This neglect is most often seen with respect to buyer-furnished equipment (BFE). If we do not adhere to the general principle that all requirements should be verified, then there is a risk that the customer equipment will not perform in the aircraft environment as it was expected to or that a safety hazard may be introduced.

11.5   Verification Sequence

Verification sequence is extremely important. If you wait until the product is built, that may be too late. Early verification using simulation and analysis will reduce this risk.

Verification takes place throughout the development process. Analyses and development tests done during the design process become the first steps in verification. Drawing examinations done for CDR are part of verification. Then qualification testing done at supplier facilities, followed by integration testing, and finally flight testing, are all part of the verification process.

Verification and system integration are inextricably linked. The manufacturer is required to develop and document a system integration plan in which the order of mating and testing is done. It is a bottom-up process: the lowest-level components of the aircraft hierarchy are mated and tested first, followed by higher-level integration and testing.

11.6   System Validation

When all requirements have been verified and the aircraft has been assured of meeting its mission objectives, this state is called *system validation*. System validation requires involvement by the customer because it is the customer who validates the system. This can be done either with flight tests or by inspecting the verification documents that have been developed.

There is an implication in the term *system validation* that the documented requirements may not satisfy all of the customer needs. If this were not true, there would be no need for system validation. If the customer is not satisfied with the aircraft as built and as flown, then further effort to satisfy these needs will be the subject of negotiations between the customer and the manufacturer.

11.4　验证客户需求

传统上，客户需求的验证往往最容易被忽略。比较常见的是忽视买方提供设备 (BFE)。如果我们不坚持所有需求都应进行验证这一普遍原则，将会导致危害产生，即客户设备可能无法在飞机环境中执行其预期功能，或可能带来安全性危害。

11.5　验证顺序

验证次序极为重要。如果等到产品制造完成再进行验证可能就太迟了。在研制早期使用模拟和分析的验证可以降低该风险。

验证工作在整个研制过程中持续进行。设计过程中进行的分析和研制试验的完成是验证工作的第一步。针对关键设计评审 (CDR) 进行图纸检查也是验证工作的一部分。然后，在供应商处进行鉴定试验，并进行集成试验和最终的飞行试验，这些都是验证过程的一部分。

验证和系统集成紧密关联。制造商需要开发并制定系统集成计划，计划中制定配合及试验的次序。这是一个自下而上的过程：飞机架构中最底层的部件首先进行配合和试验，随后进行上一层级的集成和试验。

11.6　系统确认

所有需求都得到验证，并且已经确定飞机满足任务目标，此状态被称为系统确认。系统确认需要客户参与，因为对系统进行确认是由客户完成的。此过程可以是飞行试验，或者是检查已形成的验证文件。

"系统确认"一词有一层潜在含义：成文的需求可能无法满足所有客户的需

System validation should not be confused with requirements validation as explained in Chapter 4. Requirements validation, as explained in ARP 4754A (2010, p. 13), is the assurance that the requirements are correct and complete. Although the same word is used to describe these two concepts, they are, in fact, quite different concepts.

11.7   Qualification

According to Kossiakoff and Sweet (2003, p. 452), qualification is the process of proving that a product or process meets all of its requirements. Hence it is the final decision that the aircraft is ready to be delivered and fly. There is a strong emphasis in qualification on environmental testing. That is to say, qualification assures that the aircraft has been tested in all of the environments it will experience in service.

要。如果情况并非如此，那也没有必要进行系统确认了。如果客户对制造和飞行的飞机不满意，如何进一步满足这些需要可能是客户和制造商之间谈判的主题。

不能将系统确认与第4章所述需求确认相混淆。需求确认，正如 ARP 4754A(2010, 第13页) 所解释的，是为了保证需求的正确和完整。尽管采用相同的词语来描述这两种概念，但它们实际上是两个截然不同的概念。

11.7　鉴定

根据 Kossiakoff 和 Sweet (2003, 第452页) 的观点，鉴定是证明产品或过程满足其所有相关需求的过程。因此，鉴定是飞机准备就绪拟交付或者飞行的最终决策。环境试验的鉴定需要重点关注。即，鉴定确保飞机已经在其所有预期使用环境下进行了试验。

12

Systems Engineering Management and Control

SE without good management is like an orchestra without a conductor and all the musicians are playing to a different sheet of music. In most cases, the introduction of SE into a commercial organization will result in significant changes to the management operations of a company. This section provides some guidelines to make the process of introduction easier and also some of the basic features of SE management which are essential.

SE management refers to those activities which are programmatic in nature. Control, on the other hand, refers to those activities which are conducted to assure the integrity of the SE process and the quality of the product. These include configuration management, risk management, data management, SE task scheduling, technical performance measures (TPMs), and design reviews. SE management includes the implementation of an integrated product development (IPD) program.

In spite of the fact that we have left SE management to the twelfth chapter in this book, many managers consider SE management to be *the* SE process. In fact, one major book by Kossiakoff and Sweet (2003) considers SE to be a subset of management. Therefore, its position in the book does not reflect an inferior position of importance. It is very important. Furthermore, it is recommended that within the IPD concept, integrated product team (IPT) leaders should be thoroughly familiar with the SE process and not rely on SE staff members.

12.1   Management Responsibilities

Strong management is essential for the execution of an SE program. In addition, in the commercial world management may have to do things a little differently. Following are a few guidelines for the manager:

SE planning

Traditionally, the SE manager is responsible for preparing an SE management

第 12 章　系统工程管理及控制

缺乏良好管理的系统工程就像一支没有指挥的管弦乐队，每个乐师都演奏着不同的乐谱。在大多数情况下，在商业组织中引入系统工程会对公司的管理运营产生深远的影响。为了使引入的过程更容易，本章会提供一些相关指南，同时还会介绍一些系统工程管理至关重要的基本特征。

系统工程管理是指那些带有程序化性质的活动。控制，从另一个方面来讲，是指为了确保系统工程过程的完整性和产品质量而开展的活动，包括构型管理、风险管理、数据管理、系统工程任务计划、技术性能指标 (TPM) 及设计评审。系统工程管理包括集成产品研制 (IPD) 项目的实施。

尽管本书将系统工程管理放在第 12 章，但很多管理者认为系统工程管理就是系统工程过程。实际上，Kossiakoff 和 Sweet 在他们合著的一本重要著作中将系统工程归为管理的子集。因此虽然将它安排于书中此处，但这并不表明它是次要的。这一点十分重要。而且我们建议，在 IPD 概念之下，集成产品团队 (IPT) 的领导者本人应对整个系统工程过程十分熟悉，而不是依赖于系统工程团队中的成员。

12.1　管理责任

强有力的管理对系统工程项目的执行来说至关重要。除此之外，在商业领域，管理的方式可能会略有不同。下面是针对管理者的一些指导建议：

plan (SEMP) to describe all the SE activities on a program. However, in the interest of creating fewer documents, in the commercial world the SEMP can be included in the program plan.

In addition, the planning must include the certification and the development plans shown in Table 10.1 required for certification.

Integrated Product Development (IPD) phased schedule

The IPD phased schedule, described below in Section 12.3, recommends that all program phases be scheduled and completed before a new phase can begin. Strong management is needed to make this schedule happen.

Personnel staffing

An SE program recommends up-front staffing with the expectation of down-stream savings. SE management should plan for and acquire this staffing to make it work.

Work order

As we have seen in this book, SE recommends that tasks be done in a different order. It recommends, for example, that customer requirements, functional analyses, and requirements analyses be done *before* design begins. This order means that SE should begin *really early* and that drawing development should not begin before the other steps are complete. Otherwise, the chances are high that these tasks will have to be redone when the requirements are finished.

Program integration

The SE manager should assure that the SE tasks are integrated into the balance of the program. The manager should assure, for example, that the designers are working to the developed requirements and that the identified SE trade-offs are performed and implemented. The SE manager should assure that interface meetings between all interface parties are scheduled and conducted as discussed in Section 6.6.

系统工程规划

传统上，系统工程管理者负责制订系统工程管理计划 (SEMP) 来描述项目中所有的系统工程活动。但是为了尽量减少文档编制工作，在商业领域，SEMP 可以被囊括在项目计划之中。

除此之外，规划必须涵盖合格审定的内容，以及表 10.1 所示的合格审定要求的研制计划。

集成产品研制 (IPD) 阶段性进度计划

12.3 节所述的 IPD 阶段性进度计划建议所有项目阶段都要进行进度规划，并且各阶段应在一个新阶段开始前完成。强有力的管理才能保证该进度要求得以实现。

人员配备

系统工程项目建议在前期人员配备中考虑到下游人员的节省。系统工程管理应对人员配备进行规划并配备人员，以保证系统工程管理过程的运转。

工作顺序

如本书前文所述，系统工程建议按不同的顺序来完成各项任务。比如说，建议客户需求、功能分析及需求分析应当在设计开始之前完成。这个顺序意味着系统工程确实很早就应当开始了，而且在其他步骤完成前不应开始绘图工作。否则，在需求完成时，这些任务很有可能需要返工。

项目集成

系统工程管理者应当确保系统工程的各项任务平衡地集成到整个项目中。比

Design reviews

As we will also see below, design reviews are critical elements in SE. SE-driven design reviews require planning and organizing. The manager should make sure that all the data have been developed *and reviewed* before each review. The manager should ensure that all personnel are present. For example, representatives from manufacturing and product support are essential as discussed below in Section 12.4. Most important, the *customers and suppliers* are necessary attendees at SE design reviews. Finally, the manager is responsible for assuring that the design review is closed out and that action items are complete.

Also, as will be discussed in Section 12.4, design reviews are one aspect of SE that the commercial aircraft company will find most demanding in terms of time and money. The question is: Can these reviews be conducted in more timely and cost-effective way without increasing risk? This question will be addressed in that section.

Deliverables

The SE manager should assure that all SE deliverables, such as specifications and ICDs, are complete and on schedule.

Manage Technical Performance Measures (TPMs)

SE management should manage the TPMs as described below. Many managers consider TPMs to be most useful SE tool from a management point of view.

SE principles

The SE manager has the sole responsibility of assuring adherence to the SE principles outlined in this book, the two most important being: first, the design should incorporate *all* requirements, and secondly, *all* requirements should be verified. Additionally, management should assure that requirements are identified according to the holistic approach described in Section 4.1. Finally, of all SE principles, risk is the principle for which managers should take total ownership.

Configuration management supervision

如说管理者应当确保设计人员依据已形成的需求进行设计，并保证执行和贯彻了系统工程权衡项。系统工程管理者应确保各方之间的所有接口会议均按 6.6 节所述方式如期召开。

设计评审

正如下文所述，设计评审是系统工程中的关键因素。系统工程驱动的设计评审需要计划和组织。管理者应确保所有相关数据在每次评审前都已经形成并进行复核。管理者应当确保所有人员都到场。比如说，来自制造和产品支持部门的代表就是至关重要的，正如 12.4 节中讨论的。最重要的是，客户及供应商必须出席系统工程设计评审。最后,管理者负责确保设计评审的结束及各行动项的完成。

此外，正如 12.4 节中讨论的，商用飞机公司会发现设计评审是系统工程中最费钱费时的一项内容。问题是：能否在不增加风险的前提下，以更省时更经济的方式完成这些评审？本章将提到这个内容。

交付物

系统工程管理者应当确保所有交付物，如各种规范和接口控制文件 (ICD) 的完整性和及时性。

技术性能指标 (TPM) 管理

系统工程管理应以如下文所述方式对 TPM 进行管理。许多管理者从管理的角度出发，认为 TPM 是最有用的系统工程工具。

系统工程原则

系统工程管理者对确保遵守本书所列出的系统工程原则负有独特的责任。其中最重要的两条是：第一，设计应当体现所有需求；第二，所有需求都应可验证。

Although the SE manager does not have direct control of configuration management described in Section 12.11, this is one area requiring constant oversight and reporting. It is important that *all* engineers working on a given project are working to the same configuration baseline. Within the IPD framework described in Section 12.3, the IPT has responsibility for the configuration management of its own segment of the aircraft.

Project redirection

One of the most important roles of the SE manager is to make sure all SE processes are in place and being followed during periods of project redirection. For example, a crucial time is when new requirements appear (from the customer, for example) after preliminary design review (PDR). We will see in Section 12.4 that PDR is the milestone at which all requirements are frozen. Hence, when new requirements appear after PDR, the SE manager should assure that all SE steps are taken to assure the proper incorporation of those requirements.

Supervision of risk management

One area requiring close management scrutiny is risk management (see Chapter 15). Risk management will identify many areas of program cost, schedule, and technical risk and give the manager the information with which to make educated decisions. To initiate a risk management activity, the SE manager needs to place full responsibility for the risk analysis in the hands of the program systems engineers and to assure that a non-advocate risk assessment team is in place.

Verification supervision

In order to implement the SE principle of complete verification, the SE manager should assign a single person or organization to coordinate all verification. The scope of this task is broader than design or test; hence, this person would normally be part of the SE group.

除此之外，管理层还应确保需求是根据 4.1 节所述整体论确定。最后，在所有系统工程原则中，风险是管理者负全责的一条原则。

构型管理监督

尽管系统工程管理者不直接控制 12.11 节所述的构型管理工作，但这一部分也需要持续的监督和报告。确保项目中所有工程师均依据同一构型基线开展工作十分重要。在 12.3 节所述的 IPD 框架中，各 IPT 团队对其所负责的飞机部段的构型管理负责。

项目重新定位

系统工程管理者扮演的最重要的角色之一是确保在项目重新定位期间，所有的系统工程过程到位并得到遵循。比如，以下就是一个关键阶段：初步设计评审(PDR) 之后出现了新的需求（比方由客户提出）。我们将会在 12.4 节看到，PDR是所有需求冻结的里程碑。因此，当在 PDR 之后出现新的需求时，系统工程管理者应当确保采取所有必要的系统工程步骤，以保证这些需求得到恰当的整合。

风险管理监督

系统工程中一个需要密切管理监督的方面是风险管理（见第 15 章）。风险管理将确定项目成本，进度及技术风险中的许多方面的内容，并为管理者提供信息以便其做出恰当的决定。要发起一个风险管理活动，系统工程管理者需要将风险分析的责任全权交到项目系统工程师的手上，并且要确保有一支不主张风险的评估团队参与工作。

验证监督

为了实行完全验证这一系统工程原则，系统工程管理者应当安排一个人或一个组织来协调所有的验证工作。这个任务的范围比设计或者试验都要广，该人

12.2   The Chief Systems Engineer (CSE)

One issue which has been debated in government-oriented organizations for years and is new to commercial organizations is where the person responsible for managing the SE activities should fit into the program organization. This person is normally called the Chief Systems Engineer (CSE). Ideally the program manager and/or the Chief Design Engineer (CDE) would be so familiar with SE principles that a CSE would not be necessary. If that is not the case, then where should the CSE fit? The CSE could either be parallel to or over the CDE.

The global scope of SE as outlined in this book argues that the proper position for the CSE is as the assistant program manager and hence over the CDE. This concept does not in any way diminish the importance of the CDE but rather places the CSE in a position to execute the SE activities as needed. We will see in Section

12.3 how the roles of program management and SE merge within the framework of IPD.

12.3   Integrated Product Development (IPD)

The basic idea behind IPD is that a system can be developed more efficiently and faster if the various aspects are done *together* rather than separately and in parallel. We can view IPD as the ideal management process within which SE is accomplished. There are many IPD principles. But following are the key ones, especially as they affect aircraft development.

Multifunctional teams

Multifunctional teams are especially important in aircraft development. This importance derives from the fact that aircraft development involves such a large number of technologically diverse and geographically dispersed technologies and disciplines. Multifunctional teams play the role of bringing these groups together. Hence, working together in the same room are such diverse groups as safety, human factors, maintainability, manufacturing, and aerodynamics. The presence of all these groups is recommended at design reviews, as described in Section 12.4, and in other technical coordination meetings. IPD also encourages the participation of customer representatives and suppliers.

员一般是系统工程团队的一员。

12.2　系统工程总师 (CSE)

政府主导型组织中经常讨论到的一个问题是，我们应该将那个负责管理系统工程活动的人放在整个项目组织中的什么位置上，但这对商业组织而言是全新的问题。这个人通常被叫做系统工程总师 (CSE)。理想情况下，项目经理和 / 或总设计师 (CDE) 可能十分熟悉系统工程原则，以至于不需要 CSE。但如果不是这种情况，那这个系统工程总师应该处于哪个位置？系统工程总师要么与总设计师平级，要么职位高于总设计师。

就像本书指出的那样，在通用的系统工程中，对系统工程总师来说，合适的位置是项目经理助理，因此是高于总设计师的。这种主张丝毫没有要降低总设计师重要性的意思，只是将系统工程总师放在其能根据需要执行系统工程活动的位置上。在 12.3 节我们将介绍项目管理和系统工程的作用如何融入 IPD 框架。

12.3　集成产品研制

IPD 的基本思想是：相对于单独开展或者并行开展而言，同步开展各个方面的工作，可以更加高效、迅速地完成系统研制。我们可以将 IPD 看成是一个系统工程得以实现的理想管理过程。IPD 的原则有很多，但以下所述准则最为关键，尤其是当这些准则影响到飞机研制时。

多功能团队

多功能团队对于飞机研制言而极其重要。该重要性源于以下事实：飞机研制工作涉及大量技术学科多样化，地理分布离散化的技术和学科。多功能团队起到将这些团队汇聚起来的作用。另外，各个团队，包括安全性、人为因素、可维修

Another important function of multifunctional teams is to reduce the possibility of hidden interactions that may occur between the components of different design disciplines. As discussed in Section 16.4 these types of interactions have been the root cause of accidents in many different domains. Multifunctional teams are one way of implementing the *reduce hidden interactions* rule discussed in that section.

Phased development

The key principle in phased development is that no new development phase should be initiated until the previous one is complete. The principal milestones of each phase are the design reviews described in Section 12.4. The value of this principle is that it removes much of the development risk inherent in a process in which major modifications may result from prematurely initiating tasks which are dependent on design aspects which have not yet been frozen.

Integrated Product Teams (IPTs)

IPTs are one area that has significantly altered the way large systems are developed and built. What IPT means is that parts of the aircraft are developed as *products*, such as the wing, nose, empennage, and so on, rather than as functional areas, such as avionics, mechanical, and so on. For example, the nose organization would all work together, including the electrical, mechanical, avionics, and other personnel. In addition, manufacturing and other personnel would be members of the nose IPT. This arrangement helps communications enormously.

A key aspect of the IPT is the principle of requirements ownership. That is, the IPT owns all aspects of requirements. For example, the IPT is responsible for implementing requirements allocated from higher-level IPTs. These might include, for example, weight or dispatch reliability. Secondly, the IPT is responsible for developing and allocating requirements within its own domain. Finally, the IPT is responsible for every aspect of implementing its requirements. This includes design, procurement, fabrication, installation, and verification.

Another aspect is that each aircraft segment (the product) would be integrated *and tested* completely before being joined with the other segments. This methodology is completely compatible with the bottom-up integration principle. Another way to put it is that the nose (and wing, and so on) would be *stuffed*: that is all the electrical cabling,

性、制造及空气动力学在一起开展工作。在进行 12.4 所述的设计评审，以及召开其他技术协调时，建议上述团队都参与其中。IPD 也鼓励客户代表及供应商参与研制工作。

多功能团队的另一个重要作用是减少不同设计学科之间可能存在的隐性交互。如 16.4 节所述，这种交互是导致许多不同领域事故的根本原因。本节讨论的多功能团队是减少隐藏交互的一种实施途径。

阶段研制

阶段研制的核心准则是：在前一阶段完成之前，不应启动新的阶段。各个阶段的关键里程碑节点是 12.4 节所述的设计评审。该准则的价值在于该准则可大幅减少过程中的研制风险，在这一过程中，一些基于尚未冻结的设计、过早开始的任务，可能会导致较大的更改。

集成产品团队 (IPT)

IPT 团队显著改变了大型系统的开发及制造方式。IPT 团队意味着飞机是按照产品进行研制的，如机翼、机头、尾翼等，而不是按照功能领域展开，如航电、机械等。例如，机头组织将一同工作，包括电子、机械，航电及其他人员。另外，制造及其他人员也可能是机头团队成员。这种组织方式能极大地提高沟通效率。

IPT 的最重要方面在于需求归属准则。就是说，IPT 涵盖所有方面的需求。例如，IPT 负责落实上级 IPT 团队分配的需求。例如，这些需求可能包括重量及签派可靠性。此外，IPT 负责本领域内需求的开发及分配。最后，IPT 负责需求落实的各个环节。这些环节包括设计、采购、制造、安装及验证。

另一方面是，飞机各个部段（产品）在与其他部件对接之前将进行集成及完整试验。该方法完全符合自下而上的集成准则。IPT 的另一个途径是以机头（以及机翼等）进行填充：即在机头与机身对接之前，就将所有电缆，环控管道等安

environmental ducts and so on would be in the nose before it is joined with the fuselage.

The IPT philosophy creates a problem, not insurmountable, for the SE process. The problem is that the *distributed* subsystems (electrical, hydraulics, and so on) are now resident in the various aircraft segments. Thus, the systems engineer should treat all of these subsystems as if they were separate subsystems in each segment. For example, there would be the nose electrical subsystem, the fuselage electrical subsystem, and so on. Each of these subsystems will interface with the other subsystems, thus creating additional interfaces which would not have existed as external interfaces if the whole electrical system had been treated as a single subsystem. In addition, each electrical subsystem's performance requirements will have to be separately specified and verified.

Another key issue with respect to IPTs is how SE should fit into them. Traditionally, SE was considered a staff function. The systems engineer would advise the IPT leader on various aspects of the SE process. Current SE thinking has shifted away from the staff function approach and towards a philosophy that the IPT leader should be a person thoroughly educated in SE methodology and outlook. This philosophy puts a strong responsibility on the training program.

12.4   Design Reviews

Quality design reviews are undoubtedly the most important aspect of SE management. Design reviews are where we establish that all requirements have been identified, that we have a design that meets the requirements, and that the design has been verified. We saw above that design reviews are critical to the IPD phased development and that the development program cannot proceed until each design review has been completed *and is successful*: that is, all the criteria have been met. We also saw that, using the principles of IPD, the design reviews are where the *entire development community*, from marketing, to manufacturing, to engineering, to supplier management, to the customer, gathers to put its stamp of approval on the design.

A key aspect of design reviews, from the aircraft point of view, is that *aircraft-level* reviews are an integral part of the process. A logical question is then: Is an aircraft-level review recommended for all aircraft changes. The answer is no. So what are some guidelines for conducting aircraft-level reviews? Following are a few. Aircraft-level reviews should be held for:

1. New or derivative aircraft designs.
2. Collections of changes, the accumulation of which will affect aircraft- level

装到机头内。

对于系统工程过程来说，IPT理念会产生一个问题，虽然该问题并非不可克服。这个问题是分布式子系统（电气、液压等）现在分散于飞机各个部段。因此，系统工程师应该将所有子系统视作各部段内部的独立子系统。例如，将会出现机头电气子系统，机身电气子系统等。每个子系统均会与其他子系统对接，从而产生新增接口，这些新增接口在把整个电气子系统作为一个单独子系统时不会出现。另外，每个电气子系统的性能需求将不得不单独规定和验证。

另一个关于IPT的核心问题是系统工程如何适应IPT。传统意义上，系统工程被视作一种参谋职能。系统工程师在系统工程各个环节向IPT负责人提出建议。现代系统工程思维已经从参谋职能转变为一种理念，即，IPT负责人应该在系统工程方法论及视野方面具备相当的素养。在该理念中，培训项目担负很大的责任。

12.4 设计评审

质量设计评审无疑是系统工程管理最重要的方面。通过设计评审，我们确保对所有需求定义，并且设计均满足需求，且设计均得到验证。前文已说明，设计评审对于IPD阶段性研制至关重要，在研制项目往前推进前，须保证所有设计评审完成且成功：即所有准则均被满足。我们也看到，应用IPD原则，设计评审就是整个研制团队聚在一起批准设计的地方，从市场开发、到制造、到工程、到供应商管理，再到客户。

从飞机级视角看，设计评审的核心是：飞机级评审是该过程不可分割的一部分。一个合理的问题是：是否建议对所有飞机更改都进行飞机级评审。答案是否定的。因此，进行飞机级评审的要点有哪些呢？以下为其中一些。应就下列情况进行飞机级评审：

parameters.

3. Any change affecting aircraft-level parameters.

The second logical question is: What is an aircraft-level parameter? There are many parameters measured at the aircraft level. The most obvious one is weight. Any component weight will, by definition, change the weight of the entire aircraft. If the weight change is significant, then center of gravity (c.g.) and other mass properties may also change. Another aircraft-level parameter is dispatch reliability. Any subsystem change affecting the aircraft handling capability will be a candidate for aircraft-level review.

Another important aspect of design reviews is that they are a critical requirement for certification.

Tailoring the design review process

The challenge of design review tailoring is to make the reviews both faster and less labor intensive and at the same time rigorous. But before we examine some of the options, let's look at some of the requirements that cannot be sacrificed:

First, all necessary personnel should participate, and their participation must be substantive and not just a formality. They must actually examine the material, make a professionally educated judgment and not just sign a form. In addition, the phrase "all necessary personnel" must include not just the immediate designers but members of specialty disciplines as well.

The above discussion raises the question of whom to include and whom to excuse. This question should be answered very broadly. That is, invite anyone for whom there may be a remote possibility of involvement. For example, any component with an exposure to outside air should trigger the attendance of aerodynamicists. As another example, any component emitting EMI (electromagnetic interference) or vulnerable to EMI should trigger the attendance of an EMI specialist.

Secondly, it is a common misperception that attendees at design reviews are invited to find mistakes in the material or to criticize it. On the contrary, it is the duty of the attendees to examine the material well *in advance* and provide that feedback to the designers so that they can make those corrections. So signing the design review approval forms will indeed be a formality since all comments will have been treated in advance.

One aspect of design reviews is particularly subject to adaptation. That is any aspect that has to do with requirements. If the SE organization has thoroughly reviewed

1. 全新设计或衍生型飞机设计。

2. 更改的集合，更改的累积会影响飞机级参数。

3. 影响飞机级参数的任何更改。

另一个常见的问题是：什么是飞机级参数？在飞机级测量的参数有很多。最明显的一个是重量。任何部件的重量毫无疑问会改变全机重量。如果重量更改较大，重心 (c.g.) 及其他质量属性也会随之改变。另一个飞机级参数是签派可靠性。任何影响飞机操控可靠性的子系统更改均须进行飞机级评审。

设计评审的另一个重要方面是它们是合格审定的关键要求。

裁剪设计评审过程

设计评审裁剪的主要挑战是确保评审高效进行，并降低劳动强度，同时保证过程的严格程度。但是在检查各个选项之前，我们先分析一些不能遗漏的需求：

首先，所有必要人员均应参与其中，且这种参与必须是实实在在的，而不是形式主义的。他们（参加评审的人员）必须真正地查看材料，做出专业化的判断，而不仅仅是签署表格。另外，"必要人员"不仅必须包括直接设计人员，还必须包括一些专业学科成员。

以上讨论引出了一个新的问题，即该过程参与者包括谁，由谁执行。这个问题应该从广义上回答，即，邀请任何可能关联的人员。例如，任何暴露于外部空气中的部件均应邀请空气动力学专家参与评审。另一个例子是，任何会发出电磁干扰 (EMI) 或易受电磁干扰的部件均有电磁专业人员参与评审。

其次，一种常见的错误理解是：邀请评审者就是为了找出材料中的错误，或者批判材料内容。恰恰相反，事先认真检查材料并反馈意见给设计者，以便设计者能修正错误，这才是评审人的职责。因此，签署评审批准书确实是一种形式，因为所有意见均已事先得到处理。

评审的一个方面是关于适应性的。即与需求相关的任何必要内容。如果系

383

the requirements according to the guidance of Chapter 4, much time can be saved. This is particularly true of the review of supplier specifications. The SE organization will assure, at a minimum, that all requirements satisfy the criteria of Section 4.10.

The SE organization will have to do more than just determine that the requirements are written in the proper format. They will have to determine that requirements have been allocated or derived correctly. In discussions with the design organization the SE organization will have to determine that the requirements comply with engineering design principles of the discipline at hand. In short, if the SE organization performs this task, there may be very little else to do and the formal design review can be eliminated entirely.

Finally, it cannot be overemphasized that accomplishing the two requirements above will require strong action by the program manager since these actions are not commonly adhered to. Now on to some options:

The electronic review option

In the days of electronic communications this would seem to be the logical option. In this option all material (specifications, drawing, and so forth) would be sent to the attendees electronically. Attendees could comment on the material and then sign off on an approval when that is required. Responses could be sent to all attendees for their consideration. At the risk of repetition it must be remembered that all attendees who have been selected as necessary reviewers must scrutinize the material from the point of view of their expertise. In this option the review organizer must specify exactly what materials and specific pages the attendees must review.

The video review option

This option is essentially the same as a live face-to-face meeting except that the attendees use their computers to see and hear the attendees. They still must receive the material electronically as above and respond to it.

The wall walk option

This is an option that is practiced in organizations today and is particularly useful when drawings are to be reviewed, such as for the critical design review (CDR) described below. In this option the drawings are attached to a wall in a large room in which the attendees can examine the drawings and mark them up when they apply to their area of expertise. For example, if a drawing contains both an electrical conduit

统工程团队已经按照第4章的要点完整地评审了需求，将会节约大量时间。对供应商规范进行评审时尤其如此。系统工程团队至少应确保所有需求均满足4.10节所述准则。

系统工程团队的工作绝不仅限于是否按照合适的格式编制了需求。他们必须确保需求准确地分配及衍生。在与设计部门讨论时，系统工程团队必须确定需求是否符合所属学科中现有工程设计准则。简而言之，如果系统工程团队做好这项工作，其他方面需要做的将非常有限，而正式的评审可以完全取消。

最后，完成以上两个需求要求项目管理者采取强有力的措施，因为这些措施通常并不紧密相关，这一点如何强调都不过分。以下为一些评审选项：

电子评审选项

在电子通信时代，这种评审似乎是一种合理的选项。在该选项中，所有材料（规范、图纸等）将通过电子化方式分发给参与者。参与评审者在材料上注明意见，并在需要的时候签署批准书。答复意见将发给所有参与评审者以供参考。在此需要重复说明的是，必须谨记，所有被选作必要评审者的参与者都必须从各自的专业角度仔细审查材料。在该选项中，评审组织者必须明确规定哪些材料以及哪些页面必须进行评审。

视频评审选项

该选项实质上与实时面对面会议一样，除了视频评审者通过自己的计算机看到和听到参与者，他们还需要以如上所述方式通过电子化方式接收材料，并给出反馈。

"绕墙巡检"选项

该选项是现代组织中经常使用的一种方式，在评审图纸时尤其有效，例如

attached to a structural member, engineers from both the electrical and structures department must review it and mark it up as necessary.

The following paragraphs summarize the individual design reviews.

System Requirements Review (SRR)

The SRR is the first major design review, and a very important one at that. The focus of the SRR is on the top-level requirement and not on the design. Top-level requirements are those requirements which are not derived: that is, they do not depend on design solutions as discussed in Section 4.2. The purpose of the SRR is to verify that all the top-level requirements are correct: that is, that they meet with customer approval. That is why the presence of the customer is important. At the aircraft level, if there are many potential customers, several SRRs may be recommended. Or secondarily, *surrogate customers* may be recommended, that is, members of the marketing department who have communicated with the various customers and can speak for them. In this case it is important to summarize the SRR for the various customers to obtain their concurrence. In the case of subsystem- level projects, there will normally be a single customer who can attend. Subsystem SRRs should be held after the aircraft-level SRR.

The focal point of the SRR will be the *mission statement* discussed in Section 4.3. This mission statement will capture both the key customer requirements and also describe the customer operational objectives and the environment in which the system must operate. The mission statement is a key part of the system specification, as described below.

Another function of the SRR is to present to the customer those *assumed* requirements that have been developed throughout the requirements development discussed in Section 4.4. In addition, the SRR should present to customers the capabilities of the buyer furnished equipment (BFE) they have requested. It is a common error to assume that the requirements for the BFE do not have to be examined "because the customer asked for it." Failure to examine the capabilities of the BFE and report these capabilities to the customer may result in either or both a performance shortfall (of the item) or an unexamined safety hazard.

Although the SRR will not present any solutions, it is permissible to show a preliminary concept. In addition, a *system architecture*, as shown in Figure 2.1 will be presented. The system architecture is the hierarchy into which requirements will be

在下面所述的关键设计评审 (CDR) 时。在该选项中，图纸被贴在一间大房间的一面墙上，在这个房间里，所有参与评审者对图纸进行检查，并在涉及其对应专业领域时进行标记。例如，假设图纸包括安装在某一结构件上的导线管，来自电子及结构部门的工程师都必须对其进行评审，并在必要时予以标注。

以下段落对每一个独立的设计评审进行了概述。

系统需求评审 (SRR)

系统需求评审 (SRR) 是第一项主要设计评审，也是很重要的一项评审。SRR 重点关注顶层需求而不是设计。顶层需求即那些并非衍生而来的需求：即，它们不依赖于 4.2 节所讨论的设计方案。进行 SRR 的目的在于证实所有顶层需求都是正确的；就是说，它们符合客户批准书。而这就是为什么客户参与非常重要的原因。在飞机级，如果存在很多潜在客户，那么进行若干次 SRR 是一种推荐的处理方式。或者可以推荐一些客户代言人，即已与各个客户进行过沟通并能替他们发言的市场部门人员。在这种情况下，总结概括 SRR 并取得各个客户的共识十分重要。对于子系统层级项目，通常只会有单一的客户参加。子系统 SRR 应在飞机级 SRR 之后进行。

SRR 的关键在于 4.3 节所述的任务说明。该任务说明捕获关键客户需求，并描述客户使用目标及系统工作的环境。任务说明是系统规范的关键部分，如下文所述。

SRR 的另一功能是向客户介绍 4.4 节所述整个需求开发过程中生成的假定需求。另外，SRR 必须向客户介绍客户要求的买方提供设备 (BFE) 的性能。一种常见的错误是：由于是客户要求这样的，所以 BFE 无须检查。没有对 BFE 的性能进行检查并将其性能报告给客户，会导致该设备性能缺失，或存在未识别的安全危害。

尽管 SRR 不会提出任何解决方案，但是可以展示初步概念。另外，图 2.1

allocated as discussed in Section 4.7.

System Design Review (SDR)

The SDR will be the first review of a concept which meets the top-level requirements. In addition, the SDR will show requirements which have been flowed down to a level *one level below the top level* in accordance with the principle of top-down allocation discussed in Section 4.7. Another aspect of the SDR is that it is the point at which any major trade-offs are initiated. The SDR identifies these trade-offs. Subsystem SDRs should be held after the aircraft-level SDR.

Preliminary Design Review (PDR)

The PDR has, perhaps, a misleading name because there is nothing very preliminary about it. In fact, the requirements should be defined to the lowest level of the aircraft hierarchy. In addition, the PDR should define a design to meet all those requirements. The main function of the PDR is to initiate the detail drawing process. The PDR marks the end of the front-end SE activity. Although SE continues for the life of the development process, the tasks to follow are primarily in verification. The PDR also marks the end of major customer participation. If the customer creates additional requirements after the PDR, then program management will have to alter the program schedule to accommodate them. In short, program management will have to reinitiate the requirements process. Subsystem PDRs should be held before the aircraft-level PDR.

Critical Design Review (CDR)

The CDR is not really an SE review. It is a real *design* review. That is, the purpose of the CDR is to make sure the detail drawings agree with the concept completed at the PDR. Once CDR is complete, engineering can release drawings to manufacturing, not before. If engineering does release drawings before the CDR, development risk will result.

The exception to the above rule occurs with the release of drawing for long- lead-time items. These are items whose development time exceeds the development time allowed in the program schedule. However, the release of drawings for long-lead-time

将展示系统架构。系统架构是一种组织体系，在该体系中需求按 4.7 节所述方式进行分配。

系统设计评审 (SDR)

SDR 是针对满足顶层需求的概念的首次评审。另外，SDR 将讨论已被分配到仅次于顶层的层级的需求，该需求的分配方式符合 4.7 节所述的自顶向下分配的原则。另一方面，SDR 是辨识任何重大权衡事项的时间节点。SDR 对这些权衡事项进行定义。子系统 SDR 应在飞机级 SDR 之后进行。

初步设计评审 (PDR)

PDR 是一个容易让人误解的词汇，因为关于它，没有什么非常初步的概念。实际上，需求应该定义到飞机架构的最底层。此外，PDR 应定义一个符合所有这些需求的设计。PDR 的主要功能是启动详细图纸过程。PDR 标志着系统工程前段活动的结束。尽管系统工程贯穿研制流程全生命周期，但是主要工作集中在验证。PDR 同时意味着客户大规模参与的结束。如果客户在 PDR 之后提出了新增的需求，那么项目管理部门将必须改变进度安排，以适应这种变化。简单来说，项目管理部门将不得不重启需求流程。子系统 PDR 应在飞机级 PDR 前完成。

关键设计评审 (CDR)

关键设计评审 (CDR) 实际上并不是系统工程评审，而是真正的设计评审。CDR 的目的在于确认详细图纸与 PDR 中完成的概念相符合。一旦完成 CDR，工程师即能向制造部门发放图纸，在 CDR 完成之前，该工作不能进行。如果工程部门在 CDR 之前发放了设计图纸，将会导致研制风险。

长周期项目图纸的发放可以不遵守上述规则。长周期项目是指其研制时间超过项目进度允许研制时间的一些项目。但是，长周期项图纸的发放并不意味着

items does not relieve the development risk. In this case program management has agreed to *accept* the development risk.

Subsystem CDRs should be held before the aircraft-level CDR. However, drawing release should not occur until completion of the aircraft-level CDR.

Physical Configuration Audit (PCA)

PCAs answer the question: Was the aircraft built in accordance with the drawings reviewed at the CDR? In addition, the PCA fulfills the requirements of the audit requirements of certification.

Functional Configuration Audit (FCA)

The purpose of the FCA is to verify the functionality of subsystems as they are installed on the aircraft. These reviews are also part of the reviews designed to accomplish certification.

System Verification Review (SVR)

The system verification review is the most important review prior to the delivery of the aircraft. The SVR verifies that the aircraft *as built* meets the requirements, both performance and constraints, developed in Chapter 4, Requirements. The SVR is also a key step towards certification.

First Flight Review (FFR)

This review ensures that the first test aircraft is ready for flight. All verifications should have been done to that date, and all extra requirements, such as for instrumentation, should be met. The FRR is also called the first flight readiness review (FFRR).

Engineering Safety Review (ESR)

ARP 4754A (2010, p. 104) also recommends an engineering safety review. The purpose of this review is to assure that subsystems were built in the correct configuration without flaws or errors that may affect the safety of the aircraft.

研制风险的解除。在这种情况下，项目管理部门已同意接受这种研制风险。

子系统 CDR 应在飞机级 CDR 之前进行。然而，图纸发放应在飞机级 CDR 完成后进行。

物理构型审核 (PCA)

PCA 回答如下问题：飞机是按照 CDR 中评审过的图纸制造的吗？此外，PCA 满足合格审定审核要求的需求。

功能构型审核 (FCA)

FCA 的目的是验证安装在飞机上的子系统的功能。该评审也是为完成合格审定安排的评审的一部分。

系统验证评审 (SVR)

系统验证评审是飞机交付前最重要的评审。SVR 验证制成的飞机是否在功能和约束方面均满足第 4 章"需求"中的需求。SVR 也是迈向合格审定的关键步骤。

首飞评审 (FFR)

该评审确保首架试飞飞机已做好飞行准备。在此之前，必须完成所有验证工作，且必须符合所有附加需求，例如试验设备的需求。FRR 也被称为首飞成熟度评审 (FFRR)。

工程安全性评审 (ESR)

工程安全性评审是 ARP4754A(2010，第 104 页) 推荐的评审，该评审的目的是确保各子系统按照正确构型制造，且不存在影响飞机安全的缺陷或错误。

12.5   Documentation

Documentation is the life blood of SE. Although in the commercial world there is less documentation than, say, in the military world, documentation is a necessary element for maintaining records of requirements, solutions, and verification. As we point out in Section 12.5, the word *documentation* can be used in its broadest sense to include electronic records, so that no actual paper is required.

Specifications

In the commercial world the term *specification* generally refers only to those documents which reflect agreements between the manufacturer and the customer or between the manufacturer and a supplier. Rarely are internal specifications produced to reflect the requirements for internally defined segments or subsystems. However, with the advent of SE, this internal allocation, either through a specification or another medium, such as a requirements data base, is becoming both essential and more common.

Internal specifications

In an SE environment, internal specifications will maintain a performance and constraint focus. Unlike customer specifications, the internal specifications will evolve as new requirements are refined and derived. The usual way to handle this evolution is to use TBDs (to be determined requirements) as placeholders for values of requirements to be determined at a later date. The presence of a TBD in a specification implies a task (by the manufacturer or the supplier) to determine the value. Therefore, there should be a clear linkage between the specifications and the statement of work (SOW), discussed below.

There is no standard format for internal commercial specifications. However, the Institute of Electrical and Electronic Engineering (IEEE) has published such a format, as documented in IEEE 1233 (1996). A much more common specification format is the DoD-developed MIL-STD-961D (1995) often called the *six-part* format, which contains many topics of interest only to the military community. Appendix 2 provides a version of the MIL-STD-961D format, slightly modified for commercial application,

12.5 文档管理

文档管理是系统工程的生命线。虽然商用领域的文档比军事领域里的要少，但是它们是保存需求、解决方案及验证记录的必要元素。正如我们在 12.5 节指出的，"文档管理"这个词可以用于广泛的领域，包括电子记录，所以不需要实际的纸张。

规范

在商用领域，术语"规范"一般仅指那些反映了制造商与客户之间，或制造商与供应商之间的协议文件。用于反映内部定义的部段或子系统要求的内部规范比较少见。然而，随着系统工程的推进，内部分配的重要性正日益提升，也变得更为常见，无论是通过规范或其他媒介，如需求数据库。

内部规范

在系统工程环境下，内部规范将保持对性能和约束的关注。与客户规范不同，内部规范将随着新需求的细化而派生而不断演变。处理这种演变的通常方式是使用 TBD(待定需求) 作为稍晚时候待确定需求值的占位符。在一份规范中，TBD 的存在意味着一项 (由制造商或供应商) 确定数值的任务。因此，在规范和工作说明 (SOW) 之间应该有一个明确的联系，下面讨论的即是。

内部商业规范不存在标准格式。然而，电气电子工程协会 (IEEE) 发布了这样的格式，记录在 IEEE1233(1996)。美国国防部制订的 MIL-STD-961D(1995 年) 是一种更通用的规范格式，通常被称为六部格式，其中包含了许多只有军方关注的话题。附录 2 提供了 MIL-STD-961D 格式的一个版本，该版本针对商业应用稍微作了一些修改。

393

Customer specifications

A customer specification is a contractual document rather than an engineering document. The customer specification has a total configuration focus and will reflect the aircraft *as it will be delivered to the customer*. It is not a good practice to make a single document fulfill both roles.

Procurement specifications

In the SE world, procurement specifications will cease to exist as separate documents since internal specifications will contain all the material needed for suppliers to design their components.

A "paperless" SE process

One of the greatest obstacles the systems engineer will encounter is the reluctance by design engineers to produce more paper. And they are right. Who needs more paper? But on the other hand, accurate recording of requirements is essential to the SE process. What can we do? There are several possibilities. First, the automated SE tool can organize and print out the specifications. Secondly, the systems engineer can conduct brainstorming sessions and keep all the records. The design engineer only provides verbal information.

Interface documents

As described in Chapter 6, Interfaces, the primary interface document is the Interface Control Drawing (ICD). The ICD will contain all the important information about interfaces, including the functional interfaces, a sometimes neglected topic in the commercial world.

394

Statement of Work (SOW)

In the commercial world, the SOW is often a mixed bag of requirements, solutions, and tasks. In fact, specifications often contain solutions and tasks as well. It is a basic principle of SE to make a clear distinction among these three media. The

客户规范

客户规范是一种合同文件，而不是工程文件。客户规范有一个总的构型关注点，并将反映将交付给客户的飞机。只用一份文件起到两个作用的做法并不推荐。

采购规范

在系统工程领域，采购规范将不再作为独立的文件存在，因为内部规范就包含供应商设计其部件所需要的所有材料。

"无纸化"系统工程过程

系统工程师遇到的最大障碍之一是设计工程师对创建更多纸质图纸的抵触。他们的观点是正确的。谁会需要更多的纸质文件呢？另一方面，对于系统工程过程而言，需求的准确记录又是必不可少的。我们可以做什么呢？有几种可能的方案。首先，自动化系统工程工具可以组织并打印出规范。其次，系统工程师可以开展头脑风暴并保留所有记录。相应的设计工程师只需提供口头信息。

接口文件

正如第6章所述，主要的接口文件是"接口控制文件"(ICD)。ICD将包含关于接口的所有重要信息，包括功能性接口，该接口在商用领域有时会被忽视。

工作说明 (SOW)

在商业领域，工作说明是一个包含需求、解决方案和任务的混合工作包。事实上，规范通常也包含解决方案和任务。在这三种媒介中做出明确的区分是系统工程的基本原则。工作说明仅有的功能是描述任务，无论是对飞机制造商的任务还是供应商的任务。不做明确区分的风险首先是：如果将需求放入SOW中，这

only function of a SOW is to describe tasks, either for the aircraft manufacturer or for a supplier. The risks of not making this distinction are that, first, if requirements are put into the SOW, they may not be incorporated into the design or verified; and secondly, if tasks are put into the specifications, which are technical requirements documents, they will most likely be ignored and not performed.

Baseline Concept Document (BCD)

The purpose of the BCD is to control the *physical configuration* of the aircraft. No performance data should be included in this document. The BCD will evolve during product development to capture *and control* the configuration as it is known at any time.

In addition to the description of the aircraft, another important part of the BCD is a description of the operational concept. Rather than describing how the aircraft looks, the operational concept describes how the aircraft will be used. For example, an aircraft might be designed to be used at airports with minimal support equipment.

12.6 Automated Requirements Tools

Automated requirements tools are a key method for systems engineers to develop and track functions, requirements, and solutions. As such, their support and control by SE management is valuable. We saw above that requirements documents, such as specifications, need not be hard copy items but can be electronic. Automated tools are a key way to accomplish this electronic data management. Appendix 3 provides a summary of automated tool value and characteristics.

12.7 Technical Performance Measurement (TPM)

TPM is a management tool for tracking and managing requirements compliance of selected parameters. TPM continuously verifies the degree of anticipated and actual achievement for technical parameters. TPM confirms progress and identifies deficiencies that might jeopardize meeting a system requirement. TPM assesses values which are outside established tolerances and thus indicate a need for evaluation and corrective action.

TPM is the key tool for managing the top-down allocated requirements described in Section 4.7. With TPM the program manager has total visibility of which subsystems

些需求可能不会被纳入设计或验证；其次，如果将任务放入技术需求规范文档中，任务很可能被忽略而没有得到执行。

基线概念文件 (BCD)

基线概念文件 (BCD) 的目的是控制飞机的物理构型，该文件不应包括任何性能数据。在产品研制过程中，BCD 将会演变，以随时捕获并控制这个其构型。

除了对飞机的描述，BCD 的另一个重要部分是运营概念的描述。运行概念描述如何使用飞机，而不是描述飞机外观如何。例如，飞机可能针对支持设备最少的机场的使用场景进行设计。

12.6　自动化需求工具

自动化需求工具是系统工程师开发及追溯功能、需求及解决方案的重要手段。因此，系统工程管理层对自动化需求工具的支持和控制十分重要。上文讲到，需求相关的文件，比如规范，可以是电子化的，而不必是硬拷贝件。自动化工具是完成电子数据管理的一个重要方法。附录 3 对自动化工具的价值和特点进行了概述。

12.7　技术性能指标 (TPM)

技术性能指标 (TPM) 是一个跟踪和管理符合特定参数的需求的管理工具。TPM 针对技术参数不断验证预期与实际成效的差异。TPM 确认进展并识别可能导致无法满足系统需求的缺陷。TPM 评估这些外部建立的公差值，从而指出评估及修正措施的需要。

TPM 是管理自上向下分配的需求 (4.7 节) 的关键工具。借助 TPM，项目经理可以充分了解哪些子系统可能不满足分配给它们的参数，例如，重量或签派可

may not be meeting their allocated values of, for example, weight or dispatch reliability. With this information, the program manager has three options: The first option is to ascertain whether the subsystem with an excess weight, for example, can take measures to meet the weight requirement. The other option is to reallocate the weights. If this option is taken, some other subsystem will have to live with an even more stringent weight requirement. The third option is, of course, the least desirable option, namely, that the entire aircraft will weigh more than expected. Even if this is the only possible outcome, management will be fully aware that all options have been examined.

12.8   Software Management

Although SE views software as a subsystem, like any other subsystem, management guidelines exist which govern the management of software, especially with respect to the certification. Many tasks described in Table 10.2 pertain to software development, integration, and testing.

12.9   Supplier Management

Supplier management, also called *procurement*, is one of the most critical aspects of SE management in the commercial aircraft industry because many aircraft components are manufactured by suppliers.

Supplier requirements

Requirements should be provided to the supplier in a thorough and rigorous manner. Hence, the rules for requirements development and documentation become even more important. In the traditional environment, the manufacturer develops all requirements and most often provides the supplier with only build- to specifications. In the SE environment, the supplier is responsible for the *performance* of the product, not just the physical configuration. However, the most important principle for supplier management in the SE environment is that the aircraft manufacturer owns all requirements and is responsible for assuring that the supplier's product meets those requirements.

That is, it is not good management practice to *assume* anything. The manufacturer should, first, provide the requirements to the supplier as described in the following sections. Secondly, the manufacturer should require that the supplier provide evidence

靠性。有了这些信息，项目经理有三个选项：第一个选项是，确定能否采取措施满足重量的要求，例如，对于超重的子系统。另一个选项是重新分配重量。如果采取这一选项，其他子系统将必须满足更严格的重量要求。第三个选项是整架飞机的重量将超过预期，当然，这是最不期望的选项。即使这是唯一可能的结果，管理层也要充分认识到，已经考察过所有的选项。

12.8 软件管理

虽然系统工程将软件作为一个子系统，但与其他任何子系统一样，必须制定指导准则以管控软件管理过程，尤其是合格审定方面。表10.2所示的许多任务是针对软件开发、集成及测试的。

12.9 供应商管理

供应商管理，也称为采购，是系统工程管理在商用飞机行业最重要方面之一，因为许多飞机部件由供应商制造。

供应商需求

应该以彻底和严格的方式向供应商提出需求。因此，关于需求开发和文档管理的规则就显得尤为重要。在传统的环境下，制造商制定所有需求，但往往只向供应商提供制造规范。在系统工程环境下，供应商要对产品性能负责，而不仅仅对物理构型负责。然而，系统工程环境下，供应商管理最重要原则是：飞机制造商是全部需求的所有者，并负责保证供应商的产品符合这些需求。

也就是说，假设任何事情的做法在管理上都不推荐。首先，制造商应该如下列各节所述，向供应商提出需求。其次，制造商应要求供应商提供证据证明其生产的产品能满足制造商的需求。最后，供应商应提供证据证明其制造的产品满

399

that the product will meet the requirements. Finally, the supplier should provide evidence that the product, as built, meets those requirements, that is, verifies the requirements in accordance with the principles of Chapter 11.

Requirements flow down to the supplier

In the SE environment the supplier is given only the subsystem-level requirements (that is, one level above the supplier's product). Often the subsystem-level requirements include TBDs in cases where the system-level requirements are dependent on supplier input. In addition, these requirements are in the form of performance requirements, not solutions. As part of the supplier's contract, the supplier helps develop both system-level requirements and derived supplier product requirements. The net result of this process is that the supplier can optimize the supplier product requirements and can propose solutions perhaps more cost effective than those envisioned by the aircraft manufacturer. In summary, requirements can be categorized as follows:

1. Requirements developed by the aircraft manufacturer.
2. Requirements developed jointly by the aircraft manufacturer and the suppliers.
3. Requirements which are solely the responsibility of the supplier.

The key principle is that the manufacturer *owns* all requirements, retains those which the SE process dictates, and delegates to the supplier only those below the subsystem level, that is, below the manufacturer–supplier interface. The reason for this ownership principle is that the supplier does not have the *span of knowledge* to develop the subsystem-level requirements, as discussed below.

Span of knowledge

One of the common mistakes often made by engineers not trained in the rigors of SE is the violation of the *span of knowledge* principle. This mistake is particularly damaging when dealing with suppliers. This principle states, in summary, that the information contained in the requirements for any element should be the complete information and the only information required to define the requirement. As an example, if the operation of any subsystem needs to know the altitude of the aircraft,

足这些要求，即根据第 11 章所述准则验证这些需求。

传递给供应商的需求

在系统工程环境下，提供给供应商的只是子系统级的需求（即，该供应商产品的上一级）。经常会遇到子系统级需求是待定项 (TBD) 的情况，因为系统需求依赖于供应商的输入。此外，这些需求的形式是性能需求，而不是解决方案。作为供应商合同的一部分，供应商帮助开发系统级需求及派生的供应商产品需求。该过程的最终结果是，供应商可以优化其产品需求，并可能提出比飞机制造商设想的更经济的解决方案。综上所述，需求的分类如下：

1. 由飞机制造商开发的需求。
2. 由飞机制造商和供应商共同开发的需求。
3. 完全由供应商单独负责的需求。

一条关键的原则是，制造商拥有所有的需求，保留系统工程过程规定的那些需求，委派给供应商的仅仅是子系统级下面，即制造商 - 供应商接口下面的那些需求。存在这种所有权归属原则的原因是，供应商不具有下文所述开发子系统级需求的知识面。

知识面

没有经过严格系统工程培训的工程师常犯的一个错误是违反知识面原则。在与供应商打交道时，这个错误极具危害。该原则概要说明，任何元件需求中包含的信息应该是完整的，并且只包含定义该需求所要求的信息。例如，如果任何子系统的工作需要知道飞机的高度，则该信息应该以与该子系统相关的形式出现，例如，以压力或温度的形式提供给该子系统。

then that information should be supplied to the subsystem in the manner that it matters to the subsystem, for example, in terms of pressure or temperature.

The supplier in the Integrated Product Development (IPD) process

Since the SE process can be viewed as a subset of the IPD process, the supplier will be a member of the IPD teams throughout the product development.

The supplier in the synthesis process

Except for the subsystem-level requirements, the constraints, and the interfaces provided by the manufacturer, suppliers will have a free hand to develop and build their own design. In this way a clear division of responsibility between the manufacturer and the supplier can be developed and the best design can be built.

The supplier in the verification process

If the supplier is responsible for the performance of the product, as we said in Section 12.9, then it follows that the supplier is an integral part of the verification process. The supplier's role in the verification process begins with the verification matrix shown in Section 11.1. The verification matrix will specify for *every* requirement the responsibility for the verification of that requirement. Each verification method (test, demonstration, analysis, inspection) will be incorporated into the supplier's test plan.

Supplier cost control

It is a key function of supplier management to assure the lowest cost of the procured items for the aircraft. Item cost affects both the cost of the aircraft as well as the cost of spares discussed in Section 5.7. To this end, supplier management should assure competitive bidding among suppliers and avoid sole-source acquisition. This factor should, of course, be balanced against the desire to reduce the lead time in the development process.

We saw in Section 6.1 that one method of encouraging greater competition among suppliers is by requiring common interfaces through the implementation of SAE standard AS4893 (1996).

集成产品研制 (IPD) 过程中的供应商

由于系统工程过程可以视为 IPD 过程的一个子集，在整个产品研制过程中，供应商都将是产品整个研制团队的一员。

综合过程中的供应商

除了子系统级的需求、约束及由制造商提供的接口以外，供应商可放手开发和打造自己的设计。这样，在制造商和供应商之间可以明确划分责任，而且可以创造出最优的设计。

验证过程中的供应商

如果供应商负责产品的性能，如我们在第 12.9 节所述，那么供应商就应该是验证过程一个不可或缺一部分。在验证过程中，供应商的角色从验证矩阵 (11.1 节所示) 开始。验证矩阵将对每条需求规定其验证责任。每种验证方法 (试验、演示、分析及检查) 均将被纳入供应商的试验计划。

供应商成本控制

确保飞机采购项目的成本最低是供应商管理的一项关键功能。项目成本既影响飞机的成本，也影响备件成本 (5.7 节)。为此，供应商管理应确保在供应商之间有竞争性的投标，并避免单一货源采购。当然，该因素应该与缩短研制过程中的交货时间的愿望进行平衡。

如第 6.1 节所示，鼓励更多供应商间进行竞争的一个方法是通过实施 SAE 标准 AS4893(1996) 来要求采用通用接口。

12.10   Configuration Management

Configuration management is part of the certification plan (Table 10.1) and one of the pieces of certification data in the configuration management plan. The importance of configuration management in the SE process is that it assures the continuity and the integrity of the results of the synthesis process.

Configuration management consists of configuration identification, control, verification, and accounting. Configuration identification is the baseline documentation itself. Configuration control is the formal process of controlling changes to the baseline. Configuration verification is the process of assuring that the system meets the intent of the customer. This process is conducted through reviews and audits. Configuration accounting maintains configuration data and tracks the status of the configuration.

Configuration management controls both the configuration of the aircraft as well as the data required to define the aircraft. These data include all the certification data shown in Table 10.1. The configuration management plan includes the method to show that the objectives of the configuration process are satisfied.

A key ingredient in configuration management is the configuration index. A configuration index is a catalogue of the physical elements which comprise the aircraft and its subsystems. The specification tree shown in Figure 2.1 shows a typical hierarchical structure on which a configuration index would be based. The primary index system used in the aircraft industry is the Air Transport Association (ATA) Specification 100 index. This index can be modified or adapted to the needs of a particular aircraft application. The certification data set only requires that a single index be employed in the development of an aircraft. When using the ATA index in the SE context, it is important to group the elements of the index so that they are compatible with the functions as defined in Chapter 3. The configuration index will contain the identification of each element, associated software, interconnection of elements, interfaces, and safety-related procedures and limitations.

12.11   Integration Planning

System integration is the task of assuring that all items work together individually and collectively as a group or as a whole aircraft. A system is built by taking the lowest-level components and putting them together one level at a time. The cornerstone of system integration is that it is a bottom-up process. Between each level's integration, it

12.10　构型管理

构型管理是合格审定计划的一部分（表 10.1），而且是构型管理计划中合格审定数据的一部分。在系统工程过程中，构型管理的重要性在于，它保证了综合过程结果的连续性和完整性。

构型管理包括构型标识、控制、验证及纪实。构型标识本身是基线文档管理。构型控制是控制基线更改的正式程序。构型验证是确保系统满足顾客意向的过程，该过程是通过进行评审及审核进行的。构型纪实维护构型数据，并追溯构型的状态。

构型管理既控制飞机的构型，也控制定义飞机所需要的数据。这些数据包括表 10.1 所示的所有合格审定数据。构型管理计划包含表明满足构型过程目标的方法。

构型管理的一个关键因素是构型索引。构型索引是构成飞机及其子系统的物理元件的分类目录。如图 2.1 所示的规范树（目录树）是一个可以作为构型索引的制定依据的典型架构。用于飞机行业的主要的索引体系是美国航空运输协会 (ATA) 规范 100 索引。该索引可以按特定的飞机应用进行修改和适用。合格审定数据集只是对飞机研制中某一索引的应用要求。在系统工程背景下使用 ATA 索引时，对集合索引中的元件进行分组十分重要，以使其与第 3 章定义的功能保持一致。构型索引将包含各元件的名称、相关软件、各元件的互联、接口及与安全有关的程序和限制。

12.11　集成规划

系统集成的主要任务是确保所有项目不论是作为一个群组，还是作为一整架飞机，都可以单独地或是以协同的方式完成一项任务。系统是通过将最低层级

is necessary to test the lower levels to make sure they work together. The verification process discussed in Chapter 11 will show the verification at all levels of aircraft integration. If lower-level components subsystems are tested first, problems can be uncovered before higher-level assemblies are integrated and tested. The system integration will determine the verification sequence of the system build-up.

As an example, take the verification and installation of the environmental control system (ECS). Before delivery of the product, the supplier will conduct verification (test, demonstration, analysis, or inspection) on each component made by that supplier. For example, the supplier will verify the performance of the air supply unit under prescribed electrical loads and input air sources. However, this verification will not assure that the ECS meets the aircraft-level requirements, namely, to deliver a specified air flow from all ducts and to maintain a given air temperature in the cabin. To verify the aircraft-level requirements, testing should be conducted at the aircraft level, with *passengers on board*. The importance of this principle is that only at the aircraft level can the requirements be verified under the operational conditions.

Hardware and software integration documents the entire process of integration of an aircraft and its assemblies. This process may include breadboards, prototypes, computer emulations, and laboratory or flight-worthy items. Documentation of the integration process is a certification requirement as shown in Table 10.1.

Another important part of integration planning is interface management discussed in Section 6.5. It is a role of program management to assure that the different parties to interfaces are brought together to agree on interface responsibilities and to assure that interface requirements (both functional and physical) are met.

的部件同时放在同一层级而建立起来的。系统集成的基础是它是一个自下而上的过程。在每个层级集成工作之间，必须测下层级以确保其能一起工作。在第 11 章中讨论的验证过程将表明在飞机集成的所有层级上进行的验证。如果首先试验下层级的部件子系统，在上层级的部件集成和试验前便能发现问题。系统集成将决定系统构建的验证顺序。

以环控系统 (ECS) 的验证和安装为例。在产品交付之前，供应商将对其生产的每个部件进行验证（试验、演示、分析或检查）。例如，供应商将验证在规定电气负载和输入气源条件下空气供应单元的性能。但是，这种验证不能保证 ECS 满足飞机级的需求，即，所有导管均输出规定的空气流量并保持座舱给定的空气温度。为验证飞机级的需求，试验应在飞机级进行并且试验要有乘客在飞机上。该原则的重要性在于只有在飞机级，才能验证运行条件下的需求。

硬件及软件集成文件记录了飞机及其组件集成的整个过程。该过程可包括电路试验板、原型机计算机仿真及实验室或试飞项。集成过程文档是表 10.1 所示的合格审定要求。

集成规划的另一个重要组成部分是 6.5 节所述的接口管理。保证接口的相关各方就接口职责达成一致，保证接口的需求（功能接口和物理接口）得到满足，这是项目管理的职责。

13

Adapting Systems Engineering to the Commercial Aircraft Domain

A theme of this book is *adaptation*, that is, how should an organization adapt the SE process to make it both effective and affordable. Hence this book can be seen as more of a *how* and a *why* book, rather than a *what* book. The first part focuses on the adaptation process and the importance of risk in that adaptation. The second part shows how an existing organization can be adapted to incorporate SE.

The organization itself can be viewed as a system to which the principles of SE would apply. In fact, the *NASA Systems Engineering Handbook* (1995) describes total quality management (TQM) as "the application of systems engineering to the work environment." The NASA handbook also points to other similarities between TQM and SE, such as the emphasis on customer satisfaction.

13.1   Adapting the Process

One of the principal themes of this book is that SE does not have to be practiced *in its entirety* to be effective. So the question is: If SE is not practiced in its entirety, how does the program decide what not to do? The answer lies in two rules: ① perform what has to be done to achieve the objective on any project, and ② delete what has been thoughtfully considered to be low risk. Figure 13.1 illustrates these rules.

This figure is admittedly notional, but it makes an important point, namely, that the more SE you perform, the lower the risk. However, usually more SE comes with two consequences: time and money. So the question is: How do you know what to keep and what to delete?

Experience will tell you what to keep. Most projects boil down to one or two really important requirements. For lightning projects, the question is how much current can flow through any section of the structure without creating a spark to ignite fuel. For brake design the question is how much energy the brake pads have to absorb to stop the

第13章　在商用飞机领域应用系统工程

本书的主题是应用，即组织如何应用系统工程的过程，以保证项目既高效又经济。因此，本书应该被看作是介绍"如何做"及"为什么做"的书，而不是"做什么"的书。本章第一部分重点关注适应过程及这种适应过程中风险的重要性。第二部分讨论现有组织在整合系统工程时如何适应。

组织本身能被视为一个可应用系统工程原理的系统。实际上，《NASA系统工程手册》(1995)将全面质量管理(TQM)描述为"将系统工程应用到工作环境中"。NASA手册也指出了TQM与系统工程的其他相似之处，比如它们都强调客户满意度。

13.1　过程应用

本书的一个主要主题是不必为了有效而原封不动地践行系统工程原理。那么问题是：如果不是原封不动地践行系统工程，如何决定项目中哪些是不用做的？回答这个问题需要把握两条规则：①做那些为了达到项目目标而必须做的事情；②去掉一些经仔细考虑过并认为风险较低的事项。图13.1说明了这些规则。

不可否认，这张图是理想化的，但是它说明了重要一点，即系统工程工作做得越多，风险越低。但是，较多的系统工程工作通常会带来两个后果：时间和经费。因此，问题现在变成：如何知道哪些该保留，哪些该删除？

经验会告诉你哪些该保留。大多数的项目可以归结为一到两条真正重要的

aircraft on the runway. For a project involving the replacement or incorporation of a new software module, it is absolutely essential that the interfaces with surrounding subsystems be correct. Chapter 9 describes the typical parameters for various subsystems.

Deleting steps is a much harder process. The important point is that steps should not be deleted without consulting experts in all disciplines. For example, any step critical to the safety of the aircraft should not be deleted. Beyond safety, every project

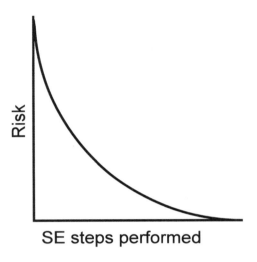

Figure 13.1   The SE adaptation diagram

should be examined to determine what disciplines would be affected, for example, human factors, reliability, and so forth. These deletions can be determined with a brief review of the project plan by all disciplines. If this review is conscientious and thoughtful, these experts can be excused from further participation.

Chapter 4, Requirements, presents a prime example of the application of the concept shown in Figure 13.1. The question at hand here is: what requirements can be deleted without incurring excessive risk? Chapter 4 shows how the number of constraints can be so overwhelming that the inclusion of all of them would put a great burden on the project. So Chapter 4 suggests that key people can make the decisions on which requirements to delete. These decisions will be based on risk.

Following are a set of guidelines for adapting SE to the commercial aviation domain. This list is summarized from a paper by Jackson (1996).

需求。对于雷电防护项目，主要问题是在不产生点燃燃油的火花的情况下，结构能够通过的电流有多大；对于刹车设计，主要问题是刹车片必须吸收多少能量，才能使飞机停止在跑道上；对于一个涉及替换或合并新软件模块的项目来说，与周围子系统的接口正确至关重要。第9章描述了不同子系统的典型参数。

　　删除步骤是一个更加困难的过程。要点是这些步骤不能在没有经过各学科专家咨询的条件下就予以删除。例如，不应删除对飞机安全至关重要的任何步骤。除了安全性，每个项目都应进行检查，以确定哪些学科会受到影响，比如说，人为因素及可靠性等。这些删除项可以通过按所有专业对项目计划进行简要评审来确定。如果这个评审是尽责并且深思熟虑的，那么这些专家便可无须参与未来的工作。

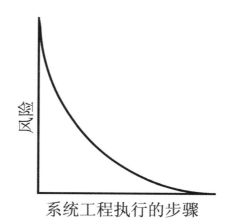

图 13.1　系统工程应用图示

　　第4章"需求"展示了概念应用的基本例子，如图13.1所示。现在，手头的问题是：什么需求能够被删掉而不导致额外的风险？第4章表明了项目约束条件的数量是如何的惊人，以至于把这些约束全部包含在内会给项目带来沉重的负担。所以，第4章推荐了一些关键人员，这些人能够在哪些需求能被删除的问题上做出决策。这些决策都是基于风险做出的。

　　接下来是一组将系统工程应用于商用航空领域的指南。这个清单是根据Jackson(1996) 的一篇论文汇总的。

Tribal knowledge

All organizations exist to some degree on *tribal knowledge*. It is not the role of SE to replace that tribal knowledge but to complement it. As an example, let's say that this organization has been procuring pitot tubes from the same supplier for many years and that the pitot tubes have been of extraordinary quality. The pitot tubes have lasted the entire life of the aircraft without replacement and never fail; after all pitot tubes are an essential component on an aircraft. The lesson is simple: stick with this supplier.

Existing processes

If the commercial organization has been in business for a long time, there may be many processes in existence. These processes may cover a variety of topics: for example, customer relations, procurement, inspection, testing, and so forth. In fact, many organizations may have processes that are recognized SE processes, such as configuration management, integration, or integrated product development (IPD). But these organizations may not implement SE as a whole. Thus, it is the best policy to make the SE process complement the existing processes. It may be necessary to change the existing processes. But it is never wise to ignore the existing processes.

Existing SE

SE may already exist in the commercial aircraft company, either because it is a result of good engineering judgment or because it has been mandated by other outside agencies. SE may be reflected in current documentation. Existing documentation may include change requests, engineering work orders (EWOs), specifications (both customer and supplier), interface documents, supplier packages, test plans, test reports, and many other documents. Mission statements (partial or complete) may appear in the change requests, EWOs, and other documents. Likewise, these same documents may provide partial or complete functional analyses. Requirements will appear in specifications (although specifications in the commercial world tend to emphasize the product design characteristics). Test plans and reports will contain much needed verification information.

We saw before Chapter 10 that the Society of Automotive Engineers (SAE) and

部落知识

所有的组织在某种程度上都存在部落知识。系统工程的作用不是取代这些部落知识，而是要对它进行补充完善。例如，假如一个组织从同一家供应商采购皮托管已经很多年了，并且这种皮托管的质量非常好。这个皮托管在整个生命周期内都无须更换，而且从来不会失效；毕竟皮托管是飞机上一个重要的部件。那么结论很简单：一直使用这家供应商。

现有的过程

如果某个商业组织已经开展业务很长时间了，那么许多的过程应该已经存在了。这些过程可能会涵盖许多不同的主题：比如客户关系、采购、检查、试验等。实际上，许多组织可能存在许多被认为属于系统工程过程的过程，比如构型管理、集成或集成产品研制 (IPD)。但是这些组织没有完整地执行系统工程。因此，最好的办法是利用系统工程补充现有的过程。可能需要对现有过程进行修改，但是绝不应忽视现有的过程。

现有的系统工程

不管是基于良好的工程判断，还是由其他一些外部机构强制要求，系统工程可能已经在商用飞机公司中存在。系统工程理念会反映在当前的文档中。现有的文档可以包括更改请求、工程工作指令 (EWO)、规范 (包含客户和供应商的)、接口文件、供应商工作包、试验计划、试验报告及其他文件。任务说明 (部分或全部) 会出现在更改请求、EWO 及许多其他文件中。同时这些文件也提供部分或全部的功能分析。需求会在规范中出现 (虽然在商用飞机领域，规范趋向于强调产品的设计特征)。试验计划和报告还包含大量需要的验证信息。

根据第 10 章之前的介绍，汽车工程师协会 (SAE) 和联邦航空局 (FAA) 联合

the Federal Aviation Agency (FAA) have jointly developed a guideline ARP 4754A (2010) for the development of commercial aircraft with a focus on certification. This document lays out a process for commercial aircraft development and certification which is based largely on SE processes. Comprehensive functional analysis, requirements analysis, and verification are all recommended. Chapter 3 explains that functional analysis can be deleted if the physical architecture is already defined. The lesson is that the existing SE processes should not be ignored or abolished.

The value-added syndrome

The systems engineer will be continually challenged to justify SE on a *value-added* basis. One common justification is that SE will make the design process so smooth that fewer designers will have to fix the mistakes that would have occurred if SE had not been employed. It is always the best policy to justify SE on the basis of value-added.

The importance of content, not form

Another challenge of SE in the commercial world is to keep it lean. As an example, instead of specifications, simple forms or spreadsheets can be substituted. These can be either in hardcopy or electronic form. In any case, in the attempt to keep the process lean, sacrificing content should be avoided.

The terminology conundrum

The one area where time can be needlessly wasted worrying is terminology. For example, well-established SE terminology will have a long-established meaning in the commercial organization. The use of the term *system* to refer to the entire product and not just the electrical system, for example, will be completely unacceptable. The best course of action is to accept the prevailing terminology.

414

Buy-in from management

SE cannot be incorporated solely from the bottom. Buy-in from top company management is absolutely essential because engineers do not have the authority to

编制了 ARP 4754A，从合格审定的角度指导商用飞机的研制。这份文件介绍了商用飞机研制及合格审定的过程，这些过程主要基于系统工程过程。该文件推荐进行全面的功能分析、需求分析及验证。根据第 3 章介绍，如果物理架构已经确定，则可删除功能分析过程。这里的结论是，现有的系统工程过程不应被忽视或抛弃。

增值症候

系统工程师会受到持续的挑战，被要求证明系统工程能为项目增值，一个通常的理由是，系统工程可以使设计过程变得更为顺畅，以至于少量一些设计员便可以修正由于没有采用系统工程带来的错误。这也是基于增值观点证明系统工程的最好方法。

重要的是内容，而不是形式

商用领域系统工程的另一个挑战是保持精益。例如，可以使用简单的表格或电子表单来替代规范。这些可以是硬拷贝或电子形式。任何情况下都应该避免为了保持过程精益而牺牲内容的做法。

语义双关的术语

术语是一个不必浪费时间去担忧的方面。比如说，明确定义的系统工程术语在商业组织中会有一个长期形成的含义。比如，术语"系统"用来代表整个产品而不仅仅是电气系统，这种说法是完全不被认可的。在这方面，最好的做法就是接受约定成俗的术语。

来自管理层的认同

系统工程不能独自从底层导入。来自公司最高管理层的认同也绝对重要，因为工程师没有改变他们工作优先级、预算或修改工作流程的权力。

change their work priorities or budget or modify their work processes.

Buy-in from management is particularly important with respect to one issue: is SE a technical or a managerial process? The correct answer is that it is both. As explained in Section 13.2 most companies are organized along technical and managerial lines. So how do you implement a process that crosses the technical vs. managerial lines? The recommended approach is to create a SE and integration team (SEIT) that covers both technical and managerial functions. To do this will require strong buy-in from management.

The importance of incremental progress

The most frustrating aspect of being a systems engineer in the commercial environment is the lack of willingness on the part of others to accept all aspects of SE at once. This aspect results from the lack of resources, the *show me* attitude, and the lack of adopted processes. Thus, we are often forced to be satisfied with incremental progress.

The importance of training

There are very few people in the commercial domain who understand the principles of SE. Thus, the establishment of a comprehensive training program is essential to the success of SE.

The importance of striving towards a complete but tailored SE process

This book does not advocate a diluted SE process; however, it does advocate a tailored or adapted process. To be effective, SE should strive to be complete. That is, the functional analysis should attempt to consider *every* function of the product. The requirements analysis should attempt to identify *every* requirement associated with a function. And *every* requirement should be verified. The systems engineer will encounter endless pressure to produce less than is needed. Nevertheless, incremental improvement and thorough training are essential to a complete SE process.

It might be concluded that the adaptation principle described at the beginning of this chapter conflicts with the completeness principle: it does not. The point is that if the deletion of an SE step can be determined to be low risk, then its deletion will not

关于管理层认同有一点至关重要：系统工程到底是技术过程还是管理过程？正确的答案是两者皆是。如 13.2 节所述，大多数公司都是沿着技术和管理两条线组织的。那么应该如何执行一个跨技术和管理的过程呢？推荐的途径是创建一个涵盖技术和管理功能的系统工程与集成团队 (SEIT)。这么做需要来自管理层强有力的支持。

逐步改进的重要性

作为商业环境中的一名系统工程师，最令人沮丧的是，有一部分人缺少一种马上接受系统工程所有方面的意愿。这种情况源于资源的缺乏、显示自我的态度及过程适应的缺失。因此，很多时候，我们不得不接受逐步改进的现状。

培训的重要性

在商用领域，真正理解系统工程原理的人很少。因此，全面的培训项目对系统工程的成功非常重要。

努力实现完整但是经过裁剪的系统工程过程的重要性

本书并不强推完全的系统工程过程，但是推荐采用经过裁剪或适用的过程。为了保证效率，系统工程应努力做到完整。也就是说，功能分析应试图考虑产品的每个功能。需求分析应试图确定与功能相关的每条需求，并且每条需求都要进行验证。系统工程师会遇到无尽的压力，被要求尽量少提要求，甚至少于实际需要的。然而，逐步改进和完整的培训对一套完整的系统工程过程仍然是重要的。

似乎可以得出这样的结论，即本章开头描述的系统工程的适用性原则与完整性原则相互冲突，但实际上并非如此。重点是如果系统工程某一个步骤的删除能够被确认风险较低，那么这个删除并不损害系统工程的完整性原则。但是，就像之前指出的，每项删除都应进行严格的评估。

endanger the completeness principle. However, as pointed out above, every deletion should be seriously evaluated.

The adoption of reason

It is difficult to tell a person when their conclusions are not based on reason. Nevertheless, persons in various domains have made conclusions and assumptions that were not based on reason. These conclusions and assumptions, often called paradigms, have sometime led to catastrophic consequences. Jackson (2010, pp. 91–119) provides a set of these paradigms from many sources. One of the most famous of these is quoted by Leveson (1995, p. 57) is called "the Titanic effect." This paradigm states in essence that a system is not necessarily safe just because you believe it is safe. Many disasters including the *Titanic* itself and the *Columbia* and *Challenger* disasters can be traced to the phenomenon of faulty paradigms.

There is no simple solution to resolving how questionable paradigms can be corrected. One cannot simply state to someone, "Think logically" and expect it to happen. The most that can be expected is that within a commercial aircraft organization there should be a system of checks and balances. That is, there should be a system of independent review. Chapter 15 discusses the concept of independent review more thoroughly and asks the question, "What is independence?"

13.2   Adapting the SE Process to the Existing Organization

Before we discuss organizational structures and how SE fits into an organization, we need to stress the point that the organization is a system itself and that the individual departments within a company are the elements of the system. Every department has a product, for example, data that must be transferred to another department so that the receiving department can execute its function, for example, design, test, and so forth.

A discussion of the organizational aspect of SE is rare in the literature with the exception of a discussion of IPTs. Hence, since all aircraft developers have, for the most part, similar organizational structures, and since each element, or department, within an organization has its own unique role in SE, it is hoped that the following discussion will aid the developer in implementing SE within its own organization and within the component parts of that organization.

Hitchins (1993, p. 55) states that an essential property of a system is *cohesion*. Cohesion is the property that all of the parts of a system must interact with all the other

推理的采纳

当结论没有基于合理原因时，很难把它们告诉别人。不过，仍然有很多领域的人做出毫无理由的结论和假设。这些结论和假设，经常被当作范式，但这有时候会引起灾难性的后果。Jackson(2010，第 91 至 119 页) 提供了一组来自不同来源的此类范式。其中最著名的一个是 Leveson(1995, 第 57 页) 引用的所谓的 " 泰坦尼克效应 "。该范式的本质是，一个系统未必仅仅因为你认为它安全，它就是安全的。很多灾难，包括泰坦尼号本身、哥伦比亚号和挑战者号等，都能被追溯到这种错误范式的现象。

如何修正可疑模式，这个问题没有一个简单的问题。你不能简单地对某人说 " 有逻辑地思考 " 就指望能解决这个问题。比较现实的期望是，在商用飞机组织里包含一个能检验和平衡的系统。也就是说，应该有一个独立的评审系统。第 15 章将对独立评审的概念进行更全面的讨论，并提出 " 什么是独立 " 的问题。

13.2　在现有组织中应用系统工程

在讨论组织架构，以及如何让系统工程与组织相匹配之前，需要强调一个观点，即组织本身就是一个系统，公司中的各个部门则是该系统的元素。每个部门都有产品，例如，必须传输到下一个部门的数据，使得接收部门能够执行其功能，如设计、试验等。

除集成产品团队 (IPT) 外，讨论系统工程中组织方面的文献很少。因此，由于所有的飞机研制单位都有类似的组织机构，并且组织中的每个元素或部门在系统工程中都有他们对应的角色，希望接下来的讨论能够帮助飞机研制方在其自有组织及各部门中实施系统工程。

Hitchins(1993，第 55 页) 认为系统的一个重要属性是凝聚性。凝聚性是同

parts of the system. In the context of an organization we can say that all the departments must interact with all the other departments. If there is any sort of irregularity among the interactions, then this irregularity will manifest itself in *variability* of the interactions, and the organizational system will drift towards a chaotic pattern in which the final product, the aircraft, is either deficient in quality or experiences higher cost and delayed schedules. Variability can manifest itself in many ways: poor data, lack of data, delayed data, and so forth. Chapter 14 discusses variability in more depth in the context of large-scale system integration (LSSI) and the causes and consequences of it. So making the organization and all its parts work together in harmony is the goal here.

Every commercial aircraft company has an organization. While the names of the departments within these companies may differ from company to company, certain department names will be common across companies. This section will use *typical* department names that should apply to any company. In this section we will capitalize department names as they would be capitalized within companies.

So why are we doing this? The purpose is to show that any existing organizational structure can be adapted to the SE process. The basic principle to be remembered here is that almost all departments have a role in the SE process. This fact may come as a surprise to some who have not yet understood the broader meaning of the word *engineering* as explained in Chapter 1.

The paragraphs below describe what roles specific organizations *should* have with respect to SE. The best of all outcomes is that these organizations *already* perform these roles. If this is the case, then the dictum from Section 13.1 to "respect the SE that already exists" will have been fulfilled. Otherwise, these organizations may need to consider a somewhat expanded role.

Many companies may already have department names that reflect functions that are already recognized as parts of the SE process. This fact supports the rule articulated in Section 13.1, namely the *Existing SE* rule that states that the company should respect and not dilute the SE that already exists. Following are a few typical departments that may already exist within a company:

- Configuration Management. As described in Chapter 12, this department is responsible for controlling both physical and functional configurations.
- Integration. Also, as described in Chapter 12, this department is responsible for assuring that all parts of the aircraft come together in a logical way.
- Integrated Product Development (IPD). Although many standards consider IPD

一系统的所有部分均须与其他所有部分交互的特性。就一个组织而言，所有的部门都必须与其他部门交互。如果在这些交互中有任何形式的不协调，那么这些不协调就会表现为交互中的变异性，并且组织系统也将漂向混乱模式，这样，最终的产品，也就是飞机，就会要么质量有缺陷，要么成本过高或者进度拖延。变异性可以表现为以下几种方式：数据质量差、数据缺失或数据延误等。第 14 章将更深入地讨论大规模系统集成 (LSSI) 中的变异性、产生原因及其后果。所以保证组织及其所有成员协调工作是这里讨论的目标。

每家商用飞机公司都对应一个组织。尽管在公司与公司之间，部门的名称会有所不同，但某些部门的名称在不同公司里是相同的。本节将使用典型的部门名称，它们应适用于任何一家公司。在本节中，我们会把部门名称大写 [1]，因为它们在公司内可能会被大写。

那么，我们为什么做这些呢？主要目的是表明任何现有组织机构都能适用系统工程过程。这里要记住的基本原则是，几乎所有的部门都会在系统工程中扮演某一角色。这对于那些尚未完全理解第 1 章所述"工程"一词宽泛含义的人来说，可能比较令人意外。

下面各段描述特定组织在系统工程方面应当履行的职责。最理想的情况是这些组织已经在履行相应职责。如果是这种情况，则满足了 13.1 节"尊重已有的系统工程"原则。否则，这些组织需要考虑在某种程度上扩展一些职责。

许多公司可能已经有一些部门的名称能反映出作为系统工程过程的部分功能。这个事实也支持 13.1 节所讨论的现有的系统工程规则，也就是说公司应该尊重，而不是忽视已经存在的系统工程。下面是一些可能已经存在于一家公司内的典型部门：

- **构型管理**。如 12 章所述，该部门负责控制物理构型及功能构型。

- **集成**。也如第 12 章所述，该部门是负责确保飞机的各个部件能够以合

1　中文对应的加粗字体表示。——译者注。

to be a separate and independent process, its basic elements have been part of SE from the beginning. These differences in definitions put aside, the existence of an IPD department within a company is a contribution to SE.

Having said that we do not want to dilute the SE that may exist in these existing organizations, we still have to ask some basic questions: Is the SE in these organizations complete? Do they interact with other organizations in a way consistent with SE principles? We will have to examine these questions as we go along.

Traditional organizations

The challenge of introducing SE into a traditional organization is that most organizations are organized along technical and managerial lines. Jackson (1997) encourages organizations to adopt the SE philosophy and processes while disturbing the current organizational structure as little as possible. Part of the challenge is that, as was stated above, SE includes both technical and associated managerial functions. So it is not an engineering function in the classical sense. There are at least two approaches to solving this dilemma.

The first alternative is not to modify the current organizational structure at all. SE will remain as a function within the Engineering department. This approach will place a large responsibility on SE to educate and collaborate with the managerial departments to explain to them what their roles are. This approach may receive some resistance from these departments and will require strong support from Program Management to make sure these departments understand and execute their responsibilities under this system.

The second alternative is more conventional and in agreement with the broad view of SE in the literature of today. This view would involve the creation of what is generally called the Systems Engineering and Integration Team (SEIT). This organization would include SE and other associated functions, such as Configuration Management and Safety. The important thing about this concept is that it would be at an organizational level above all engineering and management functions and report directly to the program manager. In this way the SEIT would be able to oversee all SE activities, whether technical or managerial.

Either way, almost all departments would have a role in SE one way or the other. Whichever alternative is chosen, though, it must be remembered that the primary

理的方式结合在一起。

- **集成产品研制** (IPD)。虽然许多标准认为 IPD 应视为一个分开和独立的过程，但其基本元素从一开始就已经是系统工程的一部分。抛开这些定义上的差异，一个公司中 IPD 部门的存在本身就是对系统工程的贡献。

虽然之前我们已经提到不能忽视现有组织中已经存在的系统工程，但我们仍然需要明确一些基本问题：这些组织中的系统工程完整吗？它们与别的组织系统工程原则一致吗？如果我们要继续往前走，就必须回答这些问题。

传统组织

把系统工程引入到传统组织的挑战是，大多数组织是按技术和管理两条线组织的。Jackson(1997) 鼓励在尽可能少地干扰现有组织机构的情况下适用系统工程的理念及过程。部分挑战正如前文所述，系统工程包括技术及与其相关的管理功能。所以从传统意义上讲，它不是一项工程功能。至少有两条途径可以解决这个困境。

第一种可选的方法是完全不修改现有的组织架构。系统工程仍然作为**工程**部门的一个功能。这种方法把系统工程的大部分职责当成对管理部门的教育及协作，来解释他们属于什么角色。该方法可能会遇到来自这些部门的一些阻力，此时就需要项目管理部门的大力支持，以确保这些部门理解和履行其在系统中的职责。

第二种可选的方法更加传统，并且与当今的文献中系统工程宽泛的视角一致。这种视角会涉及**系统工程与集成团队** (SEIT)。该组织包括系统工程及其他相关的功能，比如**构型管理和安全性**。这个概念的重要一点是，这个团队可能在组织级别上高于所有的工程和管理职能，并直接向项目经理汇报。按照这种方法，SEIT 能监督所有的系统工程活动，不管是技术活动还是管理活动。

不管采用哪种方法，几乎所有的部门都会以一种方式，或另一种方式在系

consideration is risk. Remember that Figure 13.1 shows that whatever shortcuts are taken, the one with the lowest risk is the right one. Following are a few of the salient organizations and their roles:

Organizations and their SE functions

The task is now to identify typical organizations and ask what SE functions they might perform. So as not to have to repeat this recommendation within the discussion of every organization, it is suggested here that each organization have at least one highly knowledgeable person in SE on their staff. This person would be responsible for advising the organization regarding how to implement SE. Another benefit of this recommendation is to make the minimum impact on the organization and on the organizational structure as a whole.

Program management

SE begins with program management which includes the chief engineer. The primary responsibilities of program management are as follows:

- First, program management must recognize that SE is not simply a subordinate function to program management but that they themselves play an important role in SE.
- Next, as has been stressed before, SE cannot succeed unless all organizations understand their own role in SE. This may be difficult for organizations that are not part of a traditional engineering discipline. It is the responsibility of program management to assure that all organizations understand their role in SE, as outlined in this chapter.
- Among all functions of program management probably the most important and difficult is to provide support for the risk management process as described in Chapter 15. This support consists of the following:
  - defer to expertise in the identification of both technical and non-technical risks and the methods to mitigate them;
  - resist the temptation to invoke managerial prerogative of overriding expert opinions on risk;
  - seek outside independent opinion on risks;
  - take decisive actions to mitigate risks even when costs are involved; these

统工程中履行职责。但是不管选用哪种方法，首先要考虑的是风险。必须主意，正如图 13.1 所示，不管选择哪条捷径，只有风险最低的才是正确的。下面是一些关键的组织及其职责。

组织及其系统工程职能

当前的任务是对典型的组织进行定义，并询问它们可能执行的系统工程职能。为了不在关于每个组织的讨论中重复，这里建议每个组织至少配备一名对相关系统工程知识较为了解的人员。该人员负责向该组织宣传如何执行系统工程。这项建议的另一个好处是它对整个组织及组织架构的影响最小。

项目管理

系统工程始于项目管理，这里的项目管理包括总工程师。项目管理的主要职责如下：

- 首先，项目管理必须认识到，系统工程不仅仅是项目管理下面一个子功能，其自身也在系统工程中也履行重要职责。

- 其次，就像之前强调的一样，只有所有的组织都能理解其在系统工程中的职责，系统工程才能取得成功。这对于那些不属于传统工程学科的组织来讲，可能存在一些困难。确保每个组织都能理解其在系统工程中的职责是项目管理的责任，正如本章所述。

- 在项目管理的所有职能中，为第 15 章所述风险管理过程提供支持可能是最重要的，同时也是最难的。这种支持包含以下内容：

——尊重专家在技术及非技术风险识别及减缓方法上的意见；

——抵挡得住调用管理特权，而不顾专家关于风险方面意见的诱惑；

——寻求来自外部的，关于风险的独立意见；

——即使涉及成本，也应该坚决采取措施降低风险，长远来看，这些决策实

decisions will actually save money in the long run.

- Periodically review requirements metrics, such as technical performance measurements (TPMs), and take action to assure that they are on track.
- As described in Section 4.11, the chief engineer will be one of the key people who decides what requirements are either necessary or unnecessary.

In short, SE will not succeed unless it has total program management support.

Design organizations

Every aircraft company has many design organizations. Typical design organizations are Electrical, Mechanical, Hydraulic, Avionics, and so forth. They have many roles in the SE process, for example:

- They have to write the requirements in the supplier specifications. To do this they must first have a firm grasp on what the requirements are. Some of these requirements will be derived or allocated from higher levels of the aircraft hierarchy as explained in Chapter 4. They will receive assistance from the SE organization regarding the criteria for a valid requirement. The SE organization may need to provide assistance on how to allocate or derive requirements.
- When a modification is being made at the lower level of the aircraft architecture, perhaps level 4 or 5, they will need to verify that the requirements for this modification are in compliance with the top-level requirements, that is, will a modification at level 4 or 5 affect aircraft range or durability, for example? See Section 2.3 for a discussion of the levels of aircraft architecture.
- They will have to assign verification methods to all the requirements as described in Chapter 11 and assign these verification tasks to the appropriate organizations, such as the Test or Quality organizations or the supplier.
- They will have to perform some verification tasks themselves. Primary among these tasks are verification by analysis (including similarity) and verification by inspection. When these tasks are performed, they will have to sign the appropriate documentation.
- When a supplier performs a test or analysis or any other type of verification, they will be responsible for reviewing the supplier's verification results and confirming that they are correct.
- They are responsible for reviewing all verification results and confirming that

426

际上会节省成本。

- 定期对需求指标进行评审，比如技术性能指标 (TPM)，同时采取措施确保其在正确的轨道上。

- 正如 4.11 节所述，总工程师将是一位关键人物，他（她）将决定哪些需求是必要，哪些是不必要的。

简而言之，只有得到项目管理部门的大力支持，系统工程才能成功。

设计组织

每家飞机公司都有许多设计部门。典型的设计部门是**电气**、**机械**、**液压**、**航电**等。它们在系统工程中承担许多职责，比如：

- 它们必须在供应商规范中写明需求。为了达到这个目的，它们首先必须非常深刻地理解需求的定义。如第 4 章所述，部分需求从飞机架构的上级需求衍生或分配而来。它们在有效需求的准则方面会得到系统工程组织的协助。系统工程组织可能需要在需求的分配及衍生方面提供帮助。

- 当在飞机架构中低层级（可能是第 4 或第 5 层）进行变更时，它们需要验证关于这项变更的需求是否与顶层需求一致。比如，这项在第 4 层或第 5 层进行的变更是否会影响飞机的航程或耐久性？关于飞机架构层级的讨论详见 2.3 节。

- 它们必须如第 11 章所述，为所有的需求指定验证方法，并将这些验证任务分配到合适的组织，比如说**试验**、**质量**部门或供应商。

- 它们自己也必须执行一些验证任务。这些任务中，最主要的是分析验证和检查验证。这些任务完成时，它们将必须签署相应的文件。

- 当供应商进行试验、分析或其他类型的验证工作时，它们需要负责对供应商的验证结果进行评审，并确认这些结果是正确的。

- 它们负责评审所有的验证结果，并确认这些结果符合飞机的鉴定要求。

they comply with the qualification requirements for the aircraft.

- They will have to conduct the required trade studies using the requirements that have been developed.
- Of course, using all of these requirements, they will have to produce a design that meets the requirements.

The Flight Operations organization

It is easy to overlook Flight Operations as a player in SE. Yet this organization employs a product that is essential to safe aircraft operations, the flight manual. The flight manual can be considered an end product to the same extent that the aircraft itself is a product. The flight manual is the product of the same requirements chain that resulted in the aircraft. It is therefore essential that this manual be produced with the same rigorous process that produced the aircraft.

The Marketing organization

Marketing is probably one of the last organizations that would agree that it has a role in SE. However, this is the organization that has direct contact with customers and potential customers. This role of direct contact requires special treatment. Chapter 4 explained the difference between customer needs and product requirements. The Marketing department has the responsibility for capturing customer needs in such a way that they can be transformed into product requirements. For new aircraft this role requires working with the Advanced Design department to achieve this transformation.

The Customer Engineering organization

Aircraft modifications require Customer Engineering to work with Design Engineering to accomplish the goal of modifying the aircraft. Over the lifetime of an aircraft model, the engineering effort of modifying the aircraft may very well exceed the effort to design a new model. As explained in Chapter 4, the important thing to remember is that it is the task of these organizations to determine the true needs of the customer rather than pre-conceived solutions.

The Advanced Design organization

428

- 它们必须基于已产生的需求开展必要的权衡研究。
- 当然，它们必须利用所有的需求完成满足这些需求的设计。

飞行运营组织

飞行运营在系统工程中往往被忽视。但是该组织会使用一个产品，它对于飞机的安全运营非常重要，这个产品就是飞行手册。如同将飞机本身视为一个产品一样，飞行手册也被视为一个最终产品。它是产生飞机的同一条需求链的产品。因此，遵循飞机制造相同严格程度的流程产生这份手册至关重要。

市场开发组织

市场开发可能是最不认同它在系统工程中存在对应职责的一个组织。但是，就是这个组织直接与客户和潜在客户直接接触。这种直接接触的作用需要特殊对待。第4章分析了客户需要与产品需求之间的差异。市场开发部门应该负责以某种形式捕获客户的需要，以便将这些客户需要转化为产品需求。对于新飞机，该项职责要求他们与**先行设计**部门一起工作，以实现这个转化过程。

客户工程组织

飞机改装需要**客户工程**部门与**设计工程**部门一起实现改装飞机的目标。在某一型号飞机的生命周期中，改装飞机的工作量可能远远超过设计一个新机型的工作量。正如第4章所述，值得强调的一点是，这些组织的任务是确定客户真正的需要，而不是一些事先设想的解决方案。

先行设计组织

先行设计部门首要的责任是基于客户需要，在顶层形成一些设计。**先行设计**部门将用到第8章所述方法论。**先行设计**部门将确定整架飞机的架构，比如发

Advanced Design has the primary responsibility for creating designs based on customer needs at the top level. Advanced Design will use methodologies such as the one described in Chapter 8. Advanced Design will determine the architecture of the entire aircraft, such as the number of engines, and so forth. One of the methodologies Advanced Design will use is functional analysis described in Chapter 3. Part of this architecture task is the determination of the number and identities of the subsystems. Advanced Design will be responsible for determining the top-level requirements that will be allocated to the subsystems or derived at the subsystem level. This information will have to be passed to the Design organizations responsible for each of these subsystems.

Since Advanced Design has the primary responsibility for the architecture of the aircraft, this responsibility will ultimately result in a strong participation of the introduction of *resilience* into the design, as described in Chapter 16. Although resilience is a new aspect of design, many authors, such as Zimmermann et al. (2011, pp. 257–296) have shown that this aspect is of paramount importance in aircraft design. Jackson (2010) has shown that the resilience of a system is primarily accomplished through its architecture.

The Safety organization

More advanced thinkers in safety, such as Leveson (2002) (2006, pp. 95–123) have seen an expanded role of the Safety organization and the safety function itself. Traditional safety standards, for example DoD (2012), focus almost exclusively on the design of the system and its safety. Leveson, for example, places greater emphasis on the organizational aspects of safety. Indeed, the first edition of this book (1997, pp. 126, 139, 153) discusses organizational safety. With respect to the resilience aspect of safety, as discussed in Chapter 16, the emphasis is more than just preventing failures; it is anticipating disruptions and recovering from them.

With this new and expanded view of safety, the role of the Safety organization will remain largely unchanged, that is, to assist the Design Engineering and Advanced Design organizations in implementing this expanded view.

Finally, it is recommended that the Safety organization be an integral part of the Systems Engineering and Integration Team (SEIT) suggested below. This is not to say that Safety is part of SE but to say that the functional binding between the two is so strong that organizing them together would be beneficial to the entire organization. Another

动机的数量等。**先行设计部门使用的方法之一是第 3 章介绍的功能分析。确定子系统及其数量是架构设计工作的一部分。先行设计部门**还负责分配到子系统或在子系统级衍生的顶层需求。这些信息必须被传递到负责这些子系统的**设计组织。**

由于**先行设计**对飞机的架构负有首要责任，这个责任会最终会把恢复力引入到设计工作中，如 16 章所述。虽然恢复力是设计的一个新的方面，许多研究者，如 Zimmermann 等 (2011，第 257 至 296 页) 认为，它是飞机设计最重要的方面。Jackson(2010) 也认为一个系统的恢复力主要通过其架构实现。

安全性组织

许多高级安全性问题专家，如 Leveson(2002)(2006, 第 95 至 123 页)，已经认识到**安全性**组织及安全性本身作用扩大的趋势。传统的安全性标准，比如 DoD(2012)，几乎只关注系统设计及其安全性。例如，Leveson 对组织上的安全性进行了着重强调。本书第一版 (1997，第 126，139，153 页) 讨论了组织的安全性。关于第 16 章所述安全性的**恢复力**方面，重点绝不仅是失效的预防，它能预测扰动，并能从失效中恢复。

从安全性这个新的，并且扩展的角度来看，安全性组织的职责依然没有大的变化，那就是协助**设计**部门和**先行设计**部门来执行这些扩展的职能。

最后，建议把安全性组织作为下文提议的**系统工程与集成团队 (SEIT)** 的一个必要部分。这并不表示将**安全性**视为系统工程的一部分，而是说两者之间的功能性结合非常紧密，把它们两者结合起来对整个组织更为有利。关于这种组织的另一个激烈争论是在一个组织中，安全性是否应该享有很高的优先级，以使其能及时向项目经理汇报。

合格审定组织

合格审定组织是一种不同类别的需求组织。该组织处理规章要求并与制定

strong argument for this grouping is to reflect the high priority have Safety should have in an organization so that instant access to the program manager is possible.

The Certification organization

The Certification organization is a requirements organization of a different sort. This organization deals with regulatory requirements and with the regulatory agency that produces them, such as the FAA or the ICAO. Their job is to interpret these requirements and communicate them to the Design organizations for implementation. This process requires some negotiation with the regulatory agencies who may be, to some extent, flexible in how these requirements can be implemented.

The Systems Engineering and Integration Team (SEIT) organization

The Systems Engineering and Integration Team (SEIT) is the only organization that is unlikely to exist in a present-day commercial aircraft company; although, this concept is so well-known that its existence is not totally out of the question. Secondly, the SEIT may appear to be the only new organization recommended here. This impression is only partially true since parts of it already probably exist within the company. Finally, it is strongly recommended that the SEIT have a direct reporting relationship to the project manager for reasons that will be discussed later. (It is assumed that the company will be divided into multiple projects for each new aircraft being developed.)

So the question to be answered is: Why should these organizations be combined at all? The answer is called *functional binding*. This means that these organizations have such a close interrelationship that keeping them together both functionally and physically enhances their effectiveness. According to Hitchins (1993), entities that are functionally bound have fewer interfaces and are therefore less complex and subject to flawed interactions. So what are the functions that should comprise the SEIT? The following list is only a notional structure which may be modified at the program manager's discretion:

- SE—This is the core organization within the SEIT. It will be discussed more at length below.
- Configuration Management—Most SE texts consider Configuration

这些规章要求的监管当局(如FAA或ICAO)打交道。它们的工作是解释这些要求，并与设计部门进行沟通，以贯彻这些要求。该过程需要与监管部门进行协商，在某种程度上，它们对如何贯彻这些要求有一定的灵活性。

系统工程与集成团队 (SEIT) 组织

系统工程与集成团队 (SEIT) 是仅有的在当今商用飞机公司不大可能设置的组织。虽然人们都知道这个概念，但它的存在并非毫无争议。其次，系统工程与集成团队(SEIT)可能是本书唯一建议成立的新部门。这种印象可能只有部分正确，因为它的部分功能在公司中或多或少地存在。最后，强烈建议**系统工程与集成团队** (SEIT) 对项目经理有直接汇报关系，原因下文将会讨论。(假设公司针对正在研制的飞机设有多个项目)。

那么需要回答的问题是：为什么要合并这些组织？答案是功能性结合。这意味着这些组织有非常紧密的联系，使它们在功能和物理上结合在一起时可以，提高其效能。根据 Hitchin(1993) 的研究，实体组织如果在功能上结合在一起，会减少它们之间的接口，从而降低复杂度及缺陷影响。那么**系统工程与集成团队** (SEIT) 应该包含哪些功能呢？下面的列表只是一些粗略的结构，可根据项目经理的意愿进行修改：

- **系统工程** — 这是 SEIT 里的核心组织，下文将更详细地介绍。
- **构型管理** — 大多数的系统工程文献都把构型管理作为系统工程的一部分，所以**系统工程与集成团队** (SEIT) 自然应包含该功能。
- **集成** — 这也是系统工程的一个传统部分。
- **集成产品研制 (IPD)**— 该功能至少在基本元素方面与传统的系统工程一致。
- **安全性** — 包括该功能主要是因为其重要性,及其与系统工程的内在联系。
- **可靠性** — 包括该功能被是因为其在整个飞机范围内的含义及其层级结

433

Management to be part of SE, so it is natural to include this as well.

- Integration—This is also a traditional part of SE.
- Integrated Product Development (IPD)—This function, at least the principal elements of it, are consistent with traditional SE.
- Safety—This function is included primarily because of its importance and its interrelationship with SE.
- Reliability—This function is included because of its aircraft-wide implications and its hierarchical structure.

So the next question to answer is: Why does the SEIT have to report to the project manager? The answer lies in the hierarchical depiction of the aircraft elements as described in Chapter 2 and secondly in the hierarchical nature of the SE approach. SE has often been described as a top-down approach, which is not totally true, but to the extent that it is true, having a bird's eye view of the aircraft systems and their requirements makes SE function as it is supposed to function. The SE organization at the SEIT level assures that the aircraft-level requirements are captured and allocated or derived down to the subordinate subsystems that comprise the aircraft. Chapter 4 describes the process of allocation and derivation of requirements.

The SE organization

The functions of the SE organization are many, but here is a summary of the functions that appear to be most important in the commercial aircraft domain:

First, the SE organization should be recognized as the definitive authority on what SE is and how to implement it. There is a view in some quarters that the "everyone-is-an-SE" is the best approach. This is not totally illogical but this philosophy may lead to an environment in which there are conflicting and even incorrect views of SE. This environment may even lead to conflicting and hence defective design solutions. So, the implementation of this authority function requires that the SE organization write or approach any internal documents pertaining to SE and to provide authoritative training on the subject.

Next, the SE organization will be responsible for providing SE training for key personnel. Among the key personnel are the project manager and the SE focal within each organization listed in this chapter. As mentioned above, part of the adaptation process is to have at least one knowledgeable person in each organization. Training all

构。

因此，接下来的问题是：为什么**系统工程与集成团队 (SEIT)** 必须向项目经理汇报？答案在于第2章所述飞机元件的层次架构，以及系统工程方法的层级特性。系统工程经常被描述为一种自顶向下的方法，这不完全正确，但在某种程度上，它又是正确的，对飞机的系统及其需求的鸟瞰使得系统工程的功能得以按预期方式执行。**系统工程与集成团队 (SEIT)** 层面的系统工程组织确保能捕获飞机级需求，并将其分配或衍生到构成飞机的下属子系统。对需求分配及衍生过程的介绍见第4章。

系统工程组织

系统工程组织的功能有很多，但这里总结的是商用飞机领域内最重要的一些：

首先，应当认识到系统工程组织在定义什么是系统工程，以及如何贯彻系统工程方面的权威性。目前有一种观点认为，"每个人都是系统工程师"是最好的办法。这种观点并非完全不合逻辑，但这一理念可导致一个存在相互冲突，甚至不正确系统工程观点的环境。这种环境甚至可能导致设计解决方案存在缺陷，或者相互冲突的内容。所以，这种权威功能的实现要求系统工程组织编写或处理有关系统工程内部文件，并提供关于该主题的权威培训。

接下来，系统工程组织将负责为关键人员提供系统工程培训。其中的主要人员是项目经理，以及本章列出的各个组织的系统工程联络员。如前文所述，适应过程的部分内容是在各组织中至少有一名对系统工程知识有较好理解的人员。对所有人员进行培训被证明可能是极其昂贵的。针对设计团队中负责编写供应商技术规范的人员的培训尤为重要。

针对项目经理及其他设计团队经理的培训需要特别关注。他们了解系统工程过程，并负责制定重要的设计决策。对这些领导人员进行彻底和深入的培训显

personnel may prove to be unreasonably expensive. Training of the authors within the design groups for the supplier specifications is especially important.

Training of the project manager and other design group managers requires special attention. These are people who know the SE process and are responsible for making important design decisions. Thorough and intensive training for these leaders is especially important. One technique actually used in practice is to assure that future project and design managers are members of the SE organization. That is to say, they would spend a period of time doing all the SE functions including training themselves. This technique is worthy of consideration.

With regard to requirements the SE organization plays a major role. First, this organization will be the custodian of the requirements tool. A typical tool is DOORS (Dynamic Object-Oriented Requirements System); however, there are others on the market. Chapter 12 discusses the desirable characteristics of such a tool. In the interest of adaptation, the SE organization will not necessarily capture all requirements on all subsystems. However, they will record the ones of interest as the need arises. However, a complete recording of the top-level aircraft requirements, for example, durability and range, is a necessity. As explained in Chapter 4, all other requirements must trace to these requirements. When a requirement becomes official, that is, it is the requirement to which the aircraft or any of its subsystems must be designed to, the Configuration Management group within the SEIT will record and control that requirement. In addition to customer requirements and other requirements derived from them, the SE data base will also include the certification requirements from the regulatory agencies discussed above.

With regard to requirements, a primary responsibility of the SE organization is to review supplier specifications for accuracy and completeness of requirements. If the SE organization does this thoroughly, then the need for requirements reviews will be minimized. The requirements qualities in Section 4.10 will guide the SE organization in their task.

Apart from requirements, the SE organization will need to be available to facilitate trade studies, quality function deployment (QFD), and other SE processes.

Finally, as described in Section 4.11, the systems engineer will be one of the key people who decides what requirements are either necessary or unnecessary.

The Test organization

得尤为重要。实践中常用的一种方法是，确保未来的项目和设计经理是**系统工程组织**的成员。也就是说，他们需要花费一段时间完成系统工程所有的功能，包括培训自己。这种方法是值得推荐的。

系统工程组织在需求方面起主要作用。首先，该组织是需求工具的保管者。一种典型的工具是 DOORS(面向对象的动态需求系统)；当然，市场上还有其他工具。第 12 章讨论了这类工具的理想特性。系统工程组织不一定要捕获所有子系统的所有需求。然而，他们将在需求出现时记录有关的那些需求。但是系统工程组织必须完整记录飞机的顶层需求，例如，耐用性及航程。正如第 4 章所述，其他所有需求都必须能追溯到这些需求。当某一需求变成正式需求，即该飞机或其任何子系统设计时必须依据的需求时，**系统工程与集成团队** (SEIT) 的构型管理组织将记录和控制这些需求。除了客户需求及从中衍生的其他需求之外，系统工程数据库还将包括上面提及的来自监管当局的合格审定要求。

关于需求，系统工程组织的首要职责是评审供应商技术规范中需求的准确性和完整性。如果系统工程组织认真完成了这项工作，那么就能够使需求评审的必要性变得最小。本书 4.10 节所述需求质量将在任务上给予系统工程组织较好的指导。

除了需求，系统工程组织还需要进行权衡研究、质量功能展开 (QFD) 及其他系统工程过程。最后，如 4.11 节所述，系统工程师将是关键人物之一，他们决定哪些需求是必要的，哪些需求是不必要的。

试验组织

试验组织，有时被称为**试验和评估**，显然在系统工程中扮演重要的角色，因为试验是第 11 章所述四种需求验证方法之一。**试验**组织可以执行超出需求规定范围的试验，但这些试验可以视为常规试验或无文档要求试验。然而，试验组织不决定哪些需求要求进行试验；如前文所述，该工作应由**设计组织**完成。**试验组**

The Test organization, sometimes called Test and Evaluation, obviously plays an important role in SE since test is one of the four types of requirements verification as explained in Chapter 11. The Test organization may perform tests beyond those demanded by the requirements, but those tests could be considered routine tests that pertain to *routine* or *undocumented* requirements. However, the Test organization does not determine which requirements need to be tested; that is done by the Design organization as described above. The Design organization also determines the pass-fail criteria for the requirements that the Test organization must use. Sometimes tests are witnessed by the Quality organization to determine whether the test passed or failed.

Of course, there are test organizations within the supplier companies as well. The function is basically the same, but it is important that the tests be witnessed by Quality personnel from the developer organization.

Reliability organization

Reliability is a specialty engineering subject of long standing. However, within an SE context it has a special role; that is why it is part of the SEIT. It is not simply the role of the reliability engineers to determine what the reliability of the aircraft it; it is the responsibility this group to specify what the reliability of each element of the aircraft should be. To do this the reliability engineers need to allocate the reliability from the aircraft level to each of the lower levels of the aircraft hierarchy. This is not to say that they don't do this already; if they do, then this is another example of the "SE that exists" and does not need special attention.

Specialty Engineering organizations

Within a commercial aircraft company there are many specialty engineering organizations. These include human factors, maintainability, lightning, electromagnetic interference, and others. The important principle to keep in mind here is that these organizations are not independent entities. They are tied together by the concept that the aircraft is a system and that the engineering disciplines that make the parts work together are a system as well and the *cohesion* between these disciplines needs to be maintained as well.

All of the specialty engineering organizations together comprise a group which is often called an AIT (analysis and integration team). An AIT is similar to an IPT except

织必须遵循的成败判定准则由**设计**组织制定。有时试验由**质量**组织进行目击，以确定试验成功与否。

当然，一些供应商公司内也存在试验组织。其功能基本上是相同的，但重要的是，这些试验须由**研制单位**的**质量人员**现场目击。

可靠性组织

可靠性是一门历史悠久的专业工程学科。然而，在系统工程环境下，它具有特殊的作用；这就是为什么它是**系统工程与集成团队 (SEIT)** 一部分的原因。可靠性工程师的职责不只是确定飞机的可靠性如何；规定飞机每个元件的可靠性是什么才是该组织的职责。要做到这一点，可靠性工程师需要把可靠性指标从飞机级分配到飞机层次架构下层的每一个层级。这并不表示他们之前没有这样做；如果他们已经这样做，那么这是" 现有的系统工程 "的另一个例子，并不需要特别关注。

专业工程组织

在商用飞机公司有很多专业工程组织。这些组织包括人为因素、可维修性、雷电及电磁干扰等。这里需要强调的一个重要原则是，这些组织不是独立的实体。它们借助一些概念而结合在一起，即 " 整架飞机是一个系统 "，及 " 使各部分协同工作的各工程学科也是一个系统 "，以及 " 需要维护这些学科之间的凝聚力 "。

所有的专业工程组织一起组成了通常被称为分析及集成团队 (AIT) 的团队。分析与集成团队 (AIT) 类似于一个 IPT 团队，只是它不表示如第 12 章所述的飞机特定元件或子系统。组成一个专业工程组织的工程师可能经常不超过一名到两名，所以在每个 IPT 中分配一名专业工程师并不现实。他们必须合理分配时间以配合多个 IPT 的工作。

专业工程组织的主要接口是负责执行及验证专业工程需求的设计组织，这

that it does not represent a particular element or subsystem of an aircraft as described in Chapter 12. A specialty organization may often consist of no more than one or two engineers, so it is impractical assign one specialty engineer to each IPT. They have to divide their time among multiple IPTs.

The main interface for the Specialty engineering organizations is the Design organizations who are responsible for implementing and verifying the specialty engineering requirements.

As described in Section 4.11, the specialty engineer will be one of the key people who decides what requirements are either necessary or unnecessary.

Supplier Management organization

Supplier Management, like Marketing, does not think of itself as an engineering organization; and it is not. Yet it plays an important role in SE. Chapter 14 discusses the importance of reducing the *variability* among the elements of the supply chain, where variability is the deviation in the quality of the contracts, specifications, and the resulting supplier products. Supplier Management is sometimes called Procurement.

Supplier Management does not write contracts nor does it write specifications. The Contracts and the Design organizations do those things: Supplier Management's role is to make sure they get done and done correctly. Supplier Management has other key functions, such as to assure that potential suppliers are financially sound and have a history of producing quality products.

The Contracts organization

In practice there may be two Contracts organizations, one for contracts with the airline customers and one for contracts with the suppliers. It doesn't matter. The important point is that a contract is one of the mechanisms that reduces the variability between the elements of the large-scale system (LSS) which includes the developer, the customer, and the suppliers.

Contracts with the airline customer, of course, guarantee what the customer will get, how far it will fly, how much it will weigh, and so forth. If the aircraft does not meet these requirements, there will be financial penalties on the aircraft developer. Overweight aircraft is one of the more frequent examples of such deficiencies. In accordance with the equations for aircraft performance, for each pound of overweight

些组织负责实施和验证专业工程需求。

如第 4.11 节所述，专业工程师将是关键人物之一，他们决定哪些需求是必要的，哪些需求是不必要的。

供应商管理组织

供应商管理组织像**市场开发**组织一样，并不认为自己是一个工程组织；它也确实不是。但它在系统工程中起到重要作用。第 14 章将讨论减少供应链元素变异性的重要性，这里的变异性是指在合同、技术规范及供应商最终产品方面出现的质量偏差。**供应商管理**有时也被称为**采购**。

供应商管理组织既不编写合同，也不编写规范，这些事情由**合同**和**设计**组织负责完成。供应商管理的职责是确保合同完成并且正确完成。**供应商管理**具有其他一些重要功能，如确保潜在供应商实力雄厚，拥有生产高质量产品的历史。

合同组织

实际上可能存在两个合同组织，一个负责与航空公司客户签订的合同，另一个负责与供应商签订的合同。这并非重点。重要的是，合同是减少大规模系统 (LSS) 各因素间变异性的机制之一，这里的大规模系统包括研制方、客户及供应商。

当然，与航空公司客户签订的合同保证了客户将得到什么、能飞多远、重量多少等。如果飞机不符合这些要求，飞机研制方将面临经济处罚。飞机超重是这种不符合性的一个常见例子。按照飞机性能公式，飞机每超重 1lb，都会使航程及商载受到影响。重要的一点是，如上文所述，这些相同的需求与顶层需求一样，需由**系统工程**组织在其需求管理工具中进行追溯。

与供应商的合同同样重要。正如在第 14 章所讨论的，这是用于减少大规模供应链系统变异性的主要机制之一。如第 14 章所述，重要的原则是保持合同需求和技术需求的独立性。在实践中，这很难做到；然而，把两个结合在一起时会

aircraft, there will be a corresponding deficiency in range and payload capability. The important point, though, is that these same requirements will be the same as the top-level requirements tracked by the SE organization in their requirement tool as described above.

Of equal importance are contracts with the suppliers. As discussed in Chapter 14, this is one of the primary mechanisms for reducing the variability in the large-scale supply chain system. As stated in Chapter 14, the important principle is to keep contractual requirements and technical requirements separate. In practice, this is difficult to do; nevertheless, there is risk in combining the two.

Also discussed in Chapter 14 is the importance of internal contracts, that is, contracts between different divisions of the aircraft company. Whether these internal contracts are written by the same Contracts organization may vary from company to company, but the important principle is that these contracts should be as rigorously written as external contracts.

The Production organization

Another organization unlikely to associate itself with SE is Production. Yet Production is an essential phase of the creation of a commercial aircraft, and the physical system actually produced should match the requirements in the SE requirements data base and the system actually designed by the Design organization.

The observer can appreciate the risks of not producing the product actually desired by looking at the many steps the requirements have to traverse to result in a physical product. We have already seen how customer needs are converted into product requirement and how the Design organization creates a design that meets these requirements. To become a product the requirements have to go through two more stages: First, they go to an organization called Planning who create the instructions for how to assemble the final product. The production mechanics then produce the aircraft using these instructions. The potential for mistakes increases as the number of steps increases. The insurance against mistakes is called verification as described in Chapter 11. At the Production level this verification is performed by the Quality organization.

As a simple example, the mechanics are required to install the data lines and electrical conduits at a prescribed spacing due to EMI requirements. If, due to insufficient training or any other reason, the mechanic fails to achieve this spacing, the requirements will not be met and equipment failures may occur. Once again, the

导致风险。

第 14 章还讨论了内部合同,即,飞机制造公司不同部门之间的合同的重要性。是否由同一合同部门编写这些内部合同,选择可能会因公司不同而有所不同,但重要的原则是,这些内部合同的编写应像外部合同一样严格。

生产组织

另一个本身不太可能与系统工程联系上的组织是**生产**组织。然而生产是商用飞机制造的一个重要阶段,实际制造出来的真实系统应与系统工程需求数据库中的需求相匹配,并应符合**设计**组织的实际设计。

通过查看制造出实际产品的每一步骤的需求,观察者可以评估生产出的产品不是实际期望产品的风险。我们已经看到了客户的需要如何转化为产品需求,以及设计组织如何创建符合这些需求的设计。要转变为一个产品,需求必须要经过两个阶段:首先,要去一个称为计划部门的组织,该组织编制如何组装最终产品的说明书。生产线工人使用这些说明来制造飞机。生产步骤数量的增加会导致错误出现的可能性增加。防止错误的保证措施是第 11 章所述的验证。在**生产**层级,这种验证是由**质量**组织执行的。

一个简单的例子是,由于 EMI 需求,要求技术工人被需求以规定的间隔安装数据线和导线管。如果由于培训不足或其他原因,技术工人未能达到这个间距要求,不满足该需求,设备可能会发生故障。再次,**质量**组织负责确保满足这些需求。生产组织的最后一项重要的工作是,当生产组织觉察到设计不可生产时,与**设计**组织进行沟通。这可能是一个简单的问题,例如一个舱门无法关闭。在任何情况下,这个问题需要立即上报,以修正这种情况,并防止该问题再次发生。

规划组织

如上所述,**规划**组织负责将反映产品级需求的指令传递给**生产**组织。在这

Quality organization is responsible for assuring that the requirements are met.

One essential last duty of the Production organization is to communicate with the Design organization when the Production organization perceives that the design is not producible. This may be a simple issue, such as a door that will not close. In any event, this issue needs to be reported immediately to correct the situation and prevent a repetition of the problem.

The Planning organization

As described the Planning organization is responsible for passing on instructions to the Production organization that reflect the requirements at the product level. Human error can occur at this level also. However, the Quality organization once again has the responsibility for assuring that the requirements are met.

The Quality organization

The Quality organization has broad responsibilities. However, these responsibilities all fall under the umbrella category of verification, in particular, verification by inspection as discussed in Chapter 11.

They can, for example, inspect drawings to assure that they meet the requirements of the specifications. They can witness qualification tests at supplier sites to assure that the supplier product meets the requirements. They can inspect the Planning documents to assure that they meet the design requirements. Finally, they can inspect the final aircraft product to assure all Quality organizations in all companies may not do the tasks listed above. These are only typical tasks that assure that the final product meets the customer requirements.

The Maintenance organization

Of course the Maintenance organization is not part of the commercial aircraft development company but rather part of the airline company or a supplier to the airline company. The maintenance manual produced by the aircraft company can be seen as a requirements document just as much as a procurement specification. The airline company is contractually obligated to conduct its maintenance operations in accordance with this manual.

个层级也可能产生人为错误。但是，**质量组织应负责确保满足这些需求**。

质量组织

质量组织职责广泛。然而，在第 11 章中讨论到，这些职责都属于验证类别，特别是检查验证。

例如，它们可以检查图纸来确保图纸满足技术规范的要求；它们可以在供应商所在地现场目击鉴定试验，以确保供应商的产品满足需求；它们可以检查计划文件，以确保它们符合设计要求。最后，它们可以检查最终的飞机产品，以保证各公司中所有的**质量组织**都完成了上面列出的任务。这些仅是确保最终产品符合客户需求的一些典型任务。

维修组织

当然，**维修**组织不是商用飞机研制公司的一部分，而是航空公司的一部分或航空公司的供应商。飞机公司编制的维修手册可以视为一份需求文档，正如采购规范一样。根据合同规定，航空公司有义务按照该手册进行维修操作。

对飞机企业而言，维修的重要性如何强调都不为过。Reason(1997, 第 88 页)认为不合格的维修是飞机事故的主要原因之一。他以 1979 年美国航空公司 191 航班事故作为众多例子之一进行了分析。

把维修外包给外部机构的做法会给航空器的运营带来额外的风险。第 14 章介绍了由于组织间沟通变异性的增加，组织接口的增加如何导致风险的增加。

The criticality of maintenance to the aircraft enterprise cannot be over emphasized. Reason (1997, p. 88) cites defective maintenance as one of the primary causes of aircraft accidents. He lists the 1979 American Flight 191 as one of many examples.

The practice of outsourcing maintenance to outside organizations has contributed an extra layer of risk to aircraft operations. Chapter 14 shows how the increased number of organizational interfaces adds risks due to the increased variability of communications between organizations.

Summary of organizational responsibilities

The astute reader will by now have detected that there is a central theme in the above description of organizational responsibilities. The theme is that all commercial aircraft organizations play a role in converting customer requirements into a physical aircraft product and verifying that the aircraft meets these requirements. The goal of this section is to show that even existing organizations that may not have totally understood this theme can fit into the pattern and accomplish this objective.

组织职责总结

　　细心的读者会现在会发现在上面关于组织职责的描述中有一个核心主题。这个主题就是，所有商用飞机组织在将客户的需求转化为实际的飞机产品，并证实该飞机满足这些需求的过程中都发挥了重要作用。这一节的目的是要表明，即使现有的组织可能没有完全理解本主题，也可以匹配到该模式，并实现此目标。

14

Large-Scale System Integration

The toe bone's connected to the foot bone,
The foot bone's connected to the ankle bone,
The ankle bone's connected to the leg bone,
Now shake dem skeleton bones!

"Dem Bones," American spiritual

There is no topic more important in modern commercial aircraft development than large-scale system integration (LSSI). The large-scale system (LSS) in this instance is the global supply chain. Every aircraft developer contracts with suppliers all over the world for components and other aircraft parts. This international procurement is commonly known as *outsourcing*, which will be discussed in this chapter. Commercial aircraft are part of other LSS, such as the global navigation system and the air traffic control (ATC) system. But this chapter will focus primarily on the global supply chain since it is the sources of the greatest risks and is in need of the greatest degree of management, both technically and organizationally.

14.1   The System of Systems View

Before we can address the integration of the global supply chain system, it is necessary to view it as a single system with interconnections and not a set of disconnected pieces and that the developer has the responsibility for assuring this integration. In SE this system is also known as a system of systems (SoS). This is because the suppliers themselves are systems that are independently developed. When a set of independently developed systems are required to function together to achieve a common goal, the resulting set of systems is called a SoS. More on systems of systems can be found in Jamshidi (2009).

第14章 大规模系统集成

趾骨连到足骨，
足骨连到踝骨，
踝骨连到腿骨，
骷髅骨架开始摇摆吧！

《枯骨》，美国黑人圣歌

大规模系统集成 (LSSI) 是现代商用飞机研制过程中最重要的主题。大规模系统 (LSS) 在这里是指全球供应链。各研制方都会与世界各地的部件及飞机其他零部件供应商签订合同。这一国际采购通常被称为外包，本章将对其内容进行讨论。

商用飞机是其他大规模系统的一部分，如全球导航系统和空中交通管制 (ATC) 系统，但本章将主要关注全球供应链，因为它是最大的风险源，无论在技术方面还是组织方面都需要严格的管理。

14.1 系统之系统观点

在介绍全球供应链系统集成之前，必须将其视为一个内部相互关联的单个系统，而不是一系列各不相连的碎片，研制方负有确保集成过程得以实现的责任。在系统工程中，该系统也被称为系统之系统 (SoS)。这是因为供应商本身也是独立形成的系统。一组独立开发的系统一起工作，以实现一个共同的目标，

14.2   Outsourcing

Outsourcing is the procurement of parts and services through organizations outside the developer's own organization. Very often this procurement of parts and services is through companies in other countries. There are four primary reasons for outsourcing:

- Suppliers can provide products or services at a lower cost. This is often the case for suppliers in countries where labor rates are lower. This type of outsourcing does not only apply to product development but also to maintenance. When maintenance is performed in low-cost distant countries, this practice increases the chance of poorer quality work and also makes it more difficult for regulatory agencies to inspect and approve the maintenance facilities and procedures. This type of outsourcing is performed by airline companies rather than the developer thus adding layers to the procurement process. These additional layers also add risk as will be shown later in this chapter.
- Suppliers can provide a unique or perhaps technically superior product. There are very few engine suppliers in the world. So developers have to rely on one of these for their engines regardless of where they are located.
- Suppliers may be located in a country who is a potential buyer for the aircraft. So the developer sells the aircraft to a country in exchange for the purchase of parts or services. The purchased products do not even have to be aircraft parts. They can be soccer balls or cans of ham. This practice is called *off-set* marketing. This practice sometimes occurs because the country of purchase may have a weak currency and may prefer to buy the planes with goods or services rather than cash. The supplier country also seeks to create jobs in their own country; this factor provides the incentive to them to negotiate an outsourcing contract.
- The developer may lack the investment capital to finance a totally internally developed product; hence they depend on other companies, generally called partners, to help finance the aircraft and hence share the cost risk.

Other possible reasons for outsourcing may include:

- Investors may *perceive* that outsourcing will reduce the risks of development. The discussion in Section 14.2 will show that the increase in the number of

由此产生的系统集合称为系统之系统 (SoS)。更多关于系统的系统内容可参见 Jamshidi(2009) 的介绍。

14.2 外包

外包是通过研制方自有组织以外的组织采购零部件和服务。很多时候，零部件和服务的采购是通过其他国家的公司。采用外包的主要原因有四点：

- 供应商可以以较低的成本提供产品或服务。在供应商所在国家的劳动力价格较低时经常出现此种情况。这种类型的外包并不只适用于产品的研制，也包括维修。当维修在低成本并且偏远的国家进行时，会增加产生质量缺陷的概率，也会增加监管当局检查和批准其维修设施和程序的难度。这种类型的外包是由航空公司执行，而不是研制方，从而增加了采购过程中的层级。这些额外的层级会导致风险的增加，本章后面内容将对此进行介绍。

- 供应商能够提供一种独特的，或者技术出众的产品。在世界上只有极少数的供应商能够生产发动机，因此，研制方不得不依靠这些发动机的某一家供应商，无论它们位于何处。

- 供应商可以位于一个是飞机潜在买家的国家内。因此，研制方把飞机卖到该国家以换取购买零件或服务。购买的产品甚至不必是飞机零部件。它们可能是足球或火腿罐头。这种做法被称为补偿贸易。这种做法之所以存在，有时是因为购买的国家可能货币疲软，更愿意用商品或服务而不是现金购买飞机。供应商所在国家还试图在自己的国家创造就业机会；这个因素激励他们为外包合同进行谈判。

- 研制方可能缺乏投资资金，无力资助一个完全自主开发的产品。因此，它们依赖于其他公司，这些公司通常被称为合作伙伴，他们给予飞机研

451

organizational interfaces will actually increase the risk.

- Sometimes tax incentives in foreign countries may have the effect of subsidizing development in those countries.
- Reducing internal labor costs also reduces the amount of other long-term costs like benefits, such as pensions.

Some of these reasons may be in conflict with each other. For example, if a product is outsourced to another country for sales purposes, that country may actually have labor rates greater than the developer's country.

This is not to say that outsourcing is necessarily bad. All commercial aircraft companies do it for one or more of the reasons cited above. The point is that outsourcing should only be done moderately. Excessive outsourcing can have serious consequences, that is, it will increase risks rather than decrease as was the expectation. The reasons will be delineated below.

From an organizational perspective the degree of outsourcing will be the responsibility of executive management. It is also their responsibility to remain fully informed about the risks of excessive outsourcing. It will be the responsibility of the supplier management organization to determine the soundness of the supplier organization. Engineering will determine the acceptability of the supplier's product. Sillitto (2010) emphasizes the need for continuous feedback in the system. Most importantly there needs to be feedback between these tiers. Additionally there needs to be feedback between the suppliers and the developer.

14.3 Complexity and How It Increases Risks

To understand the risks of outsourcing and LSSI it is necessary to understand the concept of complexity at least some of its primary tenets. There have been many books written on complexity, for example, Page (2011). We will only discuss a few of the important aspects of complexity to show how complexity increases risk, how it relates to LSSI, and how the commercial aircraft developer can decrease complexity and decrease risk.

The Large-Scale System (LSS) and complexity

Sillitto (2010) states that one of the primary attributes of a LSS is complexity. He says that these systems are "dominated by emergence and exist in a state of constant

制过程以资金支持，从而分担成本风险。

外包的其他可能的原因包括：

- 投资者可能认为外包可以降低研制的风险。14.2 节的讨论将表明，组织接口数量的增加实际上将导致风险的增加。

- 有时，其他一些国家的税收优惠可能会对该国的研制产生影响。

- 降低国内劳动力成本也降低了福利、养老金等其他长期成本。

其中一些原因可能彼此冲突。例如，如果一个产品出于销售的目的外包给另外一个国家，而实际上该国劳动力成本可能高于研制方本国的劳动力成本。

这并不是说，外包一定不好。所有商用飞机公司，基于上述的一个或多个原因都采用外包。重要的一点是，外包只应适度进行。过度的外包可能产生严重的后果，即，它会增加风险，而不是像期待的那样减少风险。其原因将在下面说明。

从组织的角度来看，外包的程度的确定将是高级管理层的责任。充分了解过度外包的风险也是他们的责任。确定供应商组织的健全性是供应商管理组织的责任。工程部门将确定供应商产品的可接受性。Sillitto(2010) 强调了系统中连续反馈的要求，最重要的是这些层之间需要相互反馈。另外供应商和研制方之间也需要相互反馈。

14.3 复杂性及其如何增加风险

要理解外包及大规模系统集成 (LSSI) 的风险，就有必要了解复杂性的概念，至少要理解它的一些主要原则。关于复杂性已经有许多文献论述，例如 Page(2011)。我们将只讨论复杂性的一些重要方面，以表明复杂性如何增加风险、它如何与大规模系统集成 (LSSI) 关联及商用飞机研制方如何降低复杂性及风险。

reconfiguration and evolution." Sillitto calls these LSSs *wicked* systems. For the uninitiated, emergence, according to Checkland (1999), is any property of a system that cannot be attributed to a single element, such as a single supplier. Emergence can be the result of a failure, schedule slip, or cost overrun. Emergence can also be beneficial, but the focus in this chapter will on detrimental emergence. It takes the whole LSS to create emergence. Emergence is a basic feature of complex systems. Commercial aircraft supply chain systems qualify as complex systems. So the basic question is: How do we manage complexity and reduce the risks associated with it?

Some features of complexity

Of course, the more system elements, suppliers, there are the more complex the system will be. So the obvious way to reduce complexity is to reduce the number of suppliers. However, there can be limitations to this process; we have already noted that outsourcing is a necessary part of aircraft development. So there is a limit to how much the developer can reduce complexity.

Humans in the system

It is axiomatic of SE that humans are not simply operators of the system or designers of the system; they are elements of the system itself. This is especially true of the supply chain system. There are many humans in the supply chain system. Humans alone do not make the system complex. However, there is general agreement that systems with many human elements are complex and suffer from the variability of human performance. This is not to say that humans should be replaced with technical components or that humans are necessarily untrustworthy. It is only to say that any interaction between humans can be variable, and it is this variability that needs to be managed as discussed below in the section called Variability between Elements. It is this variability that is the vulnerability of the entire global supply chain system.

According to Sillitto, human systems are regarded as *soft systems*. The products produced by suppliers are called *hard systems*. Sillitto says that hard systems exist within soft systems. This multiplicity of system types is the factor that makes the supply chain complex. Regardless of the tiers, both systems strive for a common purpose, that is, a successful commercial aircraft system. This success can only be achieved if the variability is reduced or eliminated between the systems.

大规模系统 (LSS) 及复杂性

Sillitto(2010) 认为大规模系统 (LSS) 的主要属性之一是复杂性。他认为，这些系统 "由涌现性主导，并以不断重构和演变的状态存在"。Sillitto 称这些 LSS 为邪恶系统。根据 Checkland(1999) 的研究，对于非专业人士来说，涌现性是无法归于一个单一元素，如一家供应商的任一系统属性。涌现性可能是故障、进度延误或成本超支的结果。涌现性也可能是有益的，但本章重点讨论有害的涌现性，它会使整个 LSS 都产生危机。涌现性是复杂系统的一个基本特征。商用飞机供应链系统称得上是复杂系统。所以，根本的问题是：我们如何管理复杂性，并减少与之相关的风险？

复杂性的一些特征

当然，系统元素及供应商越多，系统也就越复杂。所以降低复杂性最常见的方式是减少供应商的数量。然而，该过程存在一些限制，我们已经注意到，外包是飞机研制的必要组成部分。所以飞机研制方降低复杂性的幅度存在极限。

系统中的人

对于系统工程而言，有一点是毫无疑问的，即：人不单是系统的操作者或系统的设计者，他们也是系统本身的组成元素。对供应链系统而言更是如此。供应链系统中有很多人。人本身并不会使系统变得复杂。然而，普遍共识是，有许多人的系统往往是复杂的，并受到人的行为变量的影响。这并不是说人类应该被技术部件替代，或者说人类必然是不值得信赖的。而只是说，人类之间的任何相互作用都可能是可变的，而正是这种变异性，需要以如下方式加以管理，本节下文将其称为**元素间的变异性**。正是这种变异性导致了整个全球供应链体系的脆弱性。

455

Organizational interfaces

There are two more aspects of complexity that are often mentioned in the literature, for example by Marczyk (2009). One is the number of interfaces, and the other is the variability among the elements. Both of these are important to reducing the risks associated with complexity. We will discuss interfaces first. Chapter 6 has already discussed interfaces, but the emphasis there was interfaces internal to the aircraft. The interfaces here are generally organizational interfaces, that is, interfaces among the elements of the supply chain including suppliers, maintainers, and the developer.

Jackson (2010) discusses how the numerous interfaces among organizations have led to communication breakdown that was responsible for some of the major air disasters. For example, first there is the interface between the developer and the airline. Secondly, there is the interface between the airline and the maintenance organization. It does not matter whether the maintenance organization is part of the airline or a separate supplier. Either way, there is an organizational interface. For example, the American Flight 193 accident was caused primarily by improper engine installations. This was an internal organizational interface. The ValuJet accident was caused by the actions of an external maintenance supplier. In short, multiple organizational interfaces can add risk to the aircraft operation. Reason (1997) notes that insufficient maintenance is the root cause of a large percentage of accidents. This does not imply that the developer, the airline, or the maintenance organizations are negligent. It only implies that the increased number of organizational interfaces increases the risk. Figure 14.1 shows schematically how the maintenance organization is separated from the developer by multiple organizational interfaces.

This diagram shows how the regulatory agency, the Federal Aviation Authority (FAA) in the United States, adds at least three more organizational interfaces to the system. These agencies have an active and essential role in regulating the products of the developer, the airline, and the maintenance organization.

根据 Sillitto 的观点，人类系统可以视为软系统。供应商生产的产品则被称为硬系统。Sillitto 认为，硬系统存在于软系统中。系统类型的这种多重性是使供应链变得复杂的一个因素。不论存在多少层级，这两个系统都致力于实现一个共同目标，即，一个成功的商用飞机系统。只有降低或消除系统间的变异性，才能达成目标。

组织接口

例如，在 Marczyk(2009) 的文章中经常提到，复杂性由两个以上方面组成。一个是接口的数量，另一个是元素间的变异性。这两者对减少与复杂性相关的风险都十分重要。我们将首先讨论接口。第 6 章已经讨论过接口，但其重点放在飞机内部接口。这里的接口通常指组织接口，即包括供应商、维修商及研制方的供应链的各元素之间的接口。

Jackson(2010) 对组织中数量众多的接口如何导致沟通不畅进行了讨论，正是这些接口导致了一些重大空难的发生。例如，首先是研制方与航空公司之间的接口。其次是航空公司与维修组织之间的接口。维修组织是航空公司的一部分还是单独的供应商并不重要，无论哪种方式，都会存在一个组织接口。例如，美国航空 193 航班事故主要是由不恰当的发动机安装导致。这是一个内部组织接口的例子。ValuJet 事故则是由外部维修供应商造成的。总之，多个组织接口，会增加飞机运营的风险。Reason(1997 年) 指出，维修不善是很大一部分事故的根本原因。这并不意味着研制方、航空公司或维修组织疏忽大意，而是意味着，组织接口数量的增加会增加风险。图 14.1 示意性地介绍了维修组织如何通过多个组织接口与研制方分隔开来。

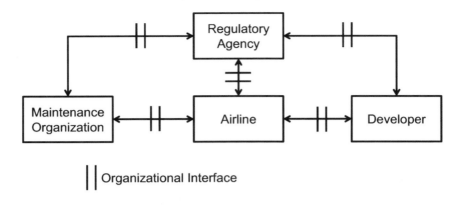

Figure 14.1　External organizational interfaces

Internal organizational interfaces

Internal to a developer's organization there are perhaps hundreds of organizational interfaces between, for example, design, test, supplier management, and so forth. Each of these interfaces is a potential source of information errors. Figure 14.2 depicts a typical set of organizational interfaces over which errors frequently occur. This set involves three functions:

- Engineering. This is where a product or a subsystem of the aircraft is designed.
- Planning. The planning function takes the product as designed in engineering and plans the steps for producing the product. This step can involve steps such as pick up drill and drill hole.
- Production. This function executes the instructions of the planner.

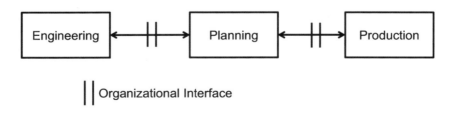

Figure 14.2　Typical internal organizational interfaces

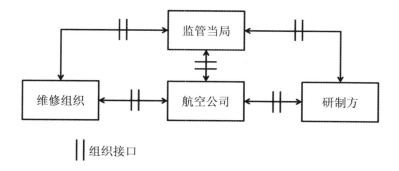

组织接口

图 14.1　外部组织接口

此图显示了监管当局，如美国联邦航空局 (FAA)，给这个系统增加了三个以上的组织接口。这些机构在监管研制方产品、航空公司及维修单位方面起到积极和重要的作用。

内部组织接口

研制方组织内部在例如设计、试验及供应商管理等组织之间可能存在数以百计的组织接口。每个这种接口都是错误信息的潜在来源。图 14.2 示出了一组典型的频繁出错的组织接口，它们包括三项职能：

- 工程。这是设计飞机产品或子系统的地方。
- 规划。规划职能是拿到工程部门设计的产品，然后规划生产该产品的步骤。该步骤可能涉及如拿起钻头并钻孔的步骤等。
- 生产。该职能执行规划者的指令。

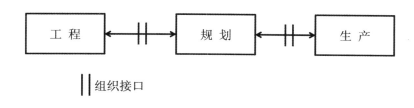

组织接口

图 14.2　典型的内部组织接口

Variability between or among elements

An important contributor to complexity is variability between or among elements as described by Marczyk (2009). An obvious question to ask is: Variability in what? A general term is variability in connections. Connections can consist of any sort of communications, either verbal or written. Variability in specified parameters, text results, contractual provisions, and so forth also count. Variability between elements can take many forms. For example, an ambiguous performance specification or contract between the developer and a supplier can constitute variability. This one aspect alone argues for intensely rigorous specifications and contracts. These will be discussed later in this chapter. Variability can consist of loose oversight of maintenance procedures. Variability can either consist of inaccurate information or an absence of information.

A potential source of variability is the differences in cultures between the developer and various suppliers. This variability can result from differences in languages and also in business practices and customs. For example, Applegate (1998) discusses the development of an aircraft in Indonesia. Factors mentioned by Applegate include the many languages spoken in that country.

The Haddon-Cave report (2009) on the Nimrod crash also discusses risks due to overly complex supply chains.

Multiple regulatory agencies

The existence of multiple regulatory agencies adds complexity and variability to the connections of the global supply chain system. Sometimes the regulations among these agencies can be in conflict. The developer is then responsible for resolving these conflicts. Of course, the multiple regulatory agencies will exist whether the developer practices outsourcing or not. However, when these multiple regulations are added on top of an already complex organizational structure, the complexity is then compounded making the risks of outsourcing even greater.

460

Intermediate stability

Rechtin (1991) emphasizes the importance of *intermediate stability*, that is, the reduction of variability, not only in the whole system, but the pieces of it as well. In this chapter the whole system is the supply chain system. In simple terms this can mean

两个或多个元素间的变异性

导致复杂性的一个重要因素是 Marczyk(2009) 介绍的两个或多个元素间的变异性。一个显而易见的问题是：变异性在哪里？一个常用的短语叫连接中的变异性。连接可以由任何形式的沟通组成，无论是口头形式还是书面形式。规定参数的变异性，如文字效果，合同条款等属于此类。各元素之间的变异性可能存在多种形式。例如，研制方与供应商之间模糊不清的性能规范或合同可能导致变异性。仅仅这个方面就说明严格的规范和合同十分重要。这些将在本章的后面进行讨论。变异性还可能由维修程序监督不严导致。变异性也可能由信息不准确或信息缺乏导致。

变异性的潜在来源是研制方与各供应商之间的文化差异。这种变异性可能是由于语言、商业惯例及习俗方面的差异导致。例如，Applegate(1998) 讨论了在印度尼西亚的飞机研制，Applegate 所提到的因素包括通行于该国的多种语言。

Haddon-Cave (2009) 对 Nimrod 坠机事件的报告也讨论了由过于复杂的供应链导致的风险

多重监管当局

多重监管当局的存在增加了全球供应链系统连接中的复杂性及变异性。有时，这些机构的监管条例可能是相互冲突的。研制方必须解决这些冲突。当然，无论研制方是否实行外包，多重监管当局都会存在。然而，当这些多重监管条例被加入到已经很复杂的组织结构的顶层时，其复杂性的叠加将使外包的风险更大。

中间稳定性

Rechtin(1991) 强调中间稳定性的重要性，也就是，不仅要减少整个系统的变异性，也要减少系统组成部分的变异性。本章所述的整个系统是指供应链系统。

that stability should be strived for between the individual suppliers and the developer. So when managing variability as described in Section 14.3, stability between individual system elements should be kept in mind.

Importance of supplier management

Rigorous supplier management is the key factor in achieving stability in the supply chain system. Within this management, rigorous contracts and specifications and their implementation and enforcement are the important aspects. This section elaborates on those aspects.

Consistent and thorough management of suppliers is central to the reduction in variability. Section 12.10 of the main body of this book discusses the function of supplier management. Supplier management is especially important in LSSI. Traditionally supplier management may have been thought of as a *managerial* function rather than a *technical* function. Nevertheless, supplier management plays an important role in SE and especially in LSSI.

Jackson (2010) provides a list of *paradigms* that are common in industry, not just commercial aircraft. Paradigms are, in this case, ways of thinking or mindsets that are risk inducing. That is, the mindset itself may lead to negative consequences if implemented. One of these is the *independent supplier* paradigm. This is the belief that suppliers can figure out what to build without any requirements from the developer. It is safe to say that such a business practice is not consistent with the rigor that LSSI would demand.

The importance of language

The importance of language between developer and suppliers cannot be underestimated. Today commercial aircraft are being built in different countries with different languages. In addition, suppliers are located all over the world. Total fluency in a common language is essential in contract negotiations and preparations and in specifications. Practitioners in these areas need to have total fluency in a common language. Although the common language is very often English, this is not always the case. Verbal competency is not sufficient; total fluency is required to detect the slightest nuances in meaning of any given thought. If the developer or the supplier needs to hire qualified interpreters or translators, this should be done.

简单来说，这可能意味着应致力于维持一些个别供应商与研制方之间的稳定性。因此，在管理 14.3 节所述的变异性时，应特别注意各个系统元素之间的稳定性。

供应商管理的重要性

严格的供应商管理是供应链系统实现稳定的关键因素。在这种管理方式中，严格的合同和规范及其实施和执行十分重要。本节将详细阐述这些方面。

对供应商进行一致和全面的管理是减少变异性的关键。本书正文 12.10 节讨论了供应商管理的职能。供应商管理在大规模系统集成 (LSSI) 中尤为重要。传统意义上的供应商管理可能被认为是一种管理职能，而不是技术职能。然而，供应商管理在系统工程，尤其是大规模集成 (LSSI) 方面发挥着重要的作用。

Jackson(2010) 提供一系列范式，这些范式在工业领域十分常见，而不仅是商用飞机领域。在这种情况下，范式是导致风险产生的思考方式或思维模式。也就是说，思维模式本身可能会导致负面后果。独立供应商范式便是其中之一。该范式认为供应商在没有来自研制方的任何需求的情况下就能判定应该制造什么。可以肯定地说，这种做法与大规模系统集成 (LSSI) 要求的严苛度极不相符。

语言的重要性

不能低估研制方和供应商之间语言的重要性。现代商用飞机往往在语言各异的不同国家制造。此外，供应商几乎遍布全球。流利的共同语言对于合同谈判、筹备工作及规范而言至关重要。这些领域的从业者需要有流利的共同语言。虽然通用语言通常是英语，但也不尽然。单凭言语能力是不够的，全面的流畅性要求能发现任何特定想法的细微含义差别。如果研制方或供应商需要聘请合格的口译或笔译员，就应尽量付诸实施。

Internal suppliers

Often large companies have many divisions located in different cities and perhaps different countries. When one of these divisions makes a product or component of the aircraft, these divisions are then assuming the role of supplier. These divisions are then internal suppliers rather than external suppliers. When products are procured internally, there has to be some internal documentation to procure these products. It is possible this documentation does not even pass through the supplier management department or the contracts department.

Here is the source of another risk. If the internal documentation does not meet the standards of external specifications and contracts, there is the risk that the product itself will not meet the same standards that an external product would meet. As a general principle this documentation should meet those standards.

14.4   Managing the Risks of a Large-Scale System (LSS)

The discussions above have identified two principal root causes of risks for LSSs. These risks are associated with the global supply chain system typical of the commercial aircraft industry. The two sources are as follows:

- the multiplicity of organizational interfaces;
- the variability of connections between the elements of the supply chain system.

Managing organizational interfaces

The above discussions point out the risks inherent in an excessive number of organizational interfaces. There are only a limited number of ways to manage this number.

The first and obvious way to reduce the number of ways is by reducing the number of suppliers, that is reducing the extent of outsourcing. As was pointed out above, there are good reasons for outsourcing. So the main goal is to minimize the extent of outsourcing. There are only two ways to do this:

1. Perform component development internally. This can only be done if the

内部供应商

大公司通常都设有很多部门，这些部门往往位于不同的城市，甚至不同的国家。当这些部门生产产品或飞机部件时，它们可以被假设为供应商的角色。此时这些部门是内部供应商，而不是外部供应商。当产品内部采购时，必须依据一些内部文件采购这些产品。这些文件可能无须供应商管理部门或合同部门批准。

这样做会带来另一种风险。如果内部文件不符合外部规范及合同的标准，就会存在产品本身不能满足外部产品所要求的相同标准的风险。一般而言，该文件应符合这些标准。

14.4 管理大规模系统的风险

上述讨论已经对大规模系统 (LSS) 风险的两种主要的根本原因进行了定义。这些风险都与商用飞机行业典型的全球供应链体系有关。这两种来源是：

- 组织接口的多样性。
- 供应链系统元素之间连接的变异性。

管理组织接口

以上讨论指出了数量过多的组织接口中蕴含的风险。只有为数不多的方法可以管理组织接口数量。

减少该数量的第一个，也是较为明显的途径是减少供应商的数量，也就是减少外包。正如上文所述，外包有其存在价值。因此主要目标是尽量减少外包。只有两种方法可以做到这一点：

1. 内部进行部件研制。这种情况只能在研制方有这种能力的情况下才是可行的。

developer has the capability.

2. Utilize internal suppliers. Similarly, this option has its risks as well. Internal suppliers can only be used if the internal specifications and cross-organizational agreements have the same rigor as external specifications and contracts.

3. Engaging one supplier to integrate an entire subsystem, for example, the avionics system or the propulsion system. This way the single supplier can assure the minimum variability among the components in that subsystem. This responsibility would have to be spelled out contractually.

4. Using multiple-source suppliers for critical equipment. A good practice is to have at least two suppliers, and even more whenever possible, in order to be able to challenge them in term of cost at least and never face a situation of a delivery interruption.

After these options have been exhausted, the developer must concentrate on reducing the variability among the elements of the global supply chain elements as described below.

Managing variability of connections

Reduction of variability through supplier management involves two principal thrusts: contractual management and management of specifications. Section 12.9 of this book discusses the role of supplier management. The main point of that section was that the supplier management process should be rigorous, both technically and managerially. Following are two rules for reducing variability of technical and contractual documents for suppliers.

Rule 1: Supplier contracts should contain only contractual information and should not contain any technical requirements

The reason for this rule is that contracts are priced on the work done, not on the performance of the product. Furthermore, if a technical requirement appears in the contract, there would be no link to the verification matrix described in Chapter 11. So the verification method would not be clear. In fact, it may not even exist. The contract must have a clear reference to the specification that will provide the information about how the product should perform.

2. 利用内部供应商。同样，该选项也存在风险。只有内部规范及跨组织协议与外部规范及合同同样严谨时，使用内部供应商才是可行的。

3. 采用一个供应商集成整个子系统，例如，航空电气系统或推进系统。通过这种方式，单一供应商可以保证该子系统中的部件变异性最小。这种责任必须在合同中明确表述。

4. 关键设备采用货源供应商。一种推荐做法是，至少选择两家供应商，有可能的话可以更多，为的是至少能够在成本上给供应商以激励，并且不会面临交付中断的情况。

这些选项之外，研制方还必须集中精力减少下述全球供应链元素的变异性。

管理连接的变异性

通过供应商管理减少变异性涉及两个主要方面：合同管理及规范管理。本书 12.9 节讨论了供应商管理的作用。该节的主要观点是，供应商管理过程在技术和管理上都应该是严谨的。以下是减少供应商技术及合同文件变异性的两条规则：

规则 1：供应商合同中应该只包含合同信息，而不应包含任何技术需求

这样做的原因是，合同上的价格是针对完成的工作，而不是产品的性能。此外，如果技术需求出现在合同中，就无法链接到第 11 章所述的验证矩阵。那么验证方法就不够明晰。事实上，它可能根本不存在。合同必须有一个明确的引用规范，该规范将提供有关产品执行方式的信息。

规则 2：供应商规范应该只包含技术需求，而不应包含任何的合同语言

提出这条规则的原因是，如果在规范中出现工作说明，将无法对其进行定价，因为所有定价说明都包含在合同中。总之，坚持这两条规则将提升严苛度，并减

Rule 2: Supplier specifications should contain only technical requirements and should not contain any contractual language

The reason for this rule is that if a work statement appears in the specification, there would be no way to price it since all pricing statements are in the contract.

In short, adherence to these two rules will add rigor to the reduction in variability of all the aspects of supplier procurement and hence reduce the complexity of the global supply chain system and the risk of significant consequences, such as technical, schedule, and cost.

Reducing the variability in supplier contracts

Continuing with Rule 1, here are a few topics that are essential to supplier contracts. These topics should apply to both external and internal contracts:

1. The contract should state what work is to be done by the supplier and over what time and how the product should be delivered to the developer.
2. The contract should list any other milestones or events in which the supplier is involved. For design reviews, the contract should state who will conduct the reviews and who will approve the outcomes, normally the developer and not the supplier.
3. The contract should state who will conduct product verification including qualification tests and who will approve the outcomes. This topic includes the five classic verification methods also described in Chapter 11:
   - Analysis—If the supplier has to verify a requirement by analysis, this should be stated in the contract. If the developer has to perform this analysis, the contract should state this fact also. Verification by analysis includes verification by similarity, a common method of verification. Other common analyses include structural analysis and aerodynamic analysis.

     There is an inherent risk in analysis by similarity. This method is only valid if the product is compared to an equivalent product in the same environment as the product being analyzed. Among verification methods similarity is the one that is least understood and most prone to a lack of rigor.
   - Inspection—Either the supplier or the developer can perform verification by inspection. The developer is responsible for final approval.

少供应商采购所有方面的变异性，从而降低全球供应链系统的复杂性及重大后果，如技术、进度及成本方面的风险。

降低供应商合同的变异性

继续规则 1，这里有几项与供应商合同有关的主题。这些主题应既适用于外部合同，也适用于内部合同：

1. 合同应规定哪些工作由供应商完成，产品应在什么时间，以何种方式交付。

2. 合同中应列出所涉及供应商的所有里程碑或事件。对于设计评审，合同中应规定谁进行审查，谁批准结果，审查及批准的执行者通常是研制方，而不是供应商。

3. 合同中应注明由谁进行产品验证，包括鉴定试验，以及由谁批准结果。本主题包括第 11 章介绍的 5 种经典验证方法：

- 分析 —— 如果供应商必须用分析方法来验证需求，则应在合同中予以规定。如果研制方必须进行这种分析，合同也应说明这一事实。分析验证包括相似性验证，这是一种常用方法。其他常见的分析包括结构分析及空气动力学分析。

相似性分析存在内在风险。这种方法只有在产品在相同的环境下，与同等的产品比较分析时才有效。在验证方法里，相似性分析是理解程度最低，最容易缺乏严苛度的方法。

- 检查 —— 无论是供应商，还是研制方，都可以通过检查进行验证。研制方负责最终审批。

- 演示 —— 无论是供应商，还是研制方，都可以进行产品演示。

- 试验 —— 一般情况下，供应商在将产品交付研制方之前进行产品试验。在任何情况下，研制方都必须批准最终结果。

- 符合性 —— 符合性是确保所有飞机部件都符合图纸。符合性过程可能十

- Demonstration—Either the supplier or developer can perform product demonstrations.
- Tests—Normally the supplier would conduct product tests before delivery to the developer. In any case the developer must approve the final result.
- Conformity—Conformity is the assurance that all aircraft parts are in compliance with drawings. The conformity process can be time consuming, but it is an essential process.

The verification method for each requirement will appear in the supplier specification. In accordance with SE principles, all requirements must be verified by at least one of the methods.

Reducing variability in specification quality

Here are a few things to keep in mind when developing supplier specifications:

1. Rigorous requirements are essential to a quality specification. Books have been written on the subject of how to write requirements. Hooks and Farry (2001) is one example of how to write good requirements. Requirements written by untrained persons tend to be ambiguous.
2. Don't forget traceability. No requirement exists in isolation. Every requirement in a supplier specification either flows down or is derived from aircraft-level requirements. See Chapter 2 for what is meant by levels.
3. Make sure your requirements are complete. Don't leave any blanks in the specification. Environmental requirements are especially vulnerable to omission. The environment inside the engine nacelle is very different from the environment inside the cargo bay.
4. Make sure your specification has a good verification matrix. See Chapter 11 of this book for a discussion of verification. From an LSSI perspective, it is important to specify what verification methods the supplier will employ and who will approve the results.

Managing multi-tier suppliers

Suppliers sometimes come in multi-tiers. That is to say, a tier 1 supplier is one that reports directly to the developer. A tier 2 supplier is one that reports to the tier 1

分耗时，但是非常重要。

每条需求的验证方法均包含在供应商规范内。依据系统工程准则，所有需求必须通过其中一个方法进行验证。

减少规范质量中的变异性

编写供应商规范时需要注意以下几点：

1. 严格的需求对质量规范而言十分重要。关于如何编写需求已有许多文献涉及。Hooks 和 Farry(2001) 给出了一个如何编写需求的例子。未经训练的人编写的需求往往是不明确的。

2. 务必保持可追溯性。没有任何需求是孤立存在的。供应商规范里的每条需求要么往下分解，要么来自飞机级需求。关于层级的定义可以参见第 2 章。

3. 确保需求是完整的。不要在规范里留下任何空白。环境需求特别容易遗漏。发动机短舱内的环境与货舱内的环境差异明显。

4. 确保规范对应一个质量较高的验证矩阵。关于验证的讨论参见本书第 11 章。从大规模系统集成 (LSSI) 的角度来看，必须明确规定供应商采用何种验证方法，以及由谁对结果进行批准，这一点十分重要。

管理多级供应商

供应商有时会存在多个层级。也就是说，一级供应商向研制方直接汇报。二级供应商向一级供应商汇报。例如，飞机发动机制造商是一级供应商，因为发动机直接安装到飞机上。发动机吊舱可能包含安装在发动机上的阀，该阀门的制造商即为二级供应商。如果需求向下分解过程严苛度不够，那么多级供应商可能会导致风险。图 14.3 是一张示意图，显示了研制方与一级供应商及二级供应商之间的关系。

supplier. For example, the aircraft engine manufacturer is a tier 1 supplier because the engine fits directly into the aircraft. In the engine pod there may be a valve attached to the engine. The manufacturer of the valve is a tier 2 supplier. The situation of multi-tier suppliers can lead to risks if the requirements flow down is not rigorous. Figure 14.3 is a schematic that shows the relation among the developer and the tier 1 supplier and the tier 2 supplier.

This schematic shows that the relation between the developer and the first tier supplier is exactly the same as between the first tier supplier and the second tier supplier. There may be third tier suppliers and fourth tier suppliers but the principle is the same.

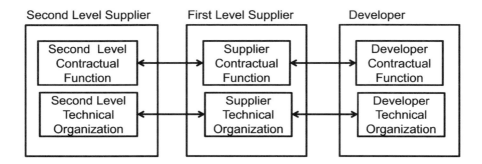

Figure 14.3   Multi-tier supplier relationships

This schematic also shows the dual relationship between the tiers, the contractual tier and the specification tier as explained in the paragraphs above. So what are the expectations about multi-tier supplier?

1. The developer can expect that the relationship between the first and second tier suppliers will be exactly the same as between the developer and the first tier supplier.

2. The developer can expect to verify that the second tier contracts and specifications meet the rigorous requirements described above.

3. The developer can expect that the third and fourth tier contracts and specifications will also meet these requirements.

As a simple example described above, the environment for the valve inside the

该示意图表明，研制方和一级供应商之间的关系与一二级供应商之间的关系完全一样。有可能存在三级和四级供应商，但原理是一样的。

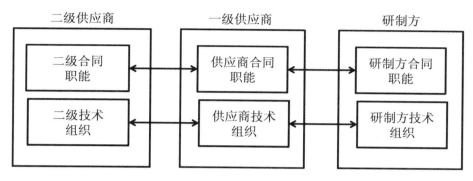

图 14.3　多级供应商之间的关系

该示意图还示出了层级之间的双重关系，即上面解释过的合同和规范层级。那么，关于多层次供应商的期望是什么？

1. 研制方可以期望一级和二级供应商之间的关系与研制方和一级供应商之间的关系是完全一样的。

2. 研制方可以期望能验证二级的合同和规范是否满足上述严格要求。

3. 研制方可以期望三级和四级的合同和规范也将满足这些要求。

如上述简单例子所示，发动机吊舱内的阀门的环境与发动机或飞机任何其他部分的环境都非常不同。研制方有责任确保这些需求都向下分解到所有层级的供应商。

确保避免这些多级供应商问题的一个可行的解决方式是，在最终的飞机交付之前完成整架飞机的集成试验。当然这种方法成本会更高，但风险将显著降低。

engine pod will be very different from the environment of the engine or any other part of the aircraft. The developer has the responsibility for assuring that requirements such as these are flowed down to all tiers of suppliers.

An accepted way to assure that these multi-level supplier problems are avoided is to perform whole aircraft integration tests before the final aircraft is delivered. This method would, of course, be more expensive, but the risks would be considerably reduced.

Supplier responsibilities

It should not be assumed that the responsibility for quality specifications belongs only to the developer. The suppliers themselves need to provide feedback to the developer regarding:

- missing requirements;
- unclear requirements;
- requirements that are not achievable;
- unnecessary requirements.

The developer needs to provide sufficient slack in the procurement schedule to allow the supplier to provide this information.

14.5 Other Large-Scale System (LSS) Principles to Apply to Commercial Aircraft

Sillitto (2010) recommends that we pay attention to the stability of the LSS and apply measures of stability to these systems. By stability Sillitto means what we have called the lack of variability. Anyone who has worked in the commercial aircraft world knows how all aspects of the system are in a state of constant flux. There are new customers with new requirements and new suppliers all the time. More importantly applicable to this chapter there may be new organizational interfaces and variability at all tiers. There may be variability in contract provisions and contracts. The introduction of new technologies, which happens on a continuing basis, adds new levels of variability and hence instability. Recent years have seen the introduction of new technologies, such as composite structures, electronic generators, and envelope control.

So how can stability be measured? On the simplest level, there are qualitative

供应商职责

不应该认为关于质量规范的责任只属于研制方。供应商本身也需要在下述方面向供应商提供反馈意见：

- 遗漏的需求；

- 不明确的需求；

- 不可能实现的需求；

- 不必要的需求。

研制方需要提供足够宽松的采购进度，以确保供应商提供这些信息。

14.5 应用到商用飞机的其他大规模系统 (LSS) 原则

Sillitto(2010) 建议关注大规模系统 (LSS) 的稳定性，并对这些系统采取稳定性措施。Sillitto 认为稳定意味着我们所说的缺乏变异性。商用飞机行业的任何从业者都知道，系统的各个方面是在不断变化的。随时都有提出新需求的新客户以及新的供应商出现。更重要的是，根据本章观点，所有层级都可能有新的组织接口及变异性。合同条款和合同中可能存在变异性。新技术的不断引进增加了变异性，因此也增加了不稳定性。近年来，新技术来断引入，如复合材料结构，电子发电机及包线控制。

那么，稳定性应该如何衡量？最简单的是定性的评价。在宏观层面上，研制方可以关注外包供应商的数量，判断它们的数量是否超过正常范围。然后，他们可以通过规范和合同看到正在发生什么变化。

在一个更复杂的层面上，Marczyk(2009 年) 开发了一种算法来衡量复杂性。他可以将该算法应用到整个全球供应链体系，并确定离危险引爆点有多远，也就是接近于解体状态的程度。

evaluations. At the macro level the developer can look at the number of outsourced suppliers to see if they exceed the normal number. Then they can look into specifications and contracts to see what changes are occurring.

On a more sophisticated level Marczyk (2009) has developed an algorithm to measure complexity. He can apply this algorithm to the entire global supply chain system and determine how close to the *tipping point* it is, that is how close to the point of disintegration.

Sillitto also recommends that we pay close attention to the node and web architecture. This architecture is a good depiction of organizational interfaces. The fewer organizational interfaces the more stable the system. There is another method for analyzing the node and web architecture and for minimizing complexity of systems that includes both humans and technological elements. This is the optimized N^2 diagram described by Hitchins (1993). In this method the analyst arranges all elements of the system on an N^2 diagram a depiction well-known to systems engineers. The method allows the analyst to rearrange the elements until all the relationships are minimized. This method results in a minimally complex system.

14.6 Summary

We have described the global supply chain system as a LSS in a SE context. This system is characterized as a complex system in which instability and resultant cost, schedule, and technical risks are incurred as a result of two major factors: an excessive number of organizational interfaces and the variability between elements of the system. In this case the elements are the developer, the suppliers, and the regulatory agency. It is shown that this variability can be minimized through rigorous control of suppliers both managerially and technically. Stability among individual elements is also emphasized. It is concluded that adherence to the rules outlined in this chapter will assist in reducing the degree of outsourcing and the associated risks.

Sillitto 建议密切关注节点及网络架构。这种架构可以对组织接口进行较好的描述。组织接口越少，系统就越稳定。还有另一种方法，用于分析节点和网络架构，并尽量降低系统的复杂性，该系统包括人和技术元素，即 Hitchins(1993) 介绍的优化后的 N^2 图。在该方法中，分析人员在 N^2 图上布置该系统上的所有元件，N^2 图是系统工程师比较熟悉的方法。该方法允许分析人员重新布置这些元件，直到所有的关联达到最小化状态。利用这种方法可以获得复杂性最低的系统。

14.6 小结

我们将全球供应链系统视为系统工程中的一个大规模复杂系统 (LSS)。该系统的特性在于作为一个复杂系统，其中的不稳定性及成本、进度及技术风险主要因两个因素而产生：组织接口过多及系统元素之间的变异性。这里所说的元素是指研制方、供应商及监管当局。它表明，这种变异性可通过供应商的严格控制实现最小化，该工作既包括管理上的，也包括技术上的。各个元素之间的稳定性也是重点。结论是：遵守本章所述的规则将有助于降低外包的比例及相关风险。

15

Risk Management

As a broad definition of risk, risk is anything undesirable that may happen as the result of either an internal or external cause. This is the consequence that occurs *if nothing is done to prevent it from happening.*

Risk management is one of the most difficult tasks to accomplish especially in a commercial aircraft environment where the demands of the marketplace and schedule constraints often place it low on the priority list. Yet its importance cannot be underestimated. This chapter will, first, provide some of the essential principles of risk management and the steps for accomplishing it. In addition, it will provide some hypothetical examples of actual risks and their possible consequences, their warning signs, and ways to mitigate the consequences. For a more comprehensive description of the risk management process, the reader can refer to, for example, Conrow (2003). However, this chapter will provide some guidance on how this process can be adapted to the commercial aircraft domain.

Not to be overlooked are the cultural obstacles to risk management. Vaughn (1997, pp. 96–111) provides the most compelling summary of this issue relative to the *Challenger* disaster in which she coined the phrase "normalization of deviance." In addition, the Columbia Accident Investigation Report (NASA 2003, p. 189) described the "broken safety culture" as a principal contributor to that accident. Finally, Conrow (2003, pp. 122–124) provides a comprehensive list of "negative attitudes" that exacerbate the risks that are already there. While none of these sources provides a clear solution to the cultural aspects, the fact remains that they remain an impediment to the effective operation of a commercial aircraft organization and should be treated seriously.

ARP5754A (2010) discusses risk with a focus on certification and safety. That is, the risks discussed in that document are almost exclusively technical. The scope of this chapter is broader. This chapter discusses risk related to the total development of the aircraft including technical, cost, and schedule risks and the interrelationships among those risks.

第 15 章　风险管理

根据风险的广义定义，风险是任何由内部或外部原因导致的，不期望发生的事情。它是如果不采取任何措施预防其发生时产生的后果。

风险管理是最困难的工作之一，尤其是在商用飞机环境下，市场的需求和进度的约束往往导致风险管理被置于优先级较低的位置。然而，其重要性不容低估。本章将首先介绍一些风险管理的基本原则及其实现路径。此外，本章内容还将给出实际风险及其后果的假设示例、风险及预警信号及减轻风险的方法。关于风险管理过程的更全面介绍可以参考 Conrow(2003) 的文章。本章将介绍如何将风险管理过程应用于商用飞机领域。

风险管理中的文化障碍是不容忽视的。Vaughn(1997，第 96 至 111 页) 对于挑战者灾难的研究结论十分具有说服力，她创造地提出了 " 越轨行为的常态化 " 这一说法。此外，《哥伦比亚号事故调查报告》(NASA 2003, 第 189 页) 认为 " 破碎的安全性文化 " 是导致该事故的一个重要原因。Conrow (2003, 第 122 至 124 页) 给出了 " 消极态度 " 的完整列表，这些消极态度加剧了已经存在的风险。虽然这些文献都没有在文化方面提供明确的解决方案，但事实上，这些文化障碍仍然会阻碍商用飞机组织的有效运行，应当认真对待。

ARP 5754A(2010) 重点讨论了合格审定及安全性方面的风险。也就是说，该文件所讨论的风险几乎完全是技术性的。本章的讨论范围更为广泛。本章讨论的风险涵盖飞机整个研制过程，包括技术、成本及进度风险及这些风险之间的相互关系。

15.1 Overview of Risk Management

There are three broad categories of risk: cost, schedule, and technical. Because of its certification implications, safety risk as discussed in Section 10.2 is treated as a separate category in the aircraft industry. The demands of the commercial aircraft industry are that all risk needs to be addressed and treated. A single risk may have all three types of consequences, or perhaps only one, or perhaps two. Hence, it is not appropriate to attempt to assign responsibilities for risk management to technical and managerial organizations. Responsibility should reside in a single technical–managerial organization, such as SE.

Risk management can minimize or eliminate many risks. It can highlight areas of uncertainty or false confidence. It provides a means of deciding the best course of action. It can provide a means of early warning of problems. It can provide a means for plan changes. Finally, and most importantly, it can increase management and airlines' customer confidence.

There are many ways of analyzing and managing risk; for example, the *NASA Systems Engineering Handbook* (1995) describes probabilistic risk assessment (PRA). The following simplified risk management process is both a manageable and effective process for commercial application:

Identify risks

The first step is to develop and document a risk statement. The risk statement should include the source of the uncertainty and the potential consequence. Each risk should be identified as one of the three categories above:

1. *Technical risk*   The uncertainty of achieving program requirements of function, performance, and operability within the planned cost and schedule.
2. *Schedule risk*   The uncertainty in achieving the program schedule if none of the technical risks should materialize.
3. *Cost risk*   The uncertainty in achieving the cost budgets if none of the technical and none of the schedule risks should materialize.

As a general rule, the best method of identifying risks is by using checklists and templates. In interviews, such phrases of uncertainty, such as "not sure," or "don't

480

15.1 风险管理的概览

风险分为三大类：成本、进度及技术。由于合格审定的需要，本书 10.2 节讨论的安全性风险在飞机工业被视为一个单独的类别。商用飞机行业的要求是所有的风险都需要加以解决和处理。单个风险可能导致三种类型的后果，或者只有一种类型，或者两种类型。因此，不应把风险管理责任分配到技术和管理组织。责任应该归到一个单一的技术 - 管理组织，比如系统工程。

风险管理可以最大限度地减少或消除许多风险。它可以帮助识别不确定性或盲目的信心。它提供了一种确定最佳行动路线的方法，也可以提供问题的早期预警手段，并为计划变更提供方法。最后，也是最重要的，它可以增加管理层及航空公司客户的信心。

分析和管理风险的方法有许多；例如，《NASA 系统工程手册》(1995) 描述了概率风险评估 (PRA)。下文所述的简化风险管理过程，对于商业应用而言既是可管理的，也是有效的。

识别风险

第一步是制定并记录风险说明。风险说明应该包括不确定性的来源及其潜在后果。每条风险均可归为上述三类风险之一：

1. 技术风险 在计划的成本和进度范围内实现项目功能、性能和操作性需求的不确定性。

2. 进度风险 因没有具体化任何一个技术风险而在完成项目进度方面存在的不确定性。

3. 成本风险 因没有具体化任何技术和进度风险而在达到成本预算目标方面存在的不确定性。

一般而言，识别风险的最好方法是使用清单和模板。进行访谈时，不确定

know" are sure clues.

Analyze risks

The grid of Figure 15.1 has been found useful in risk analysis. A common practice is to separate the risks into three categories: low, medium, and high. The general practice is to manage only those risks that are medium or high. One might ask why the two corners of the grid do not have the same level of risk. As this chart shows, one corner is low risk and the other is high risk. A logical conclusion is that high-consequence, low-likelihood risks are considered more serious than low-consequence, high-likelihood risks. This conclusion is in agreement with the fact that many disasters of high consequence are the result of low-probability events.

A key aspect of risk analysis is that the risks should be assigned and validated by persons other than the component designers or owners. An effective approach is to use an integrated product team (IPT) consisting of the designers and non-advocates, that is, impartial observers.

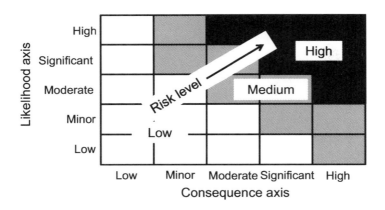

Figure 15.1   The risk grid

To determine where any given component fits into the grid of Figure 15.1 the templates of Tables 15.1 and 15.2 are useful. Every attempt should be made to base the risks on as quantitative a basis as possible. For example, if a supplier has a track record of delivering a product on time only 50 percent of the time, then the likelihood should be judged as moderate.

表述的短语必定预示着风险，如"不确定"或"不知道"等。

分析风险

实践表明，图 15.1 所示网格图在风险分析中十分有用。通常的做法是把风险分为三个等级：低、中和高。一般的做法是只管理那些中等级风险或高等级。有人可能会问：为什么在网格的两个角上的风险等级不一样？如该图所示，一个角是低风险，而另一个是高风险。一个合乎逻辑的结论是，后果严重但发生概率低的风险被认为比后果不严重但发生概率较高的风险更严重。这一结论与许多后果严重的灾难由低概率事件导致这一事实相符。

风险分析的一个关键方面是，风险的分配和确认应该由该部件设计者或所有者之外的其他人进行。采用综合产品团队 (IPT) 是一种有效的方法，IPT 成员包括设计人员和非倡导者，也就是独立的观察员。

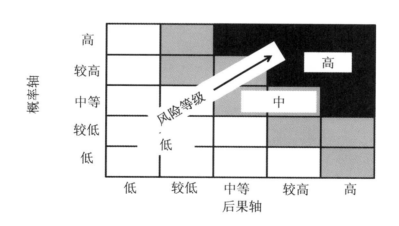

图 15.1　风险网格图

在确定将特定部件对应到哪个网络时，可以使用表 15.1 及表 15.2 中的模板。应尽可能定量地描述风险。例如，如果有记录表明某个供应商产品准时交付的概率只有 50%，则发生这种风险的概率应判断为中等。

Table 15.1   Likelihood template

Likelihood Level	Commonly Applied Probabilities	Description
Low	<1%	Desired results very unlikely to happen
Minor	<10%	Desired results have low likelihood
Moderate	~50%	Desired results likely to happen
Significant	>90%	Desired results highly likely to happen
High	>99%	Desired results near certainty of occurrence

Table 15.2   Consequence template

Consequence Level	Description
Low	Little or no impact
Minor	Problem that could easily be handled with available resources
Moderate	Problem that requires handling with additional resources
Significant	Significant change to program viability if not addressed
High	Problem that could result in program termination if not addressed

Develop risk handling plans

If steps are taken in advance, the likelihood or the level of the consequence happening or will be reduced. These steps are normally called *handling* methods.

The commonly recognized risk handling methods are as follows:

- Mitigation. A risk can be mitigated either by reducing the level of the possible consequence or by reducing its probability of occurrence.
- Acceptance. The developer can accept the risk without any action provided the probability or consequence is extremely low.
- Transfer. The risk can be transferred to another body, such as the customer, the supplier, or the regulatory agency, providing this can be done within the bounds of legal and contractual constraints.

However, as an example of risk transfer, if a customer demands a specific piece

表 15.1　概率模板

概率等级	通常应用的概率	描　述
低	<1%	预期的结果很不可能发生
较低	<10%	预期的结果发生概率较小
中等	~50%	预期的结果可能发生
较高	>90%	预期的结果很可能发生
高	>99%	预期的结果几乎肯定要发生

表 15.2　后果严重等级模板

后果严重等级	描　述
低	几乎没有或根本没有影响
较低	问题通过可用的资源很容易处理
中等	处理问题需要额外的资源
较高	如果不处理会对项目可行性产生重大的改变
高	如果不处理可能会导致项目终止

制定风险应对计划

如果事先采取措施，后果发生的概率或严重等级就会降低。这些措施通常被称为应对方法。常见的风险应对方法如下：

- **减轻**。通过降低可能后果的严重等级，或减小其发生概率来减轻风险。
- **接受**。研制方可以接受风险而不用采取任何行动，但前提是该风险的概率或后果等级非常低。
- **转移**。风险可以转移到另一主体，如顾客、供应商或监管当局，只要满足法律和合同要求。

of commercial off-the-shelf (COTS) equipment and does not provide evidence that the equipment in flight qualified, then the customer is accepting a *de facto* transfer of the risk. That is to say, if the equipment should fail, the customer is accepting the responsibility for any consequences that may result.

- Analysis. Strictly speaking one cannot handle risk by analysis. However, in the real world facts emerge that enable the analyst to reassess the risk from time to time. Of course, analysis is important as time goes on. Risk must be reassessed after each mitigation step is performed.

Track risks

Risk tracking can be accomplished by common unsophisticated computer applications. Risk tracking provides the data needed to make the decisions outlined in the risk reduction plan. A key aspect of risk tracking is to track only the moderate and high risks. This technique will keep the task manageable and effective.

Holistic Risk Management

Advanced risk thinkers have begun to understand that risk management is complex, that is, the many factors that cause risk may be interrelated. For readers interested in these advanced aspects, the book *Holistic Risk Management in Practice* by Hopkin (2002) is recommended. Among other aspects, holistic risk management integrates aspects of risk across multiple business units. As shown in Chapter 14, since the commercial aircraft system can be seen as a large-scale system (LSS) including the developer and many suppliers, this aspect is particularly relevant to the commercial aircraft domain. Although this book was intended for the financial industry, its principles still apply to commercial aircraft.

15.2   Types of Consequences

As Section 15.1 explained, there are three basic types of consequences:

- A typical schedule consequence is when the entire aircraft is delivered to the airline customer late, that is, later than the agree-to contract called for.
- A typical cost consequence is when the cost to build the aircraft is greater than

这是风险转移的一个例子：如果客户要求提供某一具体货架产品 (COTS) 设备，但未提供证据表明该产品适合飞行，那么，可以认为客户正在接受一种事实上的风险转移。也就是说，如果设备发生故障，该客户要对可能引起的任何后果负责。

- **分析**。严格来讲，人们不能通过分析来应对风险。然而，实际上，很多事实表明，分析人员对风险的重新评估一直在进行。当然，随着时间的推移，分析变得十分重要。采取减轻风险的措施后都必须重新评估风险。

风险追溯

风险追溯可以通过常用的简单计算机应用程序来实现。风险追溯为风险抑制计划提供了决策所需的数据。风险追溯的一个关键点是只追踪中高等级的风险。这种方法可以确保该任务是可管理和有效的。

整体风险管理

业界前沿的风险研究者已经开始认识到，风险管理十分复杂，也就是说导致风险的诸多因素可能相互关联。对于有兴趣了解前沿风险管理知识的读者，推荐参阅 Hopkin(2002) 所著《实践中的整体风险管理》。另外，整体风险管理集成了跨多个业务单元的风险方面。如第 14 章所述，由于商用飞机系统可以被看作是一个包括研制方及众多供应商的大规模系统 (LSS)，整体风险管理与商业飞机领域相关度很高。虽然该本书内容针对的是金融行业，但它介绍的原则也适用于商用飞机。

15.2 后果的类型

如 15.1 节所述，后果有三种基本类型：

anticipated.

- A typical technical consequence is when the aircraft does not deliver the performance that it was designed to deliver. For example, the range may not be as great as expected, or the weight is greater than expected.

These consequences may occur either at the aircraft level, as described in the above examples, or they may occur at the supplier level. For example, the supplier may deliver their product late, or it may fail the qualification test, a technical consequence.

In addition, these consequences may occur singly, or they may occur in combination. For example, if the aircraft is delivered late, there may be a cost penalty in the contract that the developer has to pay to the customer. If the aircraft does not meet its contractual performance, there may be a schedule penalty to solve the problem and a cost consequence to pay for solving the problem.

15.3   Root Causes of Risks

Risks may be external, that is, out of the control of the developer. Or they may be internal, that is, the result of decisions made within the developer's organization. It is impossible to list all of the risks that a commercial aircraft enterprise may encounter, but the following are typical.

External causes

- The basic top-level external risk is market risk. Does the customer want more range, more seats, or speed? Or has the market dried up altogether? The inability to predict these market factors can be critical.
- There are unexpected technological risks. In spite of the latest ability to establish durability requirements, cracks due to metal fatigue or composite material delamination may start appearing the fuselage. Once built these types of flaws are difficult to fix. Cost, schedule, and performance will suffer.
- Supplier risks are the most difficult. Suppliers with a spotless reputation may deliver a product that is lacking in performance or late in delivery.

Internal causes

Internally caused risks are generally the result of optimistic decisions about

- 一种典型的进度后果是，整架飞机延迟交付给航空公司客户，也就是交付时间比合同规定的时间要晚。

- 一种典型的成本后果是飞机制造成本超出预期。

- 一种典型的技术后果是，飞机交付时没有达到设计的性能要求。例如，航程小于预期，或重量大于预期。

这些后果既可能出现在飞机级，如在上述例子所述，也可能出现在供应商级。例如，供应商可能推迟交付其产品，或鉴定试验失败，即技术后果。

此外，这些后果可能会单独产生或混合产生。例如，如果飞机推迟交付，研制方有可能必须根据合同支付罚金给客户。如果飞机不满足合同的性能，可能要付出解决这个问题的进度代价和成本后果。

15.3 风险的根本原因

风险可能来自组织外部，也就是不受研制方的控制。风险也可能来自内部，就是说风险是研制方组织内部决策的结果。列出商用飞机公司可能会遇到的所有风险是不现实的，但下面列出的都属于典型风险。

外部原因

- 最基本的顶层外部风险是市场风险。客户是否希望有更大的航程、更多的座位或更快的速度？或者市场的需求已经萎缩？市场因素的不可预测性至关重要。

- 存在意想不到的技术风险。尽管建立耐久性需求是最晚考虑的能力，但因金属疲劳或复合材料脱层造成的裂纹可能开始在机身出现。一旦这种类型的缺陷产生，解决难度将会较大，成本、进度及性能都将受到影响。

- 供应商风险是最困难的部分。即使声誉很好的供应商也可能会交付性能

anything, for example, the time to develop and the capability of suppliers.

- Excessive outsourcing. Chapter 14 is devoted to the risks associated with excessive outsourcing and failure to manage the variability in contracts and specifications.
- Schedule optimism. Risks occur as a result of optimism pertaining to the development of, for example, internal components, such as electronic modules. A model that is estimated to take, for example, two years may wind up taking three years. The primary consequences are both cost and schedule.
- Lack of technical expertise. The developer may embark on a new technology, such as composite structures, without a complete understanding of some of the issues associated with it, such as delamination.
- Supplier evaluation. Risks may occur as a result of the inadequate evaluation of a supplier. A supplier may be contracted to develop a new technology that has never been built before. All three: cost, schedule, and performance risks will result.
- Flawed assumptions. Risks can occur when suppliers are given obsolete specifications to build their products to. The flawed assumption is that the ten or 15-year-old specification is still valid and accurate. This risk may occur during *modification* processes when obsolete modules are replaced with later versions.
- Technical optimism. Technical risks may appear when optimistic estimates are made about performance. A common example of this is the estimate of aircraft weight. Although efforts are made to reduce weight as time goes on, overweight on first delivery can be costly to the developer.
- Lack of internal analytical capability. This internal deficiency can result from the lack of analytical capability. For example, the developer may not have the capability to conduct the analysis of complex aerodynamics using a computational fluid dynamics (CFD) simulation or a structural analysis using a finite element analysis (FEA). Without this capability the developer may not be able to predict problems in these areas. Without structural analysis the developer may not be able to predict the physical interference between structural elements, such as control surfaces and the wing. These elements may flex substantially during flight depending on the speed, altitude, and amount of fuel on board.

不达标的产品或延迟交付。

内部原因

内部原因导致的风险一般是由对任何事物，例如研制时间及供应商能力的盲目乐观导致的。

- 过多的外包。第 14 章阐述了过度外包及合同和规范中的变异性管理失效带来的风险。

- 进度计划乐观。对电子模块等内部部件研制的乐观估计可能会产生风险。例如，一个预计两年内完成的模型可能实际需要三年。主要的后果包括成本和进度方面。

- 缺乏专业技术知识。研制方可能在没有完全了解相关细节的情况下选用某一新技术，例如，选用复合材料时对脱层现象不够了解。

- 供应商评估。对供应商的评估不足可能会导致风险，例如，供应商可能被合同要求开发以前从未有过的新技术。成本，进度及性能这三方面均会导致风险产生。

- 有缺陷的假设。当供应商用过时的规范来制造产品时就会产生风险。有缺陷的假设的一个例子是 ：10 年或 15 年前的规范仍然是有效和准确的。在过时的模块替换为更高版本的变更过程中，可能会产生风险。

- 技术上的乐观。当对性能做出乐观估计时，可能会出现技术风险。一个常见的例子是飞机重量的估计。虽然随着时间的推移，研制方面在减重上做了很多努力，首次交付时的超重还是会给研制方带来昂贵的后果。

- 缺少内部分析能力。这种内部缺陷可能是由分析能力缺乏导致。例如，研制方可能缺乏利用计算流体力学 (CFD) 进行复杂的空气动力学模拟，或使用有限元分析 (FEA) 进行结构分析的能力。如果没有这种能力，研制方可能无法预测这些领域中可能出现的问题。没有结构分析能力的研

- Lack of production quality. Production quality tends to fall due to the turnover of production workers. This deficiency can occur when one aircraft production comes to an end and workers are transferred to another aircraft. In these cases production quality can only be maintained if the new workers are given training and detailed instructions. A worker, for example, may place two electrical cables too close to one another causing electromagnetic interference (EMI) problems.
- Operator actions. The aircraft operator, that is, the airline company, may take actions that will result in risks. For example the airline company may deviate from the maintenance manual produced by the developer. This was the case with the 1979 Chicago O'Hare crash as explained by Reason (1997, p. 88).
- Lack of rigor. Failure to follow processes with rigor is a source of risk itself. This book is a catalogue of such processes that may be the source of a risk if not followed rigorously. A simple failure to specify the environment of a supplier product may lead to catastrophic consequences.
- The airline company may acquire components that are not flight qualified, that is, they do not meet the requirements of the components that were specified by the developer. They can either purchase these components, or they can lease them as explained in Section 2.6.

Warning signs

One of the most important aspects of risk management is the recognition of early warning signs. In the interest of getting the aircraft built and delivered fast, these warning signs are often ignored with predictable consequences.

- One of the most visible warning signs is test failures. These failures can happen months in advance of delivery to the developer. These failures are both a call to action and a sign of a high-risk product.
- Another warning sign is doing things that have been done before with negative consequences. The overweight aircraft is a good example. This is a warning sign that should be addressed during early advanced design.
- The next category can only be called a logical warning sign. If the electronic experts say that the module will take three years to develop, but you schedule a two-year development, then you know you have a risk.

制方可能无法预测结构件干扰，如控制面和机翼之间的物理干涉。在飞行过程中，根据速度、高度和机上燃油量，这些结构件可能会产生大幅度的弯曲。

- 生产质量不佳。由于生产工人的失误，生产质量会趋于下降。当一种飞机的生产即将结束，并且工人将转移到另一种飞机的生产时，可能发生这种缺陷。这种情况下，只有对新员工进行培训，并提供详细说明才能维持生产质量。例如，工人可能将两条电缆靠得太近，以至于造成电磁干扰(EMI)问题。

- 运营商的行为。飞机运营商，即航空公司，可能会存在导致风险的行动。例如，航空公司可能会偏离研制方编写的维修手册。1979年的芝加哥奥黑尔机场坠机的原因被 Reason(1997, 第88页) 归为此类情况。

- 严苛度不足。不严格按照流程进行工作是一个风险源。如果不按照本书的目录严格执行可能就是风险源。没有规定供应商产品的环境这样一个简单的失误就可能导致灾难性的后果。

- 航空公司拿到的部件可能不适宜飞行使用，也就是说，这些部件不符合研制方规定的部件需求。它们可以购买这些部件，或者以2.6节所述租赁。

预警信号

风险管理最重要的一个方面是识别早期预警信号。为了使飞机制造和交付速度更快，这些警告信号往往被忽视，造成可预测的后果。

- 一个最明显的预警信号是试验失败。这些失败可能在交付给研制方之前几个月发生。这些故障既需要应对，同时也是高风险产品的标志。

- 另一个预警信号是做那些以前做过，但有消极后果的事情。飞机超重是一个很好的例子。这是应该在早期设计中要解决的预警信号。

- 下一类只能被称为逻辑性预警信号。如果电气专家认为该模块需要三年

- The next category is just supplier quality. If a supplier has a record of late delivery and poor performance, and you plan to use this supplier again, then there is a risk. If, on the other hand, you are forced to use this supplier because, for example, no one else makes the product, or you are compelled to use this supplier for other reasons, then there is a risk. In this case, this becomes an external root cause. Either way it is a risk.

Unknown unknowns

The question often arises: what if you don't know the source of a risk? A serious consequence occurs, and there were no warning signs. Such was the case with TWA 800. In 1996 the fuel tank of a 747 exploded killing all 230 people on board. This accident is discussed by Jackson (2010, pp. 68–69). The flammability of aircraft fuel has always been well-known. However, identifying all ignition sources had been difficult. In 1996 an explosion occurred as a result of a worn electrical wire that came into contact with the vapor in the fuel. So the question is: how can explosions be prevented regardless of the ignition source (one of the unknown unknowns)?

The Federal Aviation Administration (FAA) (2008) addressed this problem by mandating that the vapor area in the fuel tank be filled with either nitrogen enriched air (NEA) or foam. This was a mitigation step that was independent of the source of the risk or the probability of its occurrence. The lesson of this example is that if a risk can be mitigated that is independent of the source that method should be employed or at least be considered a high-priority solution.

Off-the-table risks

Developers are sometimes reluctant even to record risks of certain types. Here are a few examples:

- If the risk source is a customer generated, the developer may not want to record the risk. For example, if the customer has a poor record of safety, maintenance, or pilot training the risk may not be recorded. Among options for handling this risk, the developer may want to discuss these items privately with the customer.
- If the risk is considered a *normal* problem, that is something that occurs all the time, the developer may consider it not worth recording. For example, if a supplier has a bad performance record, rather than recording it as a risk, the

时间研制，但实际只安排了两年的研制时间，那么这其中就存在风险。

- 下一类别就是供应商的质量。如果某供应商有延期交货和表现不佳的记录，但你还打算再次使用该供应商，那就会存在风险。另一方面，如果由于某些原因，比如没有任何其他供应商生产这种产品，或其他原因而不得不使用该供应商，那么就会存在风险。在这种情况下，这将成为一个外部根本原因，无论采用哪种方式，风险都会存在。

未知的未知

以下问题经常出现：如果你不知道风险源会怎么样？那样可能会产生严重的后果，并且没有任何预警信号。TWA800 空难就属于这种情况。1996 年，一架 747 飞机油箱爆炸导致机上 230 人全部遇难，Jackson (2010, 第 68 至 69 页) 中讨论这起事故。飞机燃油的易燃性是众所周知的，然而，识别所有点火源十分困难。1996 年发生的爆炸是磨损的电线接触燃油蒸气导致的。所以，问题是：如何才能预防爆炸，而不管点火源是什么 (未知的未知之一)？

联邦航空局 (FAA)(2008 年) 强制要求在燃油箱蒸气区域填充富氮空气 (NEA) 或泡沫，以解决这个问题。这是一种减轻风险的措施，该措施与风险源及其发生的概率无关。这个例子给我们的启示是，如果能够采取减轻风险的措施，且该措施与风险源无关，那么应该采用这种方法或至少将其作为一个高优先级的解决方案。

不公开的风险

研制方有时甚至不愿意记录某些类型的风险。下面是该类风险的几个例子：

- 如果风险源是客户产生的，研制方可能不愿记录这种风险。例如，如果客户有安全、维修或驾驶员培训方面的不良记录，则可能不记录这种风险。在处理这种风险的选项中，研制方可能希望与客户私下讨论这些问题。

- 如果风险被认为是正常的问题，即经常发生的事情，研制方可能认为该

developer may chose to deal with this problem in the way that it is usually handled. However, if the usual way is not satisfactory, there is no reason for ignoring it as a risk.

- If the risk is considered *unsolvable* or the cost is great, the developer may not record it. For example, if the solution requires reversing a major design decision, this risk may go by the wayside. On the other hand, if the risk is recognized, then the various ways of handling it can be considered.

The lesson learned from the above examples is that risks should be recorded and treated whether they are considered normal or unsolvable.

15.4   Risk Mitigation Steps

Mitigation steps are the actions taken before a consequence is realized to lessen the magnitude of the consequence or the probability of occurrence. Mitigation steps will depend on the type of risk being mitigated. Most mitigation steps will cost money, but often the cost is worth the amount.

The simple test for how much money is too much is that the cost should be less than the cost of the expected consequence. If the cost of mitigation is more than the cost of the expected consequence, then the appropriate management method is risk acceptance.

However, here are a few examples of mitigation steps:

- Schedule slip. The obvious mitigation step for a schedule risk is to slip the schedule and suffer any costs associated with it. A tempting mitigation step is to increase the work load to solve the problem and get it done on time. However, this method is more likely to increase risk rather than mitigate it.
- Redesign. This can often be the most painful mitigation step. However, if it is clear that the current design will not meet the requirements, then there may be little choice. Metal fatigue is one example. Replacing the entire metal of the fuselage will be very expensive and will also increase the weight. Often the use of stiffeners will do the job.
- Change suppliers. If caught early enough, this risk can be mitigated will little cost and schedule impact. However, if the contract is already signed, this step can wind up costing money.

类风险不值得记录。例如，如果供应商的产品性能较差，研制方可能会选择把它当作普通问题进行处理，而不是作为风险进行记录。但是，如果通常的处理方式不能令人满意，那就不应该把它忽略而不当成风险。

- 如果风险被认为是无法解决或者需要的成本很高，研制方可能不会记录下来。例如，如果解决方案需要改变重大的设计决定，这种风险可能会被搁置。另一方面，如果这种风险已被识别，则可考虑多种处理方法。

从上面的例子中可以看出，不管风险被认为是正常的，还是无法解决的，都应该进行记录和应对。

15.4　减轻风险的措施

减轻风险的措施是减轻后果严重性或降低发生概率所采取的行动。减轻措施取决于待减轻风险的类型。大多数减轻都会增加成本，但这些成本往往是值得的。

关于成本是否过高，有一个简便的方法判断，即成本应低于预期后果的成本。如果减轻风险的成本高于预期后果的成本，那么最有利的管理方式是接受风险。

不过，这里有几个减轻风险的例子

- 调整进度。减轻进度风险最常见的措施是把进度推后，并接受与之相关的任何成本。最有利的减轻措施是增加工作量来解决问题，并要求按时完成。然而，这种方法更有可能增加风险，而不是减轻风险。

- 重新设计。这往往是最艰难的减轻风险措施。但是，如果目前的设计显然不能满足需求，那么别无选择。以金属疲劳为例，更换机身的全部金属将耗费巨大，也将导致重量增加。通常情况下，该问题可以通过使用加强筋来解决。

- 更换供应商。如果及早发现，便可减轻这种风险，并且对成本及进度影响较小。但是，如果合同已经签订，则没有必要采取该措施，否则只有

15.5   Issues

Consequences that have already been realized are not risks: they are called issues. For example, if cracks appear in the fuselage as a result of metal fatigue, this is an issue. The mitigation steps listed above will still have to be performed; the difference is that they are no longer optional; they have to be done.

Issues also exist if a consequence is inevitable. Consequences are generally only inevitable when the circumstances dictate a solution that is not optional. For example, an airline customer can dictate a solution that the developer knows is not feasible without consequences. The developer must take action to mitigate the consequences.

There is a common belief that all risks for which the likelihood is 1.0 are issues. This is not always the case. It is only the case for which the consequence will occur if there is no mitigation step to prevent it. Take the case of the component development schedule that was badly estimated. The consequence will only occur if the schedule is not changed. So the schedule change becomes the mitigation step.

15.6   Independent Review

One of the risks within the risk process itself is the fact that the risk may not be impartially identified, analyzed, or managed. There is an abundance of literature that supports this assertion. Vaughn, for example, analyses the NASA culture leading up to the *Challenger* disaster. Her conclusion was that risk was "normative," that is to say risks were just considered normal and were not treated with any special attention. Similarly, the Columbia Accident Investigation Report (2003, p. 189) concluded that NASA had a "broken safety culture." This book is not going to provide a formula on how to change the culture of an organization. Many methods have been proposed on how to do that. Jackson (2010, pp. 106–112) provides a summary of frequently suggested methods.

An independent authority

Rather, we will focus on a more agreed-to approach, the appointment of an independent authority to review risks, design decisions, and any other issues that may arise. The Columbia Accident Investigation Report (2003, p. 227) recommended an Independent Technical Authority. Similarly, following the 2006 crash of the UK aircraft

浪费成本。

15.5 问题

已经变成现实的后果不属于风险：它们属于问题。例如，如果机身因金属疲劳而出现裂纹，那这就是一个问题。上面列出的减轻措施仍然必须执行，不同的是，它们不再是可选项，而是必须要做的。

如果一个后果是不可避免的，那么这也是问题。当状况只有一个不容选择的解决方案时，后果一般不可避免。例如，航空公司客户可能决定一个解决方案，该方案在研制方看来毫无作用且并不可行，那么研制方必须采取行动，以减轻后果。

普遍共识是：概率为 1.0 的风险称之为问题。但并非总是如此。它只是如果没有采取任何减轻措施以防止其发生，其后果将发生的情况。以估计不足的部件研制进度为例。如果不改变进度，将会导致后果的发生，因此，改变进度成为减轻风险的措施。

15.6 独立评审

风险过程本身的一个风险是，风险可能未被客观地识别、分析和管理。有很多文献支持这一观点。例如，Vaughn 分析认为，是 NASA 的文化导致了挑战者号的灾难。她的结论是，风险是"常态化的"，即风险被认为是正常的，并没有予以特别对待。同样，《哥伦比亚号事故调查报告》(2003, 第 189 页) 的结论是，NASA 存在一种"破碎的安全性文化。"本书无意于提供改变组织文化的公式。关于如何做到这一点的方法很多，Jackson (2010, 第 106 至 112 页) 对此类常用方法进行了总结。

the *Nimrod*, a derivative of the *Comet* aircraft, the Haddon-Cave report (2009, p. 393–495) recommended a rigorous airworthiness process. Haddon-Cave himself cited the Columbia recommendation as a model for independent review.

Independence

So having raised the topic of independent review, we have to ask, what is independence? In simple terms,

> Independence is the absence of any financial or organizational links between the reviewer and the project being reviewed.

So having defined independence, we have to ask whether independence is even possible. After all, we all work for the same government or the same enterprise. Nevertheless, some practical guidelines are possible to establish:

- An independent reviewer cannot report to the same program manager as the project being reviewed.
- Preferred reviewers report to a completely different program or if possible a completely different enterprise.
- Outside institutions, such as universities, are sometimes sources of qualified reviewers.

Regulatory agencies

Regulatory agencies, such as the FAA, are ideal independent bodies to a certain extent. However, they are generally limited to the review of safety issues. If a commercial aircraft company makes an unwise marketing decision, the regulatory agency has little authority to speak to this issue. Hence, the onus is on the company to implement the independent review process.

500

As discussed in Section 10.4, the Commercial Aviation Safety Team (CAST) of which the FAA is a member has recommended review by "third parties" of pilot training. This recommendation is based on the observation that pilots sometimes fail to execute their tasks in the manner they were trained.

独立的管理机构

相反，我们更看重更具共识性的方法：指定一个独立的管理机构来评审风险、设计决策及可能产生的任何其他问题。《哥伦比亚号事故调查报告》(2003, 第 227 页) 推荐使用**独立技术机构**。类似地，2006 年英国飞机 *Nimrod*(彗星飞机的衍生机型) 坠毁后，Haddon-Cave 报告 (2009, 第 393 至 495 页) 建议确定一套严格的适航流程。Haddon-Cave (2009, 第 393 至 495 页) 认为哥伦比亚号建议是独立评审的一个典范。

独立性

既然提出了独立评审的话题，那么问题是，什么是独立性？简单来说，

独立性是指评审者与被评审项目没有任何财务或组织上的联系。

有了独立性的定义，问题是独立性是否真有可能实现。毕竟，我们都为同一个政府或同一家企业工作。

然而，有一些实际可行的准则是可行的：

- 独立评审者不能向被评项项目的同一个项目经理汇报。
- 首选的评审者向完全不同的项目，或者如果可能向完全不同的企业报告。
- 外部机构，如大学，有时是高质量评审者的来源。

监管当局

监管当局，如美国联邦航空局，在一定程度上是理想的独立实体。然而，它们的工作一般仅限于安全性问题的审查。如果商用飞机公司做出了不明智的市场决策，监管当局几乎无权干涉。因此，实施独立评审过程是公司的责任。

如 10.4 节所述，FAA 所属的**商用航空安全小组** (CAST) 建议驾驶员训练由 "第

Creating the independent review process is no easy task. There are manifold obstacles. Yet it is of paramount importance. The main obstacle is that no program manager wants to yield any authority to an outside person. Hence, this authority must be yielded *voluntarily*. At a very minimum the program manager must be willing to accept the advice of an outside reviewer.

Sometimes the independent review may have some pretty serious and costly advice. For example, they might recommend more testing or even a major change in design or supplier. This book suggests the implementation of the independent review process may be the most important and the most difficult to implement. Yet the failure to implement it may have serious consequences.

15.7   The Risk Management Process

In the commercial aircraft domain it is unlikely that there would be a dedicated risk management organization. Risk management, as a process, would focus on the SE lead who would report directly to the program manager. Some of the responsibilities of the risk management lead are as follows:

- Create a risk management plan that would lay out how risk management is to be executed and what the roles and responsibilities are for the participants. This plan would explain the independent review process described above.
- The risk plan above would explain who the risk owner is, that is, the person responsible for explaining how each risk would be handled.
- Maintain a risk register showing the status of each risk and the handling method being employed.

15.8   Risk Management Tools

Many, but not all, commercial aircraft companies have risk management tools. Even those that do sometimes fail to use them, even for the most egregious risks. Some tools are internally created, and others are commercially available. It is not the purpose here to list or evaluate those commercially available tools.

First, having a risk management tool is not the most important aspect of the risk management process. It is a convenient way to record the risks and make them

三方"进行评审。该建议是基于驾驶员有时无法按照其受训的方式执行其任务这一观点提出的。

创建独立评审过程

创建独立评审过程不是一件容易的事，存在多方面的障碍，但其意义重大。主要的障碍是，没有任何一位项目经理愿意把任何权力授予外部人员。因此，该机构必须自愿产生，最起码，事情是项目经理必须愿意接受外部评审者的意见。

有时，独立评审可能会产生一些非常严重和成本昂贵的建议。例如，他们可能会建议进行更多的试验，甚至建议在设计或供应商方面做较大的更改。本书认为独立评审过程的实现可能是最重要的，但也最难以实施的。然而，不实现独立评审可能会导致严重的后果。

15.7 风险管理过程

在商用飞机领域，设立一个专门的风险管理组织不太可能。风险管理作为一个过程，可能关注直接向项目经理汇报的系统工程领导者。风险管理领导者的一些主要职责如下：

- 制定风险管理计划，明确如何执行风险管理，以及参与者的角色及职责。该计划将对上述独立评审过程进行解释。
- 上述风险计划将说明谁是风险所有者，也就是谁负责解释如何应对每个风险。
- 维护风险登记册，记录每个风险的状态及采用的应对方法。

15.8 风险管理工具

很多，但并非全部商用飞机公司都采用风险管理工具。即使采用风险管理工具的公司有时也不能很好地使用，甚至对于最严重的风险也是如此。工具中部

available for later review. Risks can be recorded using standard word processing or spreadsheet applications. Risk diagrams can be, and often are, created using standard chart creation applications, such as PowerPoint. These methods can be somewhat labor intensive, but if nothing else is available, it can be done.

Some risk management tool features

Here is a short list of some of the desirable features of risk management tools. Not all existing risk management tools have all these features. However, given the risk management process described above, these features would seem essential.

- The capability of recording the root cause of a particular risk. This feature is sometimes called the source. The warning signs can be considered a part of this feature.
- The tool should be capable of recording all possible consequences of a particular risk. The three major categories are technical, cost, and schedule. These are the consequences that would occur *if nothing is done to prevent them from happening*.
- The tool should be capable of recording the likelihood that the risk will occur if nothing is done to prevent it. This likelihood will be the same for all three consequences since all the consequences apply to the same risk. In the ideal case these likelihoods can be estimated using the warning signs described above. For example, if a supplier has a history of delivering their product late 50 percent of the time, then the likelihood is 0.5.
- The tool should be capable of calculating a risk level for each of the risk categories.

15.9   Opportunities

Opportunities are the counterpart to risks. Opportunities are the potential benefits that a program may realize if the consequences are better than planned for. For example:

- wheel balancing tests may reach their desired goal in fewer than planned for tests;
- the aircraft may actually weigh less than the design weight.

分是内部开发的，其他则从市场上购买。这里的目的不是列出或评估那些市场上可买得到的工具。

首先，拥有风险管理工具不是风险管理过程最重要的方面。它是记录风险，并用于后续评审的一种便捷途径。风险可以使用标准文字处理和电子表格应用程序进行记录。利用标准的图表应用软件，如 PowerPoint，便能生成，而且经常用于生成风险图。这些方法有些可能属于比较耗费精力的方法，但如果没有其他方法，那么该方法也是可行的。

一些风险管理工具的特征

下面是风险管理工具理想特征的简单列表。并非所有现有的风险管理工具都具有所有这些特征。然而，对于上述的风险管理过程，这些特征似乎是必不可少的。

- 记录特定风险根本原因的能力。此功能有时称为源。预警信号可以被认为是该特征的一部分。
- 该工具应能记录特定风险所有可能的后果。风险三种主要类别是技术、成本和进度。它们是如果不采取任何措施加以预防时，将会出现的后果。
- 该工具应该能够记录不采取任何措施时风险发生的概率。对于所有三种后果，由于它们都对应于相同风险，因此其发生概率也将是相同的。理想情况下，利用上述的预警信号可以估算其概率。例如，如果某一供应商以往交付产品有 50% 是发生延误的，那么这个概率就是 0.5。
- 工具应能计算每类风险的风险等级。

15.9 机会

机会是风险的对立面。机会是一个项目可以实现其结果好于预期的潜在利益。例如：

- 机轮动平衡试验时，可以用少于计划次数的试验就达到其预期目标；

So the question is: How will be aircraft developer take advantage of these opportunities? There may be cost or schedule benefits to be realized as a result of the uncertainties in the system. These uncertainties are the same uncertainties that might have resulted in unfavorable consequences in other circumstances. But in the alternative situations they may yield opportunities in cost or schedule, for example.

The risk process should track these opportunities as well as risks. In practice the opportunities will be less numerous than the risks. However, the aircraft developer should not fail to take advantage of them.

15.10   Challenges for Risk Management

The MIT-PMI-INCOSE Guide for Lean Enablers (2012, p. 31) provides a list of challenges for the implementation of risk management. They are as follows; these items are paraphrased for brevity:

- insufficient involvement of professionals in risk management;
- lack of understanding of program risks;
- insufficient resources and funding for risk management;
- neglect of human aspect of risk management. This item reflects the cultural barriers to risk management as reflected above. Among these is the failure of management to defer to expertise on risk and to accept the risks that have been identified, without modification or reduction;
- disconnect between risk management and other program management processes;
- failure to resolve risks quickly.

This is a concise list of challenges that management needs to address the risks that are common on commercial aircraft.

- 飞机实际重量小于设计重量。

所以问题是：飞机研制方如何利用这些机会？有可能由于系统中存在不确定性导致成本或进度优势被挖掘出来。同样的这些不确定性在其他情况下可能导致不利后果，但在其他一些情况下，它们又可能会产生机会，例如成本及进度方面。

风险过程应该跟踪这些机会与风险。实际上机会的数量会比风险少很多。不过，飞机研制方不应该放弃这些优势。

15.10 风险管理中的挑战

《MIT-PMI-INCOSE 精益使能指南》(2012, 第 31 页) 介绍了一系列实施风险管理时的挑战的列表，如下所示。为简洁起见，这些条目均为直接转述：

- 风险管理专业人员参与不足；

- 缺乏对项目风险的认识；

- 风险管理的资源和资金不足；

- 风险管理中人为疏忽方面。这一项体现了上述的风险管理中的文化障碍。其中之一是管理层的失误拖延成专业技术风险，以及不做修改和裁剪就接受已识别的风险；

- 风险管理与项目管理其他过程之间的脱节；

- 未能及时化解风险。

这是一份关于挑战的简单列表，这些挑战都是管理者处理商用飞机常见风险时需要面对的。

16
Resilience of the Aircraft System

A topic of increasing interest in SE is resilience. Both the *Systems Engineering Body of Knowledge* (SEBoK) edited by Pyster et al. (2012) and the *International Council on Systems Engineering (INCOSE) Handbook* (2006) have sections devoted to resilience.

Resilience differs from safety in that safety endeavors to prevent the failure of a system, in this case an aircraft. Resilience goes beyond safety in that it calls for mechanisms to anticipate failure and to enable the system to recover from a major disruption, such as human error or an encounter with an external threat such as a flock of birds.

In this chapter we will discuss the basic principles of resilience and show how they can be applied to the design of an aircraft.

16.1   The History of Resilience

The history of resilience begins with the basic definition of resilience and shows how this discipline has evolved into its present meaning as a field within the context of SE and as applied to commercial aircraft.

The classical definition of resilience from the *Oxford English Dictionary* (OED) (1973, p. 1807) is that resilience is "the act of rebounding or springing back." This definition described the inherent property of an entity, for example a spring, a person, or an ecosystem, to recover from a disruption.

In more recent years the emphasis on resilience has shifted to systems which can be *engineered* to bounce back, that is, some degree of human intervention is required to enable the system to bounce back. Since these systems include humans, the term *engineer* has to be interpreted in a broad way and not to just classical engineering which is a branch of the physical sciences. Resilience pioneers Hollnagel and Woods (2006) and Hollnagel et al (2011) have coined the term *resilience engineering* and have published several books on the subject. The second book has considerable emphasis on the resilience of commercial aircraft.

第16章 飞机系统的恢复力

恢复力是系统工程领域中一个日益受到关注的主题。由 Pyster 等编辑的《系统工程知识体系》(SEBoK, 2012)及《国际系统工程协会(INCOSE)手册》(2006)都有专门章节论述恢复力。

恢复力与安全性区别在于，安全性是努力防止系统的失效，这里的"系统"指的是飞机。恢复力的内涵超过安全性，它要求飞机有预测故障，并使系统从重大扰动（如出现人为错误或遭遇外部威胁，如鸟群等）中恢复的机制。

在本章中，我们将讨论恢复力的基本准则，并介绍如何将它们应用到飞机设计中。

16.1 恢复力发展历程

恢复力的发展历程需要从其基本定义说起，并介绍该学科如何演变为作为系统工程中的一个领域的当前意涵，以及如何将其应用于商用飞机。

《牛津英语词典》(OED)对恢复力的经典定义(1973, 第1807页)是"反弹或弹回的行为"。该定义描述一个实体的固有属性，例如弹簧、人或生态系统能从扰动中恢复的属性。

近年来，恢复力的研究重点已经转移到能够在工程上实现反弹的系统，即，需要某种程度的人为干预以使系统反弹。由于这些系统包括人，术语"工程师"必须广义地理解，而不能仅仅以属于物理学分支的经典工程学方式理解。恢复力理论的开创者 Hollnagel、Woods(2006)及 Hollnagel 等人(2011)创造了术语"恢

Another step in the evolution of resilience was the appearance of papers and books with an emphasis on whole systems, that is, systems that consist of humans and technological elements as well. Hollnagel and Woods place considerable emphasis on organizational systems, while later researchers, such as Haimes (2009), consider the resilience of the entire system including technological elements as well as the humans. This book has adopted the whole-systems perspective since aircraft have many technological elements that play a major role in resilience.

16.2   The Definition of Resilience

It is possible to find hundreds of definitions of resilience. However, it not necessary to review all of them since they tend to have a great deal in common. The definition by Haimes is one of the few to appear in peer-reviewed journals. It is as follows:

> Resilience is the ability of the system to withstand a major disruption within acceptable degradation parameters and to recover within an acceptable time and composite costs and risks. (Haimes 2009, p. 498)

This definition covers a great deal more territory than just bouncing back. It even includes some aspects of interest to systems engineers, such as cost and risk.

The disruption cycle

The definition of resilience is inextricably intertwined with the disruption diagram shown in Figure 16.1. This diagram depicts a system, for example an aircraft, which has encountered a threat, at a time called the *event*. The encounter of US Airways Flight 1549 as described by Pariès (2011, pp. 9–27) with a flock of geese is such an event. Before the event the aircraft will be at an initial state in a normal operating mode. Following the event the aircraft will enter a recovery state in which it is attempting to recover from the damage encountered from the threat. The final state is either a full restoration of the functionality of the aircraft or some other acceptable state. In the case of Flight 1549 the final state was the ditching of the aircraft in the Hudson River where it floated long enough for the passengers and crew to be rescued. The diagram also allows for multiple or serial threats. For example, if a human error should occur during the attempt to recover the aircraft that would be considered another threat. In the case of Flight 1549 there was no such human error.

复力工程"，并已出版了几部关于该主题的著作，其中第二本书重点介绍了商用飞机的恢复力。

恢复力学科演进的另一迈进是强调全系统的论文及书籍的出现，这类系统包括人和技术元素。Hollnagel 和 Woods 重点强调组织系统，而后来的研究者，如 Haimes（2009），则是考虑包括技术元素及人在内的整个系统的恢复力。本书采用了全系统的角度，因为飞机有很多的技术元素，它们在恢复力方面发挥重要作用。

16.2 恢复力的定义

关于恢复力的定义数以百计。但是，没有必要时它们要逐一审视，因为它们往往有共通之处。Haimes 的定义是少数出现在同行评议期刊中的一个。它的表述如下：

恢复力是系统在可接受的参数性能下降的范围内承受重大扰动，并在可接受的时间、成本及风险范围内恢复的能力。(Haimes 2009, 第 498 页)

该定义涵盖了包含了反弹在内的更大领域。它甚至包括系统工程师关注的某些方面，如成本和风险。

扰动周期

恢复力的定义与图 16.1 所示的扰动图密切相关。该图描述了一个系统，例如飞机，遭遇到威胁时的情况，这些威胁被称为"事件"。根据 Pariès (2011, 第 9 至 27 页) 的描述，美国航空公司 1549 航班遭遇鹅群就是这样一个事件。事件发生之前，飞机处于正常运行模式这一初始状态。事件发生之后，飞机将进入恢复状态，试图从威胁中遇到的损伤中恢复。最终状态是功能完全恢复或其他一些可以接受的状态。1549 次航班的最终状态是飞机在哈得逊河上迫降，并漂浮足够长时间，使乘客和船员被救出。该图还考虑了多重或一系列的威胁。例如，如果

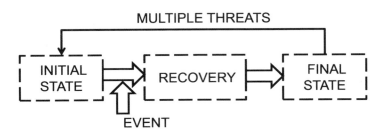

Figure 16.1   The resilience disruption cycle

Proactive vs. reactive perspectives

One issue that has dominated the discussion of definitions is whether resilience is proactive or reactive. Proactive resilience means that resilience covers the period of time before the encounter with a threat as shown as the initial state in Figure 16.1. Woods (2006, pp. 21–34) and Leveson et al. (2006, pp. 95–123) have adopted this perspective. In the proactive perspective more could be happening during the initial state of Figure 16.1; the pilot or the aircraft instrumentation could be detecting an approaching threat, for example, a flock of geese or a cloud of volcanic dust.

Reactive resilience means that resilience only applies during and after the encounter with the threat. Haimes (2009, pp. 498–501) adopted this perspective. Haimes refers to the period before the encounter as protection. The conclusion is that this distinction is one of definition and not of substance. Either perspective is valid providing analysts are consistent in their definitions, assumptions, and approach. This book has adopted the proactive perspective since this perspective is consistent with most of the literature consulted and is a simpler all-inclusive perspective.

16.3   Is Resilience Measureable?

As discussed by Jackson and Brtis (2015) whether resilience is measurable is the subject of debate. Haimes argues that it is not simply because there are too many dimensions to it; e.g., multiple threats, multiple failure modes and multiple recovery modes. However, it can be argued that while these complications make it hard to predict the resilience of a system, they do not preclude the identification of a metric, by which resilience can be gauged. This is an issue to be resolved in future research.

在试图恢复飞机原状时发生一种人为错误，这将被视为另一种威胁。在 1 549 次航班事例中，没有出现这样的人为错误。

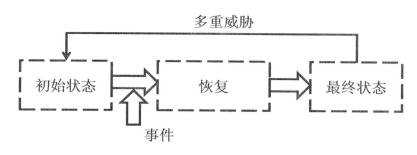

图 16.1 扰动恢复周期

主动 vs 被动

当前讨论最多的一个问题是恢复力是主动的还是被动的。主动恢复力意味着恢复力覆盖了遭遇威胁之前的时间，如图 16.1 的初始状态。Woods (2006, 第 21 至 34 页) 和 Leveson 等 (2006, 第 95 至 123 页) 都采用了这种观点。从主动的角度看，更多的事情可能在图 16.1 中的初始状态下发生；驾驶员或飞机仪器可以检测到临近的威胁，例如，一群鹅或一团火山灰。

被动恢复力是指恢复力应用于遭遇威胁期间或遭遇威胁之后。Haimes (2009, 第 498 至 501 页) 采用了这种观点。Haimes 称遭遇威胁之前的一个阶段为保护。总之，这种区别只是定义不同，而没有实质性的差异。只要分析人员在其定义、假设和方法上保持一致，无论哪种观点都是有效的。本书采用了主动的观点，因为该观点与大多数文献一致，并且是一个更简单的、兼收并蓄的观点。

16.3 恢复力可测量么？

正如 Jackson and Brtis（2015）所述，恢复力是否可测量尚无定论。Haimes 认为恢复力不可测量，因为恢复力存在太多维度，如多种威胁、多种失效模式及

16.4   Design Rules and Example Solutions

This section will first outline a set of design rules that have been collected from various sources and then provide some example solutions that have been posed in the literature, especially those solutions that pertain to the commercial aircraft domain. These design rules are, first *abstract*, that is, they are simplified replicas of concrete solutions. Since these rules are abstract they can first be applied to any domain where the solutions seem logical and practical. Secondly, there may be many concrete solutions that can be derived from a single abstract rule. Figure 16.2 provides a simplified schematic of how an abstract rule relates to a concrete solution. In this case the abstract rule of *functional redundancy*, called *design diversity* by Leveson (1995, pp. 433–437), can be implemented as two separate aircraft control systems. The feature that abstract rules and concrete solutions have in common is a common set of dominant characteristics. The dominant characteristics for *functional redundancy* are that the system must consist of at least two branches, that these branches must be physically different, and that the branches must be independent.

Jackson and Ferris (2013, pp. 152–164) have compiled a list of these rules, which they call principles, and some associated support rules. Support rules basically have the same goals as the primary rules except that they have a more limited scope. These rules are sometimes called principles (by the author), characteristics (by Woods (2006, pp. 21–34)), or heuristics (by Rechtin (1991, p. 18)).

Since these rules are abstract, there is no way of knowing which rules are better than other rules. The analyst will need to model the solutions to determine that. In other cases the best solution will seem obvious. As the previous paragraph stated, any abstract rule may be transformed into many concrete solutions. It is therefore impossible to show all possible concrete solutions. The concrete solutions to follow will be extracted from the literature or posed as hypothetical examples.

多种恢复模式。然而，虽然这些复杂情况使得难以预测系统的恢复力，找到一些衡量恢复力的指标并非毫无可能。这是今后的研究要解决的一个问题。

16.4　设计规则和示例解决方案

本节将首先对一组从不同来源收集到的设计规则进行概述，然后介绍一些已在文献中举例的解决方案，尤其是那些涉及商用飞机领域的解决方案。这些设计规则首先是抽象的，也就是说，它们是对具体解决方案的简化。由于这些规则是抽象的，它们可以应用到任何领域，只要这些解决方案在该领域是合乎逻辑和切实可行的。其次，一个抽象的规则可能会衍生出很多具体的解决方案。图 16.2 是关于抽象规则联系到具体解决方案的简化示意图。在这种情况下，被 Leveson (1995, 第 433 至 437 页) 称为设计多样性的功能冗余抽象规则可作为两个分开的飞机控制系统来实现。抽象规则和具体解决方案共同具有的特征是具有一组共同的主要特征。功能冗余的主要特征是，该系统必须至少包含两个分支，即这些分支必须是物理上不同的，且必须相互独立。

Jackson 和 Ferris (2013, 第 152 至 164 页) 编制了这些规则的清单，他们称之为准则，并编制了一些相关的支持规则。支持规则与主要规则目标相同，但它们的范围更有限。这些规则有时候被称为准则 (本书作者)、特征 [Woods (2006, 第 21 至 34 页)] 或启发 (Rechtin (1991, 第 18 页)。

由于这些规则是抽象的，无法确定哪些规则比其他规则更好。分析人员需要对解决方案进行建模，以确定优劣。在其他情况下，最好的解决方案显而易见。正如上段所述，任何抽象的规则均可以被转化为很多具体的解决方案。因此，不可能在这里展示所有可能的具体解决方案。下面的具体解决方案是从该文献中提取的，或者是假设的例子。

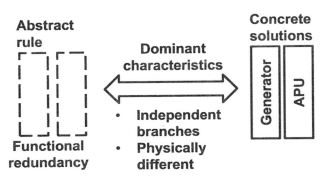

Figure 16.2   Abstract rules and concrete solutions

As we review these rules, though, it must be remembered that very rarely can the rules be implemented singly but must be implemented in combination with other rules. These combinations of rules result from the *interdependency* of the rules and the vulnerabilities of the individual rules. These interdependencies will be discussed within each section.

It is not the intent here to discuss all of the approximately 34 rules and support rules, only to provide examples for some of the more salient rules for the commercial aircraft domain.

Next, the word *design* should not be interpreted in the literal sense as in the design of hardware and software. It can imply the architectural arrangement of physical assets or of humans in the system. It can also imply the design of a system of procedures to be used by the humans in the system.

Finally, the reader should not interpret these rules to constitute a prescriptive process for designing a system. Rather, according to Hollnagel and Woods (2006, p. 348), "resilience engineering requires a constant monitoring of system performance, of how things are done."

The Absorption rule

Woods (2006, p. 23) refers to the *absorption* rule as the buffering characteristic. An obvious absorption capability of an aircraft is the design loads. One might ask: If an aircraft is designed with an adequate *absorption* capability, why would any other rules be required? The answer is that aircraft are designed knowing that abnormal conditions

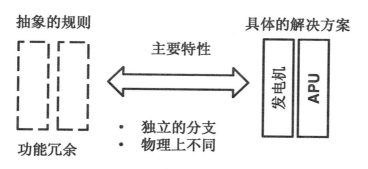

图 16.2　抽象规则与具体解决方案

　　在外面审视这些规则时，必须记住，很少有规则能单独实现，而是必须与其他规则组合在一起。这些规则的组合是由于规则的相互依赖性及单个规则的漏洞导致的。这种相互依赖性将在各节中讨论。

　　这里不打算讨论所有大约 34 条规则和支持规则，仅介绍一些商用飞机领域较为突出的规则作为示例。

　　另外，"设计"这个词不应按字面意思理解为硬件和软件的设计。它可以表示系统中实物或人的布置安排。它也可以表示系统中人所使用的程序系统的设计。

　　最后，读者不应把这些规则理解为构成系统设计的一个规定过程。相反，根据 Hollnagel 和 Woods(2006, 第 348 页) 的观点，"恢复力工程要求对系统的性能及行为方式进行持续监控。"

吸收规则

Woods (2006, 第 23 页) 把吸收规则称为缓冲特性。飞机中明显的吸收能力是设计载荷。有人可能会问：如果飞机的设计具有足够的吸收能力，为什么会需要其他规则？答案是，飞机在设计时就考虑到异常条件下飞机的载荷会超过设计载荷。另一个答案是，即使在正常条件下，实际载荷有时也会稍微超过设计载荷。

will occur that will exceed the design loads. Another answer is that even under normal conditions the design loads will be exceeded a small, but known percentage of the time.

Take the US Airways flight 1549 case discussed by Pariès (2011, pp. 9–27). All aircraft must meet stringent requirements for bird strike capability. The aircraft in question obviously met these requirements; however, on this occasion they were exceeded. To meet the resilience capabilities, the pilot was called into action using the *human in the loop* rule described below.

Another *absorption* capability is the lightning strike capability. The expected magnitude of lightning strike is 200,000 amps. This current is dispersed throughout the aircraft where it must pass through conduits without causing a conflagration. However, as the paragraph above stated, this load can be exceeded a small percentage of the time. To meet this requirement an adequate *margin* must be added to the design load.

The Limit Degradation Support rule

A pervasive threat to the *absorption* capability of an aircraft is the degradation of the capability due to poor maintenance or aging. This threat requires the invocation of the *limit degradation* support rule described by the Jackson and Ferris (2013, pp. 152–164).

Poor maintenance and aging are a constant threat to commercial aircraft. One of the more notable examples was the crash of the *Nimrod* aircraft which caught fire due to the failure of fuel seals as reported by the Royal Air Force Board of Inquiry (2007, p. 2–20).

The Margin Support rule

It is a common practice in almost all industries to add a margin to the absorption capability especially in the commercial aircraft domain. This margin allows for uncertainties in loads and in the capability of the aircraft. For most structural components the margin is 50 percent. Woods (2006, p. 23) lists margin a characteristic of resilience.

Dependency of the Absorption rule

As discussed above, most design rules should be employed in combination with

以 Pariès (2011, 第 9 至 27 页) 中讨论的美国航空公司 1549 航班为例。所有飞机必须满足鸟撞能力的严格要求，这架飞机显然满足了这些要求；然而，这一次，实际状况却超出了这些要求的范围。为了满足恢复能力，驾驶员被告知使用下面要描述的人在环规则来进行操作。

另一个吸收能力的例子是抗雷击能力。雷击的预计幅值为 20 万安培。该电流被分散在整架飞机，它必须通过导线管而不会引起火灾。然而，如上述的段落所述，可能在一段较短的时间内实际载荷超出这个载荷。为了满足这一要求，设计载荷必须有适当的余量。

限制退化支持规则

对飞机的吸收能力有普遍威胁的是由于维修不当或老化导致的性能下降。这种威胁要求调用 Jackson 和 Ferris (2013, 第 152 至 164 页) 中描述的限制退化支持规则。

维修不当和老化是对商用飞机的持续威胁。其中较为显著的例子是 Nimrod 飞机的起火坠毁，皇家空军调查委员会 (2007, 第 2 至 20 页) 报告认为该事故由燃油密封件失效导致。

余量支持规则

在吸收能力上增加一个余量，这是几乎所有行业的通行做法，特别是在商用飞机领域。该余量允许飞机载荷和能力存在不确定性。对于大多数结构部件，余量是 50%。Woods (2006, 第 23 页) 将余量列为恢复力的一个特性。

吸收规则的依赖性

如上文所述，大多数设计规则应结合其他规则一起使用，以保持有效性并弥补其漏洞。在吸收的情况下，这条规则依赖于限制退化和余量两条支持规则。

other rules in order to be effective and to compensate for its vulnerabilities. In the case of *absorption*, this rule is dependent on both the *limit degradation* and *margin* support rules. The author (2013) discusses other dependent rules the designer may want to consider.

The Physical Redundancy rule

Physical redundancy, also called *design redundancy* by Leveson (1995, pp. 433–437), is a recognized concept in engineering. The dominant characteristics of physical redundancy are that the system should have two physically identical and independent branches. However, Rijpma (1997, pp. 15–23) points out that *physical redundancy* has more vulnerabilities than is generally recognized. For example, if an aircraft has two identical software systems, and one of these has an undetected flaw, the other system will have the same flaw.

Some aircraft employ the two-load-path design philosophy. This is a form of *physical redundancy*. That is, if part of the structure, for example the skin, fails due to metal fatigue or any other reason, the rest of the structure will be able to sustain the loads that are incurred.

Dependency of the Physical Redundancy rule

To compensate for the vulnerabilities of the *physical redundancy* rule, the designer can invoke the *functional redundancy* rule discussed below or call on the services of a person employing the *human in the loop* rule.

The Functional Redundancy rule

The *functional redundancy* rule, called *design diversity* by Leveson (1995, pp. 433–437), avoids most of the vulnerabilities of *physical redundancy*. The dominant characteristics are that the system must have at least two physically different and independent branches.

An aircraft having both mechanical and fly-by-wire (FBW) control systems is an example of *functional redundancy*.

No dependent rules were identified for the *functional redundancy* rule.

笔者（2013）讨论了设计师可能要考虑的其他相关规则。

物理冗余规则

物理冗余，也被 Leveson (1995, 第 433 至 437 页) 称为设计冗余，是工程界公认的概念。物理冗余的主要特性是系统应该有两个物理上完全相同但相互独立的分支。然而，Rijpma (1997, 第 15 至 23 页) 指出物理冗余的漏洞比一般认为的更多。例如，如果一架飞机有两个完全相同的软件系统，并且其中的一个存在未检测到的缺陷，那么另一个系统将具有相同的缺陷。

有些飞机采用了双载荷路径的设计理念。这是物理冗余的一种形式。就是说，如果结构的某部分，比如蒙皮，由于金属疲劳或任何其他原因而失效，该结构的其余部分应能承受产生的载荷。

物理冗余规则的依赖性

为了弥补物理冗余规则的漏洞，设计人员可以援引下面讨论的功能冗余规则，或通过人员采用人在环规则来协助。

功能冗余规则

功能冗余规则，被 Leveson (1995, 第 433 至 437 页) 称为设计多样性，其作用是避免物理冗余的大部分漏洞。其最主要特性是系统必须至少有两条物理上不同且相互独立的分支。

功能冗余的一个例子是，一架飞机往往会有机械操纵和电传操纵 (FBW) 两套控制系统。

任何相互依赖的规则都不属于功能冗余规则。

The Layered Defense rule

Layered defense is a derived rule from Reason's (1997, p. 11) Swiss cheese model. This model states that for a failure to occur a disturbance has to penetrate a series of layers similar to layers of Swiss cheese. In this model each layer has *holes*, that is vulnerabilities which allow the disturbance to travel to the next layer. So the implication of the model is that when a system has more layers of defense, the more resilient the system will be. Pariès (2011, pp. 9–27) calls this rule "defense in depth." Like *reduce complexity* and *reduce hidden interactions*, *layered defense* is one of the key rules that enables the system to maintain a distance from its safety boundary, that is, the increased number of layers enables it to stay farther from its safety boundary.

US Airways Flight 1549 described by Pariès is an example of *layered defense*. As described by Pariès, this system had four layers of defense. Two of these layers failed and the other two enabled the aircraft to ditch successfully. The first layer is the corrective measures to rid the airport area of birds that may harm the aircraft. The second layer is, of course, the engines themselves and their certified capability to handle a bird strike to a given capacity. Obviously, this capacity was exceeded and the engines failed. The third layer was the aircraft itself. Pariès points out the various features of the aircraft that enabled the pilot to maintain control even though the engines had failed. There was, for example, the APU (auxiliary power unit) that enabled the aircraft to provide electrical power. There was the RAT (ram air turbine) which provided hydraulic pressure. Then there was the flight control system that allowed the aircraft to function normally. The fourth layer was the pilot himself. It was his extensive training and psychological preparedness that allowed him to bring the aircraft in for a controlled ditching. This fourth layer was also an example of the *human in the loop* rule discussed below. This example also serves to illustrate the interdependency of the design rules.

Another example of *layered defense* is the prevention of collisions between aircraft in flight. The first layer belongs to air traffic control (ATC) who are responsible for tracking the aircraft by radar and making sure that they are at the assigned altitudes, which should be different for each aircraft. This layer uses the *drift correction* design rule. The second layer involves the use of the Traffic Collision Avoidance System (TCAS) described below. This system warns the pilot of an impending collision with another aircraft. Hence this layer also employs the *drift correction* design rule. However, a third layer is required in the form of action by the pilot to perform a corrective maneuver. Hence, the *human in the loop* design rule is required also.

多层防护规则

多层防护规则是由 Reason's (1997, 第 11 页) 的瑞士奶酪模型衍生出来的。该模型表示的是，要发生一个故障，必须要有一个扰动穿透一系列的层级，就像穿透瑞士奶酪一样。模型中，每一层都有一些孔洞，这些洞就是所说的漏洞，它使扰动能够到达下一层。所以，该模型的含义是，一个系统的防御层越多，系统的恢复力就越强。Pariès (2011, 第 9 至 27 页) 称这个规则为"纵深防护"。与降低复杂性和降低隐藏交互一样，多层防护规则也是一条重要规则，它能使系统与其安全性边界保持一定的距离的，也就是说，增加层数能使系统远离其安全边界。

美国航空公司 1549 航班就是多层防护的一个例子。如 Pariès 所述，这个系统有四个防护层。其中的两层失效，另外的两层使飞机成功水上迫降。第一层是正确的措施以摆脱可能会对飞机造成伤害的机场有鸟区域。当然，第二层是，发动机本身及其经合格审定的应对鸟撞能力。显然，这个实际情况超过了该能力范围，并且发动机失效了。第三层是飞机本身。Pariès 指出飞机的很多特征能使驾驶员在发动机失效的情况下操纵飞机。比如，可以用辅助动力装置（APU）为飞机提供电源，用冲压空气涡轮（RAT）为飞机提供液压源。那样的话，飞行控制系统就能够控制飞机以使其功能正常。第四层是驾驶员本身，正是全面的培训和心理素质使他能够驾驶飞机实施可控的水上迫降。第四层也是下面讨论的人在环规则的一个例子。这个示例也说明了设计规则的独立性。

多层防护的另一个例子是飞机飞行中的防撞。第一层属于航空交通管制（ATC），它负责用雷达追踪飞机，并确保飞机在指定的高度飞行，并且每架飞机的高度都是不一样的。这一层运用了漂移修正的设计规则。第二层涉及下面所述的飞机空中防撞系统（TCAS）的使用。这个系统警告驾驶员与别的飞机可能发生的碰撞。因此，这一层也使用了漂移修正的设计规则。但是第三层要求驾驶员执行正确的机动。因此，也需要人在环设计规则。

Layered defense dependency

Layered defense does not itself have any dependent rules. However, layered defense is dependent on the implementation of at least two other rules. In the case of US Airways Flight 1549 there were four rules: ① *drift correction* (corrective action to keep the birds away from the airport), ② *absorption* (having engines with sufficient capacity), ③ *functional redundancy* (having alternative ways to control the aircraft), and ④ *human in the loop* (having a pilot with training and instincts to perform the task). As was stated above, it was the last two rules that enabled the aircraft to ditch satisfactorily.

The Human in the Loop rule

The *human in the loop* rule, discussed by Madni and Jackson (2009, pp. 181–191) is one of the most accepted rules in the commercial aviation domain. This rule asserts that humans should always be employed as system elements when there is a need for human cognition. The fact that human pilots continue to be the dominant mode of aircraft control attests to this assertion. Human dominance in ATC is further evidence of its validity.

The Automated Function Support rule

Billings (1991, pp. 244–245) elaborates on the *human in the loop* rule by saying that when there is a choice between the human or an automated system performing a given function, the task should always be performed by the human provided the task is within the time limitations and capability of the human.

The designer will need to balance this rule against the possible advantages of the *flight envelope protection* system discussed in Chapter 2. As discussed in that chapter, the purpose of that system is to prevent the pilot from allowing the aircraft to exceed its flight envelope.

The Reduce Human Error Support Principle

It is axiomatic that human error cannot be eliminated. However, there are numerous accepted ways to reduce and minimize human error. Reason (1990, pp. 217–250) presents a comprehensive list of these methods.

多层防护的依赖性

多层防护本身不依赖于任何其他规则。但是，多层防护至少需要两个其他规则的执行。在美国航空公司 1549 航班的例子中存在四个规则：①漂移修正（使鸟类远离机场的修正措施）；②吸收（有性能达标的发动机）；③功能冗余（有备选的飞机控制方法）；④人在环（有训练有素，且能正确执行任务的驾驶员）。如上文所述，正是最后两条规则使得飞机能安全进行水上迫降。

人在环规则

Madni 和 Jackson（2009，第 181 至 191 页）讨论的人在环规则是商用飞机领域通行最广的规则之一。该规则主张当系统需要人的认知时，人应该始终被当作系统的元素。实际上，人类驾驶员一直是飞机控制中的主要模式，这个事实正好佐证了上述观点。ATC 中人占支配地位，这也是该规则的进一步证据。

自动功能支持规则

Billings (1991，第 244 至 245 页) 详细描述了人在环规则，他认为，如果在执行一个既定的功能时可以选择由人还是由自动化系统来完成，只要时间限制及人的能力允许，那么始终应该选择由人来执行。

设计人员将需要对照第 2 章所述飞行包线保护系统可能的优点，对本条规则进行平衡。第二章讨论到，飞行包线保护系统的用途是为了防止驾驶员驾驶飞机超出其飞行包线。

减少人为错误支持规则

可以肯定的是，人为错误是无法完全避免的。但是有许多方法可以减少人为错误并使其最少化。Reason (1990，第 217 至 250 页) 全面介绍了这些方法。

Human in the loop dependency

The *human in the loop* rule is primarily dependent on the *reduce human error* support principle to compensate for its primary vulnerability, that is, human error.

The Reduce Complexity rule

Complex systems are subject to highly erratic operation and sometimes failure. The *reduce complexity* rule states that the complexity of the system should be reduced as much as possible. Like *layered defense* and *reduce hidden interactions*, the purpose of this rule is to remove the system from its safety boundary as much as possible

The phenomenon of complexity is prevalent in both the aircraft system and the larger system called the large-scale supply chain system described in Chapter 14. One of the effects of complexity is the phenomenon of hidden interactions discussed in the section on the *reduce hidden interactions* rule discussed below. According to Marczyk (2012) there are three contributors to complexity as discussed below.

The first contributor is the number of elements of the system, in our case the number of parts of the aircraft. The number of elements by itself does not create complexity; however it is an important multiplier. It is possible for a system to have many components and not be complex. Such a system is called *complicated* rather than complex. The example often given, by for example Giachetti (2010, pp. 34–36), is a clock whose parts fit together perfectly. It is the lack of a perfect fit that makes the system complex; this lack of fit is called *variability* as discussed below.

The second contributor is the number of interfaces. Some authors, for example Carson (2000), consider this to be the primary contributor. With respect to the number of interfaces, this is an aspect entirely within the purview of the aircraft designer, in particular, the aircraft architect who determines the number and relationships among the subsystems. The methodology the designer might use is called cluster analysis as described by Hitchins (1993, pp. 135–147). The premise of cluster analysis is that the parts of the aircraft with the greatest *functional binding* will be a candidate to be an identifiable subsystem and will have the fewest interfaces among those parts that constitute the subsystem. As an example, the designer may want to decide whether to make guidance and navigation two software modules or one. Cluster analysis will help

人在环的依赖性

人在环规则主要依赖于减少人为错误支持规则，以弥补其主要的漏洞，也就是人为错误。

降低复杂性规则

复杂系统容易被误操作，且有时会发生失效。降低复杂性规则表明应该尽可能减少系统的复杂性。与多层防护和减少隐藏交互规则一样，这条规则的目的是使系统尽可能远离其安全边界。

复杂性在飞机系统和大型系统中非常普遍，所谓大型系统是指 14 章描述的大规模供应链系统。复杂影响之一是本节讨论的隐藏交互现象，下面还会讨论减少隐藏交互规则。根据 Marczyk (2012) 的研究，导致复杂性的因素有三个，下面将分别讨论。

第一个因素是系统元素的数量，本书中指的是飞机零部件的数量。元素数量本身不产生复杂性，但它是一个重要的乘数。一个系统可能有很多的部件但并不复杂，这样的系统被称为繁复的而不是复杂的。一个经常被乏引用的例子（ Giachetti (2010, 第 34 至 36 页)）是一座各部件匹配完美的钟表。如果缺少完美的配合，将会使系统变得复杂，这种配合的缺乏在下面的讨论中被称为变异性。

第二个因素是接口的数量。有些作者，比如 Carson (2000)，把它作为复杂性最主要的影响因素。至于接口的数量，完全是由飞机设计人员决定的，特别是那些能决定子系统数量及相互关系的飞机架构师。设计人员可能会使用的方法被 Hitchins (1993, 第 135 至 147 页) 称为群分析。群分析的前提是将功能结合最紧密的飞机零部件当作一个确定的子系统，从而使构成这个子系统的零部件之间的接口最少。例如，设计人员可能要决定是制造导引和导航两个软件，还是一个软件模块。群分析能通过最小化接口数量进行决策。

make that decision by minimizing the number of interfaces.

The third contributor is the variability in the relationships among the parts of the aircraft. Chapter 14 describes this factor more in depth. However, in short, this factor has to do with the variability in the quality of the data or other parameters that are exchanged among the parts of the aircraft. All of the parameters listed in Chapter 6 are candidates for variability.

The Reduce Variability Support Rule

As described above, variability is one of the primary contributors to complexity. Hence, reducing complexity requires reducing variability. Variability can be seen in its simplest technical terms, such as the variability of an electrical current between two components. If there are wide fluctuations in this current, then there will be variability and resultant complexity. Of course, real complexity is a composite of all the fluctuations of different parameters all over the aircraft. Chapter 14 also discusses the variability of parameters between the various elements of the supply chain also contributing to its complexity. Hence, rigorous management of all of these parameters will minimize the complexity of both the aircraft and the supply chain.

Reduce complexity dependency

There are two categories of rules upon which the *reduce complexity* rule is dependent. The first dependency is the *reorganization* rule described below. Reorganization includes both the reduction in the number of elements and the number of interfaces described above. The second category is the *reduce variability* rule described above.

The Reorganization rule

Woods (2006, p. 23) refers to this rule as the *restructuring* rule. He states that in order for a system to be resilient, it must be capable of restructuring itself. Although it may seem unreasonable for an aircraft to restructure itself during operation, it has happened.

The most obvious example is the Sioux City DC-10 crash of 1989. In this case, the aircraft lost control when its control system was damaged. The pilot managed to

第三个因素是飞机零部件之间相互关系的变异性。第 14 章对该因素进行了更深入的介绍。简而言之，该因素必须要处理数据或其他参数质量的变异性，这些数据和参数在飞机各零部件之间进行交换。第 6 章列出的所有参数都可能是变异性的参数。

减少变异性支持规则

如上所述，变异性是复杂性主要的来源。因此，要降低复杂性，就需要减少变异性。变异性可以从最简单的技术词语中看出，比如两个部件之间电流的变异性。如果电流有比较大的波动，则会存在变异性，以及相应的复杂性。当然，真正的复杂性是飞机中不同参数波动的组合。根据第 14 章讨论，供应链不同元素之间参数的变异性也会增加复杂性。因此，对这些参数的严格管理有助于将飞机及供应链的复杂性降到最低。

降低复杂性的依赖性

有两类规则与降低复杂性有关。第一个是下面讨论的重组规则。重组包括减少上文所述的元件数量及接口数量。第二类是上面所说的减少变异性规则。

重组规则

Woods (2006, 第 23 页) 把这个规则称为重构规则。他认为，为了使系统能自我恢复，系统必须有能力重组自己的结构。虽然对于一架飞机而言，运营时的结构重组听起来无法理解，但它确实发生了。

最明显的例子是 1989 年苏城 (Sioux City) DC-10 坠机事件。在那个情况下，飞机由于控制系统损坏而失去控制。驾驶员用推进控制装置努力保持一定程度的控制。虽然不是所有乘客都幸免于难，但大多数的乘客活了下来。这种控制的改变就相当于飞机控制系统的重组。如第 2 章所述，FAA 认为强制进行推进控制

maintain some degree of control by using the propulsion controls. Most, but not all, of the passengers survived. This change of controls was tantamount to reorganizing the control system of the aircraft. As mentioned in Chapter 2, the FAA did not see fit to mandate propulsion control. Nevertheless, this incident did illustrate the principle of reorganization and its value.

Reorganization dependency

The *reorganization* rule is primarily dependent on the human in the loop rule because reorganization normally requires humans to implement the rule.

The Drift Correction rule

Drift correction, discussed by Dekker (2006, pp. 77–92) is the primary rule in the pre-event phase of disruption. It is the rule that enables the system either to anticipate or detect its drift towards an unsafe condition and makes a corrective action. *Drift* correction can either be in real-time or a long-term anticipation of a problem which may involve latent conditions not evident to the operators. Reason (1997, pp. 10–11) describes latent conditions in which flaws in the system are hidden until a catastrophic failure occurs.

The concept of latent flaws brings to mind a well-known axiom of system safety, paraphrased as follows:

Just because your aircraft has not had an accident does not mean it is safe.

One of the most well-known threats to aircraft safety is bird strikes as discussed by Pariès (2011, pp. 10–11) that was a major factor in the US Airways Flight 1549 case. So the anticipation of this threat and steps to thwart it are an example of long-term drift correction. Skybrary (2013, unpaginated) points out that in many parts of the world mammals, such as reindeer, are also threats to aircraft because they can unexpectedly walk onto runways. Many corrective actions have been taken, such as reducing the bird population. However, as was seen in the Flight 1549 case, this was not effective, hence the need for *layered defense* as discussed above.

Typical of latent flaws in the aircraft domain is the degradation of the aircraft due to poor maintenance or aging as described above for the *Nimrod* case in the discussion

是不合适的。不过，这个案例还是说明了重组的准则及其价值。

重组的依赖性

重组规则主要依赖人在环规则，因为重组一般都需要人来执行规则。

漂移修正规则

Dekker (2006，第77至92页) 所介绍的漂移修正是扰动之前阶段的主要规则。该规则能使系统预测或探测其向不安全状况的漂移，并采取修正措施。漂移修正可以是实时的，也可以是对一个问题的长期预测，这可能会牵涉到运营方不甚明了的潜在状况 Reason (1997，第10至11页) 介绍了隐藏状况的概念，即，在灾难性失效发生前，系统中的缺陷一直是隐藏的。

潜在缺陷概念让人想到一个与系统安全性相关的著名公理，该公理表述如下：

不能因为飞机没发生事故就认为着飞机是安全的。

如 Pariès (2011，第10至11页) 所述，飞机安全性威胁中最为人熟知的是鸟撞，它是美国航空公司1549航班事件的最主要原因。所以，对此类威胁的预测及解决措施是长期漂移修正的一个例子。Skybrary (2013，非分页) 指出，世界上很多哺乳动物，比如驯鹿等，也是对飞机的威胁，因为它们可能会意外地出现在跑道上。目前已经采取了很多修正措施，比如减少鸟群。但是，正如我们在1549航班例子上看到的，这往往还不够，所以我们需要上面所说的多层防护。

飞机领域典型的潜在缺陷是维修不良及老化造成的飞机退化，这在上面讨论限制退化支持规则时提到的 Nimrod 事件中已经讨论过了。这些缺陷的修正措施更加频繁，需要更加细致的检查及监督维修来探测它们。对于一些类型的缺陷，比如疲劳导致的结构裂纹，可能需要特殊的设备来探测。

of the *limit degradation* support rule. Corrective actions for these flaws are more frequent, and detailed inspections and oversight of the maintenance operations are required to detect them. For some types of flaws, for example structural cracks due to fatigue, special equipment may be necessary to detect them.

Regarding real-time *drift correction*, some existing methods already fall into that category, for example Terrain Avoidance Warning System (TAWS) as described by Skybrary (2012, unpaginated). This system warns of an approaching mountain or any other geological feature that may be a hazard to the aircraft and allows the pilot to perform a corrective maneuver.

Another example of *drift correction* is the Traffic Collision Avoidance System (TCAS). This is mandated system that uses a transponder to warn the pilot of the possible collision with another aircraft.

Another interesting example of real-time *drift correction* is the volcanic ash detector, called AVOID, which is in the planning stage, as reported by the BBC News (2010). This is a joint project between the airline EasyJet and the aircraft company Airbus. This device would warn the pilot if the volcanic ash density is becoming too dense and thus allow the pilot to take an alternative route. This development hinged on the adoption of a criterion for the maximum particle size of volcanic ash through which an aircraft could fly as also reported by the BBC (2010). This development followed the eruption of the Icelandic volcano Eyjafjallajökull in 2010, disrupting air traffic. It is not known whether this device would be mandatory on aircraft, but in any case, some airlines may voluntarily choose to install it.

Drift Correction Support rules

The *drift correction* rule has two important support rules: the *detection* support rule and the *corrective action* support rule. The *detection* support rule can be interpreted broadly. It can mean, for example, the detection of volcanic ash, as described above or the detection of hidden flaws in a software system. *Corrective action* can be any action to correct these problems.

Drift correction dependency

The drift correction rule is dependent on the two support rules listed above; it is also dependent on the *inter-node interaction* rule to convey the detection information to the control center of the system. This control center may be the pilot.

关于实时漂移修正，已有一些方法实施，比如 Skybrary (2012, 非分页) 描述的地形提示和告警系统（TAWS）。该系统对靠近山脉，或其他可能对飞机产生危害的地形特征进行警告，以使驾驶员做出修正的机动动作。

另一个漂移修正的例子是空中防撞系统（TCAS）。这是一种强制性的系统，它使用应答机来警告驾驶员可能要与其他飞机相撞。

实时漂移修正的另一个有趣的例子是火山灰探测器，也被称为 AVOID，据 BBC 2010 年的新闻报道，这个设备目前处于规划阶段。这是易捷航空公司与空客公司的合作项目。这个设备会在火山灰密度太大时警告驾驶员，以使得驾驶员选取替代航线。该研究涉及判断飞机能够飞越的最大的火山灰颗粒大小，这部分内容在 BBC（2010）也有报道。冰岛火山 Eyjafjallajökull 在 2010 年喷发并干扰了空中交通，从而催生了该研究项目。尚不明确在飞机上安装这个设备是否属于强制要求，但无论如何，有些航空公司可能会自愿选装。

漂移修正支持规则

漂移修正规则有两个重要的支持规则：探测支持规则和修正措施支持规则。探测支持规则可以广义地理解比如，它可以表示对上述的火山灰探测或者软件系统中隐藏缺陷的检测。修正措施可以是修正这些问题的任何措施。

漂移修正的依赖性

漂移修正规则依赖于上面列出的两个支持规则，它还依赖于节点间交互规则，以传递探测的信息到系统控制中心，这个控制中心可能是驾驶员。

The Inter-Node Interaction rule

The *inter-node interaction* rule is an adaptation of the *cross-scale interaction* concept described by Woods (2006, p. 23). This rule finds its roots in systems theory in which a basic property of a system is *cohesion* as described by Hitchins (1993, p. 55). That is to say, for a system to be a system—and an aircraft is a system—there must be a relationship among all the parts. We have seen much evidence of the relationships among the parts of the aircraft in all the interfaces discussed in Chapter 6. Of primary interest is the relationship between the pilot and the aircraft automation. Billings (1997, pp. 232–262) lays out his set of rules (which he calls *requirements*) for all of these relationships. All of Billings' rules are in fact *heuristics*, that is, rules based on his own experience and observations.

The Informed Operator Support rule

The *informed operator* support rule as formulated by Billings (1997, p. 240) states that the operator, that is the pilot, should be completely knowledgeable about the operations of the automated system. In other words, the automated system should not perform any operation that the pilot does not understand. This support rule is particularly important in the implementation of *flight envelope protection*, as described in Chapter 2. Billings argues that this rule is most important when the aircraft is not operating in normal operating conditions.

The value of this rule is evident in those cases in which the failure to observe it resulted in a catastrophic event. For example, according to Zarboutis and Wright (2006, pp. 359–368), the aircraft in the Nagoya incident of 1994 was in a go-around mode while the pilot desired to land. This miscommunication between the pilot and the automated system led to the ultimate crash and loss of life.

The Knowledge between Nodes Support rule

Another Billings (1997, p. 243) rule, the *knowledge between nodes* support rule is similar to the *informed operator* support rule except broader in scope. This rule pertains to any combination of elements on the aircraft, not just the pilot and the automated system. However, the Nagoya case study satisfies this rule as well.

节点间交互规则

节点间交互规则是 Woods (2006, 第 23 页) 介绍的跨尺度交互的一个用例。该规则的来源是系统理论，根据该理论，系统的基本属性是 Hitchins (1993，第 55 页) 介绍过的凝聚性，即，系统之所以能成为一个系统（飞机也是一个系统），其所有部分之间必定存在联系。第 6 章已经讨论了飞机所有接口部件之间存在联系的证据。对这些交联关系中最引人关注的是驾驶员与飞机自动化设备之间的关系。Billings (1997, 第 232 至 262 页) 列出了所有这些关系的一组规则（他称为需求）。Billing 的所有规则实际上就是试探法，也就是说这些规则都是基于他自己的经验和观察。

知情操作者支持规则

知情操作者支持规则由 Billings (1997, 第 240 页) 提出，它说明了操作者，即驾驶员，应该对自动化系统的操作有全面的了解。换句话说，自动化系统不应该执行任何驾驶员不理解的动作。该支持规则对第 2 章所讨论的飞行包线保护的执行尤为重要。Billings 表示，该规则在飞机没有在正常情况下运行时是最重要的。

当这些规则没有得到执行时，将会产生灾难性后果，由此可见该规则的价值。比如，根据 Zarboutis 和 Wright(2006, 第 359 至 368 页) 的研究，1994 年名古屋事件中的飞机处于复飞模式，而驾驶员希望着陆。驾驶员与自动化系统之间的沟通错误导致了飞机坠毁和生命损失。

节点间知识支持规则

Billings (1997, 第 243 页) 的另一个规则，即节点间知识支持规则与知情操作者支持规则相似，只是它范围更广。这个规则涉及飞机元素的所有组合，不只是驾驶员与自动化系统这两个元素。但是，名古屋事件的研究也能满足该规则。

The Human Monitoring Support rule

This rule is the reverse of the *informed operator* rule; it states, according to Billing (1997, p. 240) that the automated system should monitor the human pilot. According to Billings, one of the most important aspects of human monitoring that the automated system should perform is to know when the human has made an incorrect data entry.

The Automated System Monitoring Support rule

This rule, according to Billings (1997, p. 243) goes beyond the *informed operator* rule; it states that the human should know the *intent* of automated system actions. Once again, this rule seems to have been absent in the Nagoya incident. If the automated system does not behave in accordance with pilot intentions, this rule will have been violated.

The Inter-Node Impediment Support rule

This support rule can be inferred from various case studies suggest this support rule which was documented by Jackson and Ferris (2013, pp. 152–164). It says that there should be no administrative or technical impediments to communication or cooperation among the nodes of a system. Among the more notable examples of when these impediments existed was the 9/11 Commission's (2004, p. 11) concern that the FAA could not communicate directly with the pilot of United 93 that a terrorist threat was imminent. This lack of communication was the result of protocols in the FAA at that time.

The Reduce Hidden Interactions rule

Perrow (1999, pp. 79–86) states that many failures result from hidden interactions that result from excessive complexity. It is from this observation that the *reduce hidden interactions* rule is inferred.

There are many ways that interactions can occur on aircraft and cause failures. Prominent among these is the effects of electromagnetic interference (EMI) when electrical devices, such as solenoids or generators, are located near data lines.

A dramatic example of hidden interactions was the Helios 522 aircraft in which a

人类监控支持规则

该规则与知情操作者规则正好相反，根据 Billing (1997，第 240 页) 观点，它表示自动化系统应该监控人类驾驶员。Billings 说，人类监控最重要的一方面是，自动化系统应该知道人在什么时候输入了错误的数据。

自动化系统监控支持规则

根据 Billings (1997，第 243 页) 观点，该规则比知情操作者规则要求更高，他认为人应该知道自动化系统动作的意图。这条规则看起来已经存在于名古屋事件中了。如果自动化系统的行为不符合驾驶员的意图，则已经违反了这个规则。

交互节点间障碍支持规则

Jackson 和 Ferris (2013，第 152 至 164 页) 的书中提到了这条规则，该规则在很多研究案例中都可以见到，它是指系统的节点之间不应存在沟通和合作方面的障碍，这个障碍包括行政和技术方面。这些障碍存在的一个著名的例子是 9·11 事件委员会报告 (2004，第 11 页)，由于恐怖主义威胁逼近，FAA 不能与联合航空公司 93 航班飞机的驾驶员直接通话，这种交流的缺乏是当时 FAA 内部协议的结果。

减少隐藏交互规则

Perrow (1999，第 79 至 86 页) 指出，许多失效源于隐藏交互，而隐藏交互又源于极度的复杂性。由此可见减少隐藏交互规则的义要性。

飞机上存在多种交互，它们会导致失效。其中最突出的例子是电磁干扰，它是由于线圈或发电机等电气设备在数据线附近产生的。

根据跟 Dekker 等 (2008) 介绍，Helios 522 飞机是隐藏交互的一个生动的例子，

malfunctioning pressurization system and the automated flight control system combined to cause the crew and all the passengers to die of hypoxic hypoxia as described by Dekker et al. (2008).

So how can hidden interactions be reduced? At first, it might seem like a formidable task, and it can be. The most tedious method would be a mapping of all possible interactions as suggested by Jackson (2010, p. 39). This method can be expedited by exploring specific categories of interactions one at a time, such as EMI.

A more global approach is to reduce the complexity of the entire aircraft in accordance with the reduce complexity rule discussed above. However, even this approach would involve examining the variability of the parameters between the individual elements.

Finally, Leveson et al. (2006, p. 97) suggests that the problem and the solution might be managerial in nature. That is to say, closer cooperation between different design groups would reduce the chance that there would be any inconsistencies in the design approaches that would cause hidden interactions. Within SE such a managerial approach exists; it is called integrated product teams (IPTs) as described in Chapter 12. As Chapter 12 explains, an IPT has members from all technical disciplines working on the same product, for example, the wing or fuselage. The premise is that this close working relationship would result in fewer inconsistencies and hence hidden interactions.

16.5 Other Rules

If readers have access to Jackson and Ferris's paper (2013, pp. 152–164), they will notice that there are rules other than the ones listed above. These other rules may have some application to the commercial aircraft domain, but perhaps in a more limited way. For this reason, the following list will briefly summarize these rules and some potential applications.

The Repairability rule

There is no doubt that aircraft need to be repaired from time to time. Chapter 5 has devoted some space to the subject of maintainability. With respect to resilience the type of maintenance that is probably of interest is field repairs, such as, damage due to volcanic ash.

该例子中，增压系统和自动飞行控制系统同时出现功能异常，导致机组和所有乘客因为缺氧死亡。

那么如何能减少隐藏交互呢？初看来，它看起来是一项艰难的任务，它可能确实如此。最烦琐的方法是根据 Jackson (2010, 第 39 页) 的建议，把所有可能的交互都画出来。这种方法能通过每次探测一类特定的交互来提高效率，比如先探测 EMI。

减少全机复杂性更通用的方法是通过上面所述的降低复杂性规则。但是即使采用这种方法，也可能需要检验所有单个元件间参数的变异性。

最后，Leveson 等 (2006, 第 97 页) 提出，问题和解决方案在本质上是可管理的。就是说，不同设计团队之间更紧密的合作能减少设计方法的不一致，这些不一致可能会导致隐藏交互。在系统工程中，这样的可管理的方法是存在的，那就是 12 章讨论的集成产品团队（IPT）。正如 12 章所介绍的，一个 IPT 成员来自所有技术专业，他们为同一个产品工作，如机翼或机身。当然，前提是这种紧密的合作关系能够减少不一致和隐藏交互。

16.5 其他规则

如果读者已经看过 Jackson 和 Ferris 的论文 (2013, 第 152 至 164 页)，就会注意到有许多上面没有列出的规则。这些规则在商用飞机领域会有一些应用，但是比较有限。出于这个理由，下列清单只简要概述这些规则，以及一些潜在的应用。

可修理性规则

毫无疑问，飞机需要时不时进行修理。第 5 章对可维修性进行了讨论。关于恢复力，比较相关的维修类型是现场修理，例如，由于火山灰造成的毁坏。

The Localized Capacity rule

Localized capacity has to do with the localization of elements of the aircraft so that a failure in one does not cause a failure in the other. For example, one would not like a fault to cause a failure of all the engines.

The Loose Coupling rule

Loose coupling has to do with slack between elements of the system, so that a failure in one element does not propagate to other elements. In commercial aircraft probably the most applicable area of study is the cluster analysis described by Hitchins (1993, pp. 135–147). This analysis assures that subsystems are internally functionally bound, while connections between subsystems are loosely coupled. For example, there would be very little, if any coupling between the environmental control system (ECS) and the flight control system.

The Neutral State rule

Neutral state is a rule of interest mostly to pilots and to control systems. This rule allows the aircraft following a disturbance to enter a neutral state to allow the pilot to make important decisions.

16.6   A Final Word on Interdependency

The reader should not underestimate the importance of interdependency which is discussed above within the discussion of each design rule. The reason is that if the designer employs only the design rules one at a time without their dependencies, the resultant design will most likely be brittle, that is, not resilient.

The second thing that the reader might notice is that the design rules form a chain, that is, when design rule A is dependent on design rule B, and design rule B is dependent on design rule C, a chain of design rules will form that will comprise a whole system of design rules. It is this whole system the designer will want to implement and not just the individual rules.

能力局部化规则

能力局部化是要把飞机的元件局限在某个特定位置，那么一个元件的失效就不会导致另一个元件失效。比如你不会希望一个故障导致所有发动机都失效。

松耦合规则

松耦合是要让系统元素间的关系呈松弛状态，那么一个因素的失效不会传递给其他元素。在商用飞机领域最适用的研究是 Hitchins (1993, 第 135 至 147 页) 描述的群分析。该分析保证子系统内部功能紧密结合，而子系统之间的联系是松耦合的。比如，环控系统 (ECS) 和飞行控制系统之间即使有耦合也是很少的。

中间状态规则

中间状态是驾驶员和控制系统最为关注的。该规则使飞机在经历一个扰动之后进入中间状态，以使驾驶员做出重要的决策。

16.6 关于相互依赖性的最后几句话

读者不应低估相互依赖性的重要性，该规则在讨论每个设计规则时有所涉及，原因是，如果设计人员只使用这些设计规则，而不考虑其相互依赖关系，这样得到的设计往往是比较脆弱的，也就是说不易恢复的。

读者应该注意的另外一点是设计规则形成了一条链条，即，当规则 A 依赖于规则 B，并且规则 B 依赖于规则 C 时，设计规则的链条就会形成，并构成整个设计规则体系。设计人员将要执行的是整个规则体系，而不仅仅是单条规则。

Final Comments

In the first edition I tried to emphasize the common sense approach to systems engineering and in particular its application to the commercial aircraft domain. I still believe that common sense is important, but there is more to SE than common sense. It is a question of mindset; mindset is the collection of assumptions and philosophies that rule your life.

The Systems Mindset

A mindset of importance here is the systems mindset. If your education is in aerodynamics, for example, you may view the whole world in terms of Mach number or Reynolds number. (This was my undergraduate and master's field of study.) As time goes on, you may have a broader view of a system. It may even result in the realization that the aircraft is a system. Sometime later you may conclude that this system is not really complete until it includes the pilot. Your view of a system will get broader and broader. It may eventually include the airline, the regulators, and the passengers. My point is that once you have the systems mindset, then figuring out what to do next is almost automatic. Your job is to make sure the whole thing, the system, hangs together, works together, and performs the desired functions.

The Risk Mindset

The subject of risk permeates this entire book even though only Chapter 15 treats it as a separate subject. The point is that if the principles of SE are not adhered to, at least as adapted in this book, there will always be a risk of the system failing to meet its objectives or failing in its entirety. The risk mindset is a positive mindset that helps you be aware of possible risks and things you can do to manage them as described in Chapter 15.

There is a negative counterpart to the risk mindset; it is sometimes called the risk denial mindset. The risk denial mindset is self-explanatory; it is the mindset that says that there are no risks or at least risks that cannot be handled until it happens. In my previous book (2010, pp. 91–119) I made a collection of these negative mindsets

最后评论

在第一版中，我试图强调系统工程的常识性方法，特别是其在商用飞机领域的运用。我仍然相信常识是很重要的，但与常识相比，系统工程的内涵要丰富得多。这是思维模式的问题，思维模式是主导生活设想及哲学的集合。

系统思维模式

这里重要的思维模式是风险思维模式。例如，如果你的专业是空气动力学，你可能会用马赫数或者雷诺数来观察世界（这是我的本科及研究生阶段的学习领域）。随着时间的推移，你可能会有一个更广阔的系统视野，这可能引导出飞机是个系统的认识。一些时间之后，你可能会得出结论，系统如果不包括驾驶员就不是真正完整的。你的系统观越来越开阔，定义的系统最终可能包括航空公司、监管机构及乘客。我的观点是一旦你有了系统思维模式，几乎就能自动识别出接下来该做什么。你的工作就是确保整个事物、系统结合在一起、工作在一起，并实现想要达到的功能。

风险思维模式

虽然只有第 15 章将风险作为一个单独的主题，但实际上风险的主题从头到尾贯穿全书。问题在于，如果不坚持系统工程基本原则，至少是书中改编的原则，就存在系统不能满足其目标，或者全部失败的风险。第 15 章所述风险思维模式是一个积极的思维模式，它帮助你意识到可能的风险，以及管理风险可以做的事情。

which I called paradigms. Prominent among these is a famous one called the Titanic Effect which is the belief that your system is safe when it is not. The Titanic Effect is tantamount to risk denial. In my opinion, and there is abundant evidence to support the existence of this mindset. That is, there is a widespread cultural disorder of risk denial. *Challenger, Columbia, Concorde, Katrina*, and *Deepwater Horizon* are all products of risk denial.

The problem is that there are no straightforward solutions to risk denial. Telling decision makers to shape up is not a solution since they are usually the victims of risk denial. As explained in Chapter 15, the only concrete solution on the table is an independent authority which was recommended by the Columbia Accident Investigation Board. This solution is not easy either since there is little agreement on what independent means and how to achieve independence. In commercial aircraft the regulators supply a certain degree of independence. In the end, however, the aircraft developers need to embrace the concept of independence and implement it.

The Resilience Mindset

Resilience is a new concept that is gaining favor in many domains including commercial aircraft. As explained in Chapter 16, resilience goes beyond safety in that safety is concerned with preventing failures, while resilience deals with methods to anticipate, withstand, and recover from disruptions of many kinds. However, resilience requires a mindset of its own, which may be difficult for many engineers to adopt. As pointed out in Chapter 16, many authorities claim that resilience is , at the very least, difficult to measure. This is a fact that will be difficult for many engineers to accept. The net result is that in order to find a satisfactory solution, the designer will have to use an iterative, that is, trial and error solution.

Also, as pointed out in Chapter 16, many of the resilience concepts, such as *layered defense*, are already present, to some extent, in the commercial aircraft domain. The "Miracle on the Hudson" case illustrates that point. Nevertheless, acceptance of resilience principles and the adoption of the resilience mindset promise to make safer skies.

风险思维模式有负面的对应物，它有时被称为风险否认思维模式。风险否认思维模式很容易理解，它认为不存在任何风险，或者至少风险在发生之前不可应对。在我之前的书中（2010, 第 99 至 119 页）收集了这些负面的思维模式，我称之为范式。范式中最突出的是一个称为泰坦尼克效应的著名思维模式，泰坦尼克效应是指系统被认为是安全的，但其实并非如此。泰坦尼克效应相当于否认风险。我认为，有大量的证据可以支持这种思维模式的存在。也就是说，风险否认是一个普遍的文化病症。挑战者号、哥伦比亚号、协和飞机、卡特里娜飓风及深水地平线号灾难事件都是风险否认的产物。

问题在于，对于风险否认没有直接的解决措施。敦促决策者表现更好并不是一个解决方案，因为他们通常就是风险否认的受害者。正如第 15 章讨论的，唯一的具体解决方法是由哥伦比亚号事故调查委员会推荐的独立管理机构。这个方案实现起来并不容易，因为在独立的含义，及如何获得独立方面几乎无法达成任何一致。在商用飞机领域，一些监管机构有一定程度的独立性。然而，飞机的研制方最终需要接纳独立性的概念并予以实施。

恢复力思维模式

恢复力是一个在很多领域受到重视的新概念，包括商用飞机领域。正如第 16 章所述，恢复力在某些方面超过了安全性，安全性关注的是防止失效，而恢复力则关注多种扰动的预测、承受及恢复的方法。然而，恢复力需要自己的思维模式，可能很多工程师难以采纳。如第 16 章所述，很多权威人士声称，恢复力至少是很难被测量的。这点将造成很多工程师难以接受这个概念。最终结果就是为了得到一个满意的解决方案，设计者必须使用迭代的方法，也就是试错法。

而且，如第 16 章所述，很多恢复力概念，例如多层防护，在一定程度上已经出现在商用飞机领域了。"哈德逊河奇迹"的例子可以证明以上观点。然而，恢复力准则的接受和恢复力思维模式的采用注定将使飞行更加安全。

Appendix 1
The Mathematics of Reliability Allocation

The purpose of this appendix is to provide an explanation of how multiplicative parameters, such as reliability, are allocated. In this appendix we are using the third definition (see Glossary) of the word *allocation*, namely, a breakdown of a top-level requirements parameter into its component parts.

A1.1 Basic Reliability

The basic reliability equation is:

$$R = R_1 \times R_2 \cdots R_i!$$

What this equation says is that, if several components have reliabilities R_1 through R_i, then the reliability of all the components working together is R. For example, for the case of only two components, if component 1 has a reliability of 0.99 and component 2 has a reliability of 0.99, then the reliability of the whole subsystem (consisting of the two components) is about 0.98.

A1.2 Allocation for Generically Similar Components

Allocation does it in reverse: that is, it sets a required value of R and determines what R_1 and R_2 must be to keep the reliability at least equal to R (that is, "allocates" values to R_1 and R_2). For example, let's say that R must be at least 0.99. Then one possible allocation is $R_1 = 0.995$ and $R_2 = 0.995$. (This can be checked by multiplying them together.) This would be a good allocation if both components were ① of the same generic class (mechanical or electrical) and ② these reliabilities were considered to be achievable. In our specification the allocation would look like:

附录1 可靠性分配的数学计算

本附录提供可靠性等乘法参数如何分配的说明。在本附录中，我们使用词汇"分配"的第三个定义（见术语表），即顶层需求参数分解到其组成部分中。

A1.1 基本可靠性

基本的可靠性公式是：

$$R = R_1 \times R_2 \cdots R_i!$$

该公式表明，如果几个部件的可靠性是 R_1 到 R_i，那么所有部件的可靠性就是 R。例如，对于只有两个部件的情况，如果部件 1 有 0.99 的可靠性，部件 2 有 0.99 的可靠性，那么整个子系统（由这两个部件组成）的可靠性大约是 0.98。

A1.2 属类相似部件的分配

分配可以逆向进行：也就是说，它设定了一个要求值 R，确定 R_1 和 R_2 必须是什么值才可以保证可靠性至少为 R（也就是说，"分配"值给 R_1 和 R_2）。例如，我们假设 R 必须至少达到 0.99，一个可能的分配就是 R_1=0.995，R_2=0.995。（可以通过将两个值相乘来检查）。如果这两个部件都是①同一属类（机械类或者电气类）且②这些可靠性被认为是可以实现的，那么这个分配就是合理的分配。在我们的规范中，分配如下表所示：

(Continued)

Total reliability	0.99
Component 1	0.995
Component 2	0.995

A1.3 Allocation for Generically Different Components

Another possible solution occurs if the reliability specialist knows, for example, that the components are different: one is mechanical and the other is electrical. He or she knows that electrical components are more reliable than mechanical; the reverse is true for some types, but this is only an example. The first step is to set R_1 (electrical component) to 0.999 and R_2 (mechanical component) to 0.991. The resulting solution yields an acceptable value of $R = 0.99$. The table would look like

Total reliability	0.99
Component 1	0.999
Component 2	0.991

A1.4 Redundancy

Redundancy becomes necessary when the reliability specialist knows that the total reliability of 0.99 is not achievable with the available components. Let's say that the mechanical component still has a reliability (R_2) of 0.991. However the electrical component has a reliability (R_1) of 0.99. The net reliability (R) would then be only 0.981 (by the equation, above). This is obviously too low. What can be done? We can use redundant electrical parts, that is, if one of the parts fails, the other will still do the job. The equation for redundancy is a little more complicated:

$$R_1 = 1 - (1 - R_e)^2$$

where R_e is the reliability of an individual electrical part and R_1 is the reliability of the two redundant electrical parts working together. Thus in this case, if R_e is 0.99, then R_1 would be 0.999 9. Much better, right? Hence, the total reliability of the subsystem would be about 0.991, that is $R_1 \times R_2$, and we would have met the requirement of at least 0.99. The allocation table would look like:

总可靠性	0.99
部件 1	0.995
部件 2	0.995

A1.3　属类不同部件的分配

如果可靠性专家发现，例如部件是不同属类的：一个是机械类，另一个是电气类，这时候就会出现另外一个可能解。他或她知道，电气部件比机械部件更加可靠；对于某些类型来说，反过来也是正确的，这仅仅是一个例子。第一步是设 R_1（电气部件）为 0.999，R_2（机械部件）为 0.991。最终方案产生一个可接受的 R 值 0.99。表格如下所示：

总可靠性	0.99
部件 1	0.999
部件 2	0.991

A1.4　冗余

当可靠性专家发现无法达到 0.99 的总可靠性时，必须使用冗余。例如，假设机械部件仍然有 0.991 的可靠性（R_2）。但是电气部件的可靠性（R_1）为 0.99。此时，其净可靠性可能只有 0.981（通过上面公式计算得出）。这显然太低了。我们能做些什么呢？我们可以使用冗余的电气组件，也就是说如果其中一个部件失效了，另一个部件可以继续工作。冗余公式有一点复杂：

$$R_1 = 1 - (1 - R_e)^2$$

其中，R_e 表示单个电气部件的可靠性，R_1 是两个冗余电气部件一起工作的可靠性。因此，在这种情况下，如果 R_e 是 0.99，那么 R_1 将是 0.999 9，结果好很多，是吗？所以，这个子系统的可靠性将是大约 0.991，也就是 $R_1 \times R_2$，我们达到了至少 0.99 的要求。分配表如下所示：

Total reliability	0.99
Electrical assy	0.999 9
Component 2 (mechanical)	0.995

Actually, in this case the achievable reliability will be greater than 0.99, namely, about 0.995 since the electrical assembly is now so reliable due to redundancy. However, the requirement will still be met. The reliability specialist may now choose to reduce the mechanical reliability allocation to 0.99 if there are some major cost savings to be gained.

A1.5   The Whole Airplane

Of course, our examples above are for only two components. In reality there may be thousands of components. The reliability specialist approaches this problem in the same hierarchical way as the weight specialist does, that is, by product center. The equation is a little longer, but not much, for example:

$$R_{airplane} = R_{wing} \times R_{fuselage} \times R_{empennage} \times R_{systems} \times R_{propulsion}$$

The specialist then proceeds as shown in the above examples for only two components.

总可靠性	0.99
电气部件	0.999 9
部件 2（机械的）	0.995

事实上，在这种情况下可以达到的可靠性将超过 0.99，即差不多 0.995，因为电气部件由于冗余变得更加可靠了。然而，仍可以达到这个要求。如果可以获得重大的成本节约，可靠性专家则会选择降低机械可靠性分配值至 0.99。

A1.5 整架飞机

当然，我们上述例子只有两个部件。实际上，可能存在成千上万个部件。可靠性专家处理这个问题的方式和重量专家一样，运用同样层级的方式，也就是通过乘积的方式。这个公式有点长，但不是太长，例如：

$$R_{飞机} = R_{机翼} \times R_{机身} \times R_{尾翼} \times R_{系统} \times R_{推进}$$

按照上面例子，专家只对两个部件进行计算。

Appendix 2
Example Commercial Specification Outline

This example specification is adapted from MIL-STD-961D (1995), tailored, and simplified for commercial applications. For many years, the standard specification format was MIL-STD-490A (1985) which was recently superseded by MIL-STD-961D. While somewhat differently organized, these two documents cover substantially the same information. Military considerations, such as vulnerability, have been deleted. Terminology, such as *logistics*, not often used in commercial practice has been replaced with terms, such as *support*, more commonly used in commercial practice.

The reader may notice some differences in terminology between this example specification and the body of this book. For example, MIL-STD-961D places reliability and other constraints in the functional and performance requirements category rather than in a separate constraints category. These differences reflect some of the differences in theory and terminology among systems engineers. However, these differences are not significant from an implementation point of view: Most systems engineers are in basic agreement regarding the factors which should be considered.

This format cannot be considered rigid, but rather adaptable to the individual system being developed. This document can be used, with modification, for aircraft system level and aircraft-level and subsystem-level requirements in accordance with the principles of Chapter 4. This format can be used, at management discretion, first, as an airline customer document to clarify the requirements of the aircraft which the aircraft will be designed to meet. Secondly, it can be used as an internal document for the tracking and verification of requirements. Finally, the format can be used as a medium for establishing supplier requirements and conducting supplier management as explained in Section 12.9. This document can be part of the contractual documentation between the manufacturer and both the customer and the supplier.

A2.1 Scope

This section defines the scope of the document. It names the system (aircraft system, aircraft, or subsystem) being addressed and the purpose (establish requirements and

附录 2　商业规范概要示例

本案例规范改编自 MIL-STD-961D（1995），对内容进行裁剪，并针对商用领域进行了简化。多年来，该标准规范格式是 MIL-STD-490A（1985），此规范最近由 MIL-STD-961D 取代。虽然编排方式有所不同，但这两份文件包含的信息却基本相同。一些基于军事方面的内容已被删除，比如漏洞。一些在商用领域不经常使用的术语已被替换为商用领域更常用的术语，如后勤替换为支持。

读者可能会注意到这个示例规范与本书正文所用术语存在一些差异。例如，MIL-STD-961D 将可靠性及其他方面的约束归到功能及性能需求，而不是一个单独的约束类别。这些差异反映了一些系统工程师之间关于理论及术语的不同理解。然而，这些差异从实施角度看并非重点：大多数系统工程师在应该考虑哪些因素方面基本上是一致的。

该格式不能被认为是刚性的，而是适应于正在研制的各个系统。根据第 4 章的原则，本文件经修改可用于飞机系统级、飞机级及子系统级需求。根据管理层的意愿，这种格式首先可以作为航空公司客户文件，澄清需要满足的飞机设计需求。其次，它可作为对需求的追溯及验证的内部文件。最后，这种格式可以用于建立供应商的需求，并进行供应商管理 (12.9 节) 的媒介。这份文件可以作为制造商与客户及供应商之间合同文档的一部分。

A2.1　范围

本节定义了该文件的范围。它指明所述系统（飞机系统、飞机或子系统）及该系统的用途（规定需求及验证方法）。本节是纯粹的介绍，不包含任何可验证

verification methods) for that system. This section is purely introductory and does not contain any verifiable requirements. It describes the major components of the system, presents some diagrams, if applicable, and describes, in general terms, the external interfaces of the system. It describes any mandatory components, that is, non-development items (NDIs), which are particularly important for derivative aircraft as described in Section 2.2.

A2.2   Applicable Documents

A2.2.1   General

This section should list *all* the documents and *only* the documents which constitute part of this specification. These include company, industry, and regulatory documents, including applicable certification documents. These documents will be made available to the supplier if necessary.

A2.2.2   Certification basis

This section specifies the basis on which the aircraft will be certified as described in Chapter 10. Specific FARs will be cited.

A2.3   Requirements

All requirements in this section, with the exception of Sections A2.4.1 and A2.4.2, should be verifiable in accordance with the principles of Chapter 11 and should contain a *shall* as part of the requirements statement.

As explained in Section 4.8, various requirements may be in conflict. That section also explains several ways to resolve those conflicts. The results of that resolution should be documented in Section A2.8 below.

A2.4   Functional and Performance Requirements

A2.4.1   Missions

This section describes the missions of the aircraft or subsystem as necessary to describe requirements. These missions can include passenger or cargo missions as

的需求。它描述系统的主要部件，如果适用，则介绍一些图形，并使用术语描述外部借口。描述任何必需的部件，即非研制项（NDI），这些部件对衍生型飞机 (2.2节) 尤为重要。

A2.2　适用文件

A2.2.1　概述

本节应列出所有文件和仅构成本规范一部分的文件。这些文件包括公司、行业及监管性文件，包括适用的合格审定文件。如有必要，这些文件将提供给供应商。

A2.2.2　合格审定基础

本节规定了飞机合格审定 (第 10 章) 的基础。将引用特定的 FAR。

A2.3　需求

除 A2.4.1 节和 A2.4.2 节之外，根据第 11 章所述原则，本章中所有需求均应可验证，并应将"应"作为需求陈述的一部分。

正如在 4.8 节所介绍的，不同需求之间可能会产生冲突。该节还介绍了几种解决这些冲突的方法。解决方案的结果应载入下面的 A2.8 节中。

A2.4　功能及性能需求

A2.4.1　任务

本节描述对于需求描述所必需的飞机或子系统任务。这些任务包括 3.3 节介绍的客运或货运任务。

described in Section 3.3.

A2.4.2   Operational phases and modes

This section describes the operational phases and modes of the system, as described in Section 3.3. These phases or modes may differ for individual subsystems or components.

A2.4.3   Product capability

This is the primary paragraph in the specification. It should not be deleted without good reason. This is the paragraph that dictates how the product being delivered should perform.

All requirements in this section pertain to the performance of the system and must be derived from functions as described in Section 4.2. Each requirement should contain a *shall* to indicate that it is mandatory and must be verified. The term *entity* refers to the system being specified, whether at the aircraft system level, aircraft level, subsystem level, or component level. The term *capability* refers to performance requirements as defined in this book in Section 4.2.

A2.4.4   Reliability

This section specifies both the dispatch and operational reliabilities for the system as described in Section 5.4. It does not contain the safety related probabilities of failure discussed in Section A2.4.13 below.

A2.4.5   Maintainability

This section specifies the maintainability requirements on the aircraft as described in Section 5.7. It does not specify requirements for maintenance equipment as in Section A2.4.18 below, except for the specification for the support system as described in Section 2.3.

A2.4.6   Environmental conditions

This section describes all the natural and induced environments which the system

A2.4.2　运行阶段及模式

本节介绍系统的运行阶段和模式，如 3.3 节所述。这些阶段或模式对于单个子系统和部件或有所不同。

A2.4.3　产品能力

这是本规范的主要段落，无特殊理由不应删除。该段落对交付的产品应如何工作进行定义。

本节中的所有需求都属于系统性能，而且这些需求必须源于 4.2 节讨论的功能。每条需求应包含一个"应"，以表明它是强制性的，并且必须进行验证。术语"实体"，是指被规定的系统，无论是在飞机系统级、飞机级、子系统级或部件级。术语"能力"指的是性能要求，如本书 4.2 节所定义的。

A2.4.4　可靠性

本节规定 5.4 节所述的系统签派可靠性及运行可靠性。它不包括在下面 A2.4.13 节讨论的与安全相关的故障概率。

A2.4.5　可维修性

本节规定如第 5.7 节所述的飞机维修需求。除了第 2.3 节所述支持系统的规则之外，本节不对下面 A2.4.18 节所述关于维修设备的需求进行规定。

A2.4.6　环境条件

本节描述系统必须承受的所有自然及诱发环境 (5.6 节所述)。如果产品在飞机内部，该环境应反映该产品在飞机上所处的位置。这些环境可能千差万别。

如果产品在飞机外部，本节也应反映其所处环境。

must endure as discussed in Section 5.6. If the product is internal to the aircraft, the environment should reflect the section of the aircraft in which the product will be placed. These environments may vary widely.

If the product is external to the aircraft, this section should reflect that environment also.

A2.4.7 Transportability

This section specifies any transportability constraints on the system as discussed in Section 5.12.

A2.4.8 Materials and processes

This section specifies any limitations on materials, processes, and parts, as specified in company, industry, or government design standards as described in Section 5.8. Limitations specified by the customer may also be included. It may also specify any manufacturing processes which are needed to produce the item as described in Section 5.14.

A2.4.9 Electromagnetic radiation

This section specifies the electromagnetic radiation a component is permitted to *emit* as described in Section 5.10, as opposed to the electromagnetic environment of Section A2.4.6, above.

A2.4.10 Nameplates or product markings

This section specifies how the entity should be marked.

A2.4.11 Producibility

This section specifies any process requirements, such as tolerances, related to the manufacturing process as discussed in Section 5.14.

A2.4.12 Interchangeability

A2.4.7　可运输性

该节规定 5.12 节所述的关于该系统的任何可运输性约束。

A2.4.8　材料及工艺

本节规定在公司、行业或政府设计标准 (5.8 节) 内规定的关于材料、工艺及零部件上的任何限制条件，客户规定的限制条件也可能包括在内。它可能还规定生产零部件需要的任何制造程序 (5.14 节)。

A2.4.9　电磁辐射

本节规定允许部件发生的电磁辐射 (5.10 节)，它与上面 A2.4.6 中的电磁环境正好相反。

A2.4.10　铭牌及产品标记

本节规定如何对实体进行标记。

A2.4.11　可生产性

本节规定与制造过程 (5.14 节) 相关的任何过程需求，如公差。

A2.4.12　互换性

This section imposes interchangeability requirements on the design of the aircraft components as described in Section 5.7.

A2.4.13 Safety

This section imposes both quantitative safety requirements as described in Section 10.2 as well as the qualitative safety factors described in Section 5.5 and 10.2.

A2.4.14 Human engineering

This section specifies all requirements related to human factors as described in Section 5.5. This section is normally used in situations in which people are not part of the system, but interface with the system. This section is used in those situations to specify only the aspects of the system which may be affected by that interface. If people are part of the system, then specific requirements relating to people, such as, number, training, and physical characteristics, will be specified in Section A2.4.19.

A2.4.15 Security requirements

This section specifies any requirements related to the security of the system, either electronic or physical.

A2.4.16 Software requirements

This section specifies the top-level software requirements described in Section 10.3 Firmware requirements are also included in this section.

A2.4.17 Design and implementation constraints

A2.4.17.1 Physical characteristics
Physical characteristics include mass properties as described in Section 5.2 and dimensions as described in Section 5.3.

A2.4.17.2 Design-to-cost requirements
Although cost constraints are not covered in MIL-STD-961D, this section

本节规定关于飞机部件设计的互换性需求 (5.7 节)。

A2.4.13 安全性

本节规定定量安全性需求 (10.2 节) 及定性安全性因素 (5.5 节及 10.2 节)。

A2.4.14 人因工程

本节规定与人为因素相关的所有需求 (5.5 节)。本节通常在人不是系统一部分，但与系统存在接口的情况下使用。本节在这些情况下使用，并仅规定系统一些可能受到该接口影响的方面。如果人是系统的一部分，则与人有关的特定的需求，比如人数、培训及物理特性等，将在 A2.4.19 节中规定。

A2.4.15 安保需求

此节规定关系到系统的安保需求，无论是电气类还是物理类。

A2.4.16 软件需求

本节规定顶层软甲需求 (10.3 节)。本节还包含固件要求。

A2.4.17 设计及实现约束

A2.4.17.1 物理特性

物理特性包括质量特性 (5.2 节) 及尺寸特性 (5.3 节)。

A2.4.17.2 按成本设计需求

虽然 MIL-STD-961D 不包括成本约束，此节仍包含关于系统的任何成本约束 (5.11 节)。包括飞机系统级成本 (8.1 节，第 17 项)、直接运营成本 (8.6 节)

contains any cost constraints on the system as described in Section 5.11. Included are aircraft system-level costs as described in Section 8.1, Item 17, direct operating costs (DOC) described in Section 8.6, and indirect costs described in Section 8.6. These costs are normally not treated in military specifications.

A2.4.17.3 Noise

Noise is also not specifically mentioned in MIL-STD-961D. This section specifies the noise levels the system may emit as described in Section 5.9, as opposed to the noise environment of Section 5.6, above. This is an important constraint for commercial aircraft.

A2.4.17.4 Flexibility and expansion

In commercial aircraft, flexibility and expansion are important from a total aircraft point of view. MIL-STD-961D mentions flexibility and expansion in a software context. This section specifies features the aircraft should have in order to accommodate future growth options described in Section 5.13.

A2.4.18 Support requirements

This section specifies specific requirements for maintenance and other support. For the aircraft system, this section will simply refer to the support subsystem to be described in Section A2.9, below.

A2.4.19 Personnel and training

This section specifies the number and qualifications of personnel which will operate, maintain, or service the aircraft as described in Section 8.1. It also specifies the levels of training required for each personnel category.

A2.4.20 Requirements traceability

This section may consist of a table to provide traceability from each requirement to higher-level requirements, as discussed in Section 4.2.

及间接成本 (8.6 节)。军用规范通常不处理这些成本。

A2.4.17.3 噪声

噪声也没有在 MIL-STD-961D 具体提到。本节规定可能发出的噪声水平 (5.9 节)，与上述噪声环境 (5.6 节) 相反。这是商用飞机的重要约束。

A2.4.17.4 灵活性及可扩展性

在商用飞机领域，灵活性及可扩展性从整个飞机角度看十分重要。MIL-STD-961D 介绍了软件方面的灵活性及可扩展性。本节规定飞机应具备的特征，以适应未来发展选项 (5.13 节)。

A2.4.18　支持需求

本节规定关于维修和其他支持的具体需求。对于飞机系统，本节将简要介绍以下 A2.9 节所述的子系统支持。

A2.4.19　人员及培训

本节规定飞机运行、维修或服务人员的数量及资质 (8.1 节)。同样还规定每一类人员需要的培训级别。

A2.4.20　需求的可追溯性

本节可能是一份表格，可使每一条需求追溯到该需求的上层需求，如第 4.2 节中所述。

A2.5 Interface Requirements

This section specifies the functional and physical interfaces between this entity and any other entity, as discussed in Chapter 6.

A2.6 Design and Construction

This section specifies design standards discussed in Section 5.8, workmanship requirements, and special production inspection requirements as required by the producibility requirements discussed in Section 5.14.

A2.7 Documentation

This section specifies any documentation which must be delivered as part of the deliverable package to the airline, such as training and maintenance manuals.

A2.8 Precedence of Requirements

The purpose of this section is to specify the order of relative importance of the requirements. For example, requirements related to safety would supersede any other requirements.

Section 4.8 explains how various requirements may be in conflict. That section also describes various ways of resolving those conflicts. This section should explain the resolution of that analysis and the resulting driving requirements.

A2.9 Major Component Characteristics

MIL-STD-961D does not have a section pertaining to the requirements of major component characteristics. However, some systems engineers find it useful. This section will provide the top-level requirements for the elements one level below the system addressed in this specification. For the aircraft system, these elements are the training equipment, support equipment, and facilities. For the aircraft level, these elements are the subsystems of Figure 2.1.

A2.5　接口需求

本节规定本实体与任何其他实体之间的功能及物理接口，如第 6 章所述。

A2.6　设计与建造

本节规定设计标准 (5.8 节)、工艺水平需求及可生产性需求所要求的特殊生产检验需求 (5.14 节)。

A2.7　文档管理

本节规定作为向航空公司提交的交付物包的一部分，必须交付的任何文档，例如《培训手册》及《维修手册》。

A2.8　需求优先次序

本节的目的是规定需求的相对重要性顺序。例如，涉及安全的需求时，其优先级将高于任何其他需求。

本书 4.8 节介绍了不同的需求可能如何发生冲突。此节还将介绍解决冲突的各种方法。本节应介绍分析的解决方案，以及产生的推动需求。

A2.9　主要部件特征

MIL-STD-961D 没有关于主要部件特征需求的章节。然而，一些系统工程师发现主要部件特征十分有用。本节将提供在本规范中提及的系统的下一级元素的顶层需求。对于飞机系统，这些元素是培训设备，支持设备及设施。对于飞机级，

A2.10   Verification

This section contains the verification matrix and any other material related to the verification of requirements for the system as explained in Chapter 11. The verification matrix should contain all the elements required for certification discussed in Section 10.1 Section 10.17. It also contains the verification descriptions for all requirements.

A2.11   Preparation for Delivery

This section provides guidance on how specific elements of the aircraft are to be delivered to the manufacturer.

A2.12   Notes

Any information which should be made known as background information or as instructions to the suppliers may be included.

A2.13   Appendices

This section may be used for any information deemed useful. It may contain requirements which are of a temporary or limited nature, such as for a test aircraft or requirements for a specific customer. Only those requirements which are necessary to describe the temporary features of the item are listed in this appendix.

这些元素是图 2.1 所示的子系统。

A2.10　验证

本节包含第 11 章讨论的与系统需求有关的验证矩阵，以及所有其他材料。验证矩阵应该包含合格审定 (10.1 节及 1.17 节) 要求的所有元素，同时还包含了所有需求的验证说明。

A2.11　交付准备

本节提供关于飞机的特定元件如何交付给制造商的指南。

A2.12　备注

作为背景信息或作为对供应商的指令应了理解的任何信息都应包括在内。

A2.13　附录

此节可以用于任何认为有用的信息。它可能包含临时的或有限属性的需求，例如，关于试验飞机的需求，或对某一具体客户的需求。只有那些对于描述项目临时特征必需的需求才被列入此附录中。

Appendix 3
Systems Engineering Automated Tools

Frequently SE is facilitated by an automated SE tool. In fact, the vast number of functions and requirements make automated tools a virtual necessity. This appendix summarizes the features and benefits of automated tools.

A3.1   Features of Automated SE Tools

Following is a summary of the basic features an automated SE tool usually has. The SE tool should be able to:

1. Support functional analysis.
2. Assign a requirement (or requirements) to each function as the requirements are developed. In some tools, the product of the requirements task is the requirements allocation sheets (RASs).
3. Allocate the requirements to specific components, subsystems, or systems. This allocation is also recorded on the RASs.
4. Create a traceability record of requirements so that the source of each requirement can be identified.
5. Sort for missing data in order to assure completeness.
6. Assign consistent design constraints in keeping with regulatory requirements and other sources.
7. Create specifications automatically from the data base in standard or custom specification formats.

Many tools have other specialized capabilities. These include configuration management and simulation capabilities. However, the above seven features are considered to be the most fundamental capabilities.

附录3 系统工程自动化工具

通常来讲，自动化的系统工程工具可以促进系统工程运行。但事实上，数量众多的功能及需求使得自动化工具成为实质上的必需品。本附录对自动化工具的特征及优点进行了概述。

A3.1 自动化系统工程工具的特征

以下是自动化系统工程工具通常具有的基本特征。系统工程工具应能：

1. 支持功能分析。

2. 在需求开发时，将需求（或一些需求）分配至各项功能。在一些工具中，这项需求任务的产物是需求分配表(RAS)。

3. 分配需求给特定的部件、子系统或系统。这种分配也记录在需求分配表内。

4. 给这些需求创建一个可以追溯的记录，以辨识每项需求的来源。

5. 给丢失数据排序，以保证完整性。

6. 指定一致的设计约束，以符合监管要求和其他来源。

7. 根据数据库，按标准格式或客户规定的格式自动生成规范。

很多工具具有其他一些专门的功能。其中包括构型管理及仿真能力。然而，上述7项特征是最基本的能力。

A3.2   Benefits of Automated SE Tools

Automated requirements tools have the following benefits:

1. They are fast and can accomplish the analysis in a small fraction of the time required by hand.
2. They provide the traceability of requirements discussed above.
3. They provide increased program technical control by assuring that all organizations, including suppliers, if required, are using the same data base. The tool becomes an integral part of the configuration management process.
4. They enhance data management by providing a single source of data.
5. They provide an efficient tool for fast change control.
6. They provide a medium for requirements data exchange among groups and with suppliers.

A3.2　系统工程自动化工具的优点

自动化需求工具主要优点如下：

1. 它们很快捷，完成分析所用时间相比手动操作大幅减少。

2. 它们可以为上述需求提供可追溯性。

3. 可以提供额外的程序技术控制，以保证包括供应商在内的所有组织在需要的情况下使用相同的数据库。该工具是构型管理过程一个不可或缺的部分。

4. 可以通过提供单一数据源以加强数据管理。

5. 为快速变更控制提供有效工具。

6. 为团队间及团队与供应商之间的需求数据交换提供媒介。

符　号

AR	Aspect ratio	展弦比
C_D	Drag coefficient	阻力系数
C_{D0}	Lift-independent drag coefficient	零升阻力系数
C_L	Lift coefficient	升力系数
$CL^2/\Pi ARe$	Induced drag	诱导阻力
C_{LIC}	Initial cruise lift coefficient	初始巡航升力系数
C_{Lmax}	Maximum lift coefficient	最大升力系数
C_{Lto}	Take-off lift coefficient	起飞升力系数
ΔC_{Dc}	Compressibility drag coefficient	压缩阻力系数
e	Lift efficiency	升力效率
Fb	Block fuel	轮挡燃油量
L/D	Lift to drag ratio	升阻比
$\Lambda_{C/4}$	Sweepback angle at quarter chord	1/4 弦处后掠角
Ma_{div}	Divergence Mach number	发散马赫数
R	Total range	总航程
R_cl	Climb range	爬升距离
R_{cr}	Cruise range	巡航距离
SFC	Specific fuel consumption	单位耗油率
$t/c)_{ave}$	Average thickness to chord ratio	平均厚度比
U	Aircraft utilization factor	飞机利用率系数
V_b	Block speed	轮挡速度
$W/S)_{IC}$	Initial cruise wing loading	初始巡航翼载
$W/S)_{to}$	Take-off wing loading	起飞翼载
W_0	Initial cruise weight	初始巡航重量
W_1	$W_0 - W_f$	中间重量
W_f	Weight of fuel	燃油重量
W_{to}	Weight at take-off	起飞重量
W_{wing}	Wing weight	机翼重量

W_{fuselage}	The fuselage weight	机身重量
$W_{\text{landing gear}}$	The landing gear weight	起落架重量
$W_{\text{nacelle \& pylon}}$	The nacelle and pylon weight	短舱及吊挂重量
W/T	Thrust loading	推重比
W_{TS}	The tail section weight	尾翼重量
W_{fuel}	The fuel weight	燃油重量
W_{payload}	The payload weight	商载重量
$W_{\text{fixed equipment}}$	The fixed equipment weight	固定设备重量

缩略语

术语	定义	中文
AC	Advisory circular	咨询通告
AC	Alternating current	交流（电）
ACARS	ARINC communication addressing reporting system	ARINC 通信寻址与报告系统
ADI	Attitude direction indicator	姿态方向指示器
AFM	Airplane flight manual	飞机飞行手册
AIT	Analysis and integration team	分析与集成小组
ALAR	Approach and landing accident reduction	减少进近及着陆事故
APU	Auxiliary power unit	辅助动力装置
ARINC	Aeronautical Radio Incorporated	航空无线电公司
ARP	Aerospace recommended practice (SAE)	航空推荐项目实践 (SAE)
ASA	Airplane state awareness	飞机状态感知
ASAP	Aviation safety action plan	航空安全行动计划
ATA	Air Transport Association of America	美国航空运输协会
ATC	Air traffic control	空中交通管制
AVOID	Airborne volcanic object identifier and detector	机载火山灰识别及探测装置
BCD	Baseline concept document	基线概念文件
BFE	Buyer furnished equipment	买方提供设备
BITE	Built-in test equipment	机内检测装置
BWB	Blended wing-body	翼身融合体
CAST	Commercial Aviation Safety Team	商用航空安全小组
CCA	Common cause analysis	共因分析
CDE	Chief design engineer	总设计师
CDR	Critical design review	关键设计评审
CEO	Chief executive officer	首席执行官
CFD	Computational fluid dynamics	计算流体力学

术语	定义	中文
CFIT	Controlled flight into terrain	可控撞地飞行
CM	Configuration management	构型管理
CMR	Certification maintenance requirement	审定维修要求
CONOPS	Concept of operations	运行概念
COTS	Commercial off-the-shelf products	货架产品
c.p.	Center of pressure	压力中心
CRM	Cockpit (or crew) resource management	驾驶舱（机组）资源管理
CSD	Constant-speed drive	恒速传动
CSE	Chief systems engineer	系统工程总师
DC	Direct current	直流（电）
DCAS	Digital core avionics system	数字核心航电系统
DF	Development fixture	研制样机
DFMA	Design for manufacture and assembly	面向制造及装配的设计
DME	Distance measurement equipment	测距器
DOC	Direct operating cost	直接运营成本
DOORS	Dynamic Object-Oriented Requirement System	面向对象的动态需求系统
DOS	Director of safety	安全性主管
ECS	Environmental control system (or subsystem)	环控系统（或子系统）
EDF	Electronic development fixture	电子研制样机
ELB	Emergency locator beacon	应急定位信标
ELT	Emergency locator transmitter	应急定位发射机
EMI	Electro-magnetic interference	电磁干扰
ESE	Enterprise systems engineering	企业系统工程
ESR	Engineering safety review	工程安全性评审
EWO	Engineering work order	工程工作指令
FAA	Federal Aviation Administration	（美国）联邦航空局
FAR	Federal aviation regulation	（美国）联邦航空条例
FBL	Fly-by-light	光传操纵
FBW	Fly-by-wire	电传操纵

术语	定义	中文
FCA	Functional configuration audit	功能构型审核
FDA	Food and Drug Administration	食品药品监督管理局
FEA	Finite element analysis	有限元分析
FFBD	Functional flow block diagram	功能流框图
FFRR	First flight readiness review	首飞成熟度评审
FFR	First flight review	首飞评审
FHA	Functional hazard assessment	功能危害性评估
FMS	Flight management system	飞行管理系统
FOD	Foreign object debris	机场跑道异物
FOQA	Flight operational quality assurance	飞行运行质量保证
FTA	Fault tree analysis	故障树分析
G&A	General and administrative [costs]	行政及管理 [成本]
GMT	Greenwich mean time	格林威治标准时间
GPS	Global positioning system	全球定位系统
GPWS	Ground proximity warning system	近地警告系统
HBPR	High by-pass ratio	高涵道比
HF	High frequency	高频
HIRF	High-intensity radiation field	高强度辐射场
HSCT	High-speed civil transport	高速民用运输机
HSI	Horizontal situation indicator	水平状态指示器
HUD	Heads-up display	平视显示器
ICA	Initial cruise altitude	初始巡航高度
ICD	Interface control drawing (or document)	接口控制文件 (或文档)
IDEF0	Integrated definition for function modeling—type 0	功能建模
IDG	Integrated drive generator	综合传动发电机
IEEE	Institute of Electrical and Electronic Engineering	电气电子工程师学会
IFR	Instrument flight rules	仪表飞行规则
ILS	Instrument landing system	仪表着陆系统
IMACH	Improved methods for aircraft cargo handling	空运货物装卸改进方法

术语	定义	中文
INCOSE	International Council on Systems Engineering	国际系统工程协会
INS	Inertial navigation system	惯性导航系统
IPD	Integrated product development	集成产品研制
IPT	Integrated product team	集成产品团队
IVATF	International Volcanic Ash Task Force	国际火山灰工作团队
JAA	Joint Aviation Authorities	（欧洲）联合航空局
JAR	Joint airworthiness requirements	（欧洲）联合适航要求
LRU	Line replaceable unit	航线可更换单元
LSS	Large-scale system	大规模系统
LSSI	Large-scale system integration	大规模系统集成
MAP	Maximum allowable probability	最大允许概率
MCBF	Mean cycles between failures	平均无故障周期
MCBUR	Mean cycles between unscheduled removals	平均非计划拆换间隔
MEL	Minimum equipment list	最低设备清单
MEW	Manufacturer's empty weight	制造空机重量，制造空重
MMEL	Master minimum equipment list	主最低设备主清单
MMH/1 000FH	Maintenance man-hours per 1 000 flight hours	每 1 000 飞行小时维修工时
MN$/1 000FH	Maintenance cost per 1 000 flight hours	每 1 000 飞行小时维修成本
MSAW	Minimum safe altitude warning	最低安全高度警告
MT$/1 000FH	Material cost per 1 000 flight hours	每 1 000 飞行小时材料成本
MTBF	Mean time between failures	平均无故障时间
MTBUR	Mean time between unscheduled removals	平均非计划拆换间隔时间
MTTR	Mean time to repair	平均修理时间
MVA	Minimum vectoring altitude	最小引导高度
NDI	Non-development item	非研制项
NEA	Nitrogen enriched air	富氮空气
NOX	Nitrous oxide	氮氧化物
NTSB	National Transportation Safety Board	国家运输安全委员会

术语	定义	中文
OEM	Original equipment manufacturer	原始设备制造商
OSHA	Occupational Safety and Health Administration	职业安全与卫生管理局
PBW	Power-by-wire	电传动力
PCA	Parametric cost analysis	参数成本分析
PCA	Physical configuration audit	物理构型审核
PCA	Propulsion controlled aircraft	推进控制飞机
PDR	Preliminary design review	初步设计评审
PRA	Probabilistic risk analysis	概率风险分析
PSAC	Plan for software aspects of certification	软件合格审定计划
PSE	Product systems engineering	产品系统工程
PSSA	Preliminary system safety assessment	初步系统安全性评估
QFD	Quality function deployment	质量功能展开
RAS	Requirements allocation sheet	需求分配表
RATs	Ram air turbine	冲压空气涡轮
RI	Runway incursion	跑道进近
RNAV	Area navigation	区域导航
RNP	Required navigation procedures	按需导航程序
RTCA	Radio Technical Commission for Aeronautics (former name of RTCA, Inc.)	航空无线电技术委员会
SAE	Society of Automotive Engineers	（美国）汽车工程师协会
SATCOM	Satellite communications	卫星通信
SCM	Software configuration management	软件构型管理
SCI	Software configuration index	软件构型索引
SDR	System design review	系统设计评审
SE	Systems engineering	系统工程
SEBoK	Systems Engineering Body of Knowledge	系统工程知识体系
SEIT	Systems engineering and integration team	系统工程与集成小组
SELCAL	Selective calling	选择呼叫

术语	定义	中文
SEMP	SE management plan	系统工程管理计划
SFC	Specific fuel consumption	单位耗油率
SOP	Standard operating procedure	标准操作程序
SoS	System of systems	系统之系统
SOW	Statement of work	工作说明
SSE	Service systems engineering	服务系统工程
SQA	Software quality assurance	软件质量保证
SRR	System requirements review	系统需求评审
SSA	System safety assessment	系统安全性评估
SVR	System verification review	系统验证评审
TAWS	Terrain avoidance warning system	地形提示和告警系统
TBD	To be determined (for a requirement)	待定（针对需求）
TCAS	Traffic collision avoidance system	空中防撞系统
TEAM	Technology evaluation and adaptation methodology	技术评估与适应方法论
TPM	Technical performance measure	技术性能指标
TQM	Total quality management	全面质量管理
TRL	Technology readiness levels	技术成熟度级别
UER	Unscheduled engine removals	非计划发动机拆换
VFR	Visual flight rules	目视飞行规则
VGSI	Visual glide slope indicator	目视下滑道指示器
VHF	Very high frequency	甚高频
VMC	Visual meteorological conditions	目视气象条件
VOR	VHF omni-directional radio	甚高频全向无线电信标
VSCF	Variable-speed constant frequency	变速恒频

术 语 表

Absorption 吸收	The capability of withstanding a design-level disruption	承受设计级破坏的能力
Airworthiness 适航性	The condition of an item (aircraft, aircraft system, or part) in which that item operates in a safe manner to accomplish its intended function (ARP 4754A, 2010)	产品（飞机、飞机系统或部件）以安全的方式完成其预定功能的状态（ARP 4754A, 2010）
Allocation [def. 1] 分配[定义1]	The assignment of a performance requirement to a function	性能需求向功能的分配
Allocation [def. 2] 分配[定义2]	The assignment of a requirement to a system element	需求向系统元件的分配
Allocation [def. 3] 分配[定义3]	The breakdown of a top-level requirement into its subordinate components, for example, weight	顶层需求分解到其组成部件中；如：重量
Analysis 分析	A type of verification. Any kind of mathematical, computational, or logical task performed to verify a requirement which cannot be verified in any other manner. Includes in-service evaluation and similarity analyses	一种验证类型。通过任一一种数学的、计算的或逻辑的任务来验证需求，并且无其他验证方式。包括在役评估和相似性分析
Anthropocentric 人类中心主义	Pertaining to the human view, for example, the human view of a system	关于人的观点，例如，人对系统的观点
Anthropometry 人体测量学	An applied branch of anthropology, concerned with the measurement of the physical features of people (Chapinis, 1996)	人类学的一个应用分支，关于人体特征的测量（chapinis, 1996）
Architecting 构建	The process of determining the arrangement of the parts of a system	确定系统内各部件布置的过程
Architecture 架构	The arrangement of the parts of a system or subsystem; this can apply either to the physical architecture or the functional architecture	系统或子系统中部件的布置，可以应用于物理架构或功能架构
Arousal 激励	A form of stress where activation is resolved following termination of the stressor or perturbing event (Satchell, 1993)	一种压力形式，伴随着压力源或扰动事件终止后产生的激活。（Satchell, 1993）

Assurance 保证	The planned and systematic actions necessary to provide adequate confidence that a product or process satisfies given requirements (DO-178B)	为确保产品或过程满足给定的要求而必需的有计划的和系统化的行动（DO-178B）
Blended wing-body (BWB) 翼身融合体	An aircraft configuration in which the wing and body are integrated into a single unit	一种机翼和机体集成到一体的飞机构型
Canard 鸭式布局	A horizontal surface located on the forward portion of the fuselage for improved control. Canards have rarely been used on modern aircraft, but were part of the Wright brothers' Kitty Hawk	为了改进操控性能而将水平安定面布置在机体前部的布局方式。该布局方式在现代飞机中很少用到，但在怀特兄弟的猎鹰飞机上使用了
Certification 合格审定	The legal recognition that a product, service, organization or person complies with the applicable requirements. Such certification comprises the activity of technically checking the product, service, organization or person, and the formal recognition of compliance with the applicable requirements by issue of a certificate, license, approval or other document as required by national laws and procedures (ARP 4754A, 2010)	法律认可产品、服务、组织或人员符合要求。这种合格审定包括：对产品、服务、组织或个人的技术性检查活动，以及通过颁发国家法律和程序要求的证书、执照、批准书或其他文件正式认可与适用要求的符合性 (ARP 4754A,2010)
Certification basis 合格审定基础	The set of particular standards and FARs on which the certification of an aircraft is based	据以进行合格审定的特定标准和联邦适航规章的集合
Certification maintenance requirement (CMR) 审定维修要求	A required periodic task established during the design certification of the aircraft as an operating limitation on it	飞机设计合格审定过程中所制定的周期性任务要求，也是对飞机运行的一种限制
Change-based aircraft 改装飞机	An aircraft for a specific customer which may have a large number of requested changes	针对特殊客户的一类飞机，可能需要大量的需求
Cluster analysis 群分析	An analysis by which different parts of a system are shown to have functional affinity and hence constitute a potential subsystem; this technique has been shown to be useful in the architecting of a system	一种分析方式，通过分析发现系统的不同部分间存在功能联系，因此各个部分都可以构成一个潜在的子系统。这种技术在系统架构分析中很有用

Cockpit resource management (CRM) 驾驶舱资源管理	Class of programs designed to reduce the number of incidents and accidents which are behavioral in origin. Focus on such activities as training and pilot selection processes	一类用于减少由于（机组）初始行为造成的事故数量的程序，重点关注驾驶员培训及筛选过程等活动
Common cause 共因	Event which bypasses or invalidates redundancy or independence; that is, an event which causes the simultaneous loss of several redundant or independent items	一种绕过或导致冗余性和独立性失效的事件；该事件可导致多个冗余或独立设备的同时丧失
Common cause analysis 共因分析	Generic term encompassing zonal safety analysis, particular risk analysis, and common mode analysis. (ARP 4654A, 2010)	包含区域安全性分析、特定风险分析及共模分析的通用术语。（ARP 4654A，2010）
Complexity 复杂性	The state of a system characterized by many components, many interfaces, and variability in the relationship among the components; this term is also used to describing the difficulty in understanding such a system	指一个系统的状态，该系统的特征是组件众多，接口繁多，组件之间的关系多变，这个术语也被用来描述理解这样一个系统时的难度
Component 部件	Any self-contained part, combination of parts, subassemblies, or units, which perform a distinctive function necessary to the operation of the system	任何独立的部件，零件、部件，或单元的组合，可以执行系统运行所必需的特有功能
Configuration index 构型索引	A catalogue of the physical elements which comprise the aircraft and its subsystems	组成飞机及其子系统的物理单元目录
Configuration management 构型管理	A process for controlling both the configuration of the aircraft and the data required to define the aircraft	控制飞机构型和定义飞机所需数据的过程
Constraint 约束	Any non-performance requirement. Constraints include weight, dimensions, environments, and any other factor which may constrain the design	非性能需求。约束包括重量、尺寸、环境及其他可能约束设计的任何因素
Control 控制	Technical SE management activities which occur during all program phases and at all levels of the aircraft hierarchy, such as configuration and risk management	系统工程技术管理活动，覆盖（飞机）项目的每个阶段及飞机的各个层级，诸如构型和风险管理
Demonstration 演示	A method of verification, similar to test except does not require sophisticated instrumentation	一种验证方法，类似于试验，区别是演示不需要复杂的仪器（仪表）

Derivative aircraft 衍生型飞机	An aircraft which utilizes major components of existing aircraft as the basis for the development of an aircraft which meets new requirements	以现有飞机大部件为基础研制的满足新需求的飞机
Derived requirements 衍生需求	Requirements that are dependent on the design solution	依赖于设计方案的需求
Design 设计	The result of the design process, as distinct from the requirements process	设计过程的结果，有别于需求过程
Design requirement 设计需求	The design characteristic which is the product of the synthesis process	一种设计特征，是综合过程的产品
Development fixture 研制样机	A mock-up of the aircraft used during development to assure that the spatial allocation for all components is correct and that they fit correctly	研制过程中使用的一种飞机模型样机，用以确保所有部件的空间位置正确，与所分配的位置相吻合
Disruption 扰动	Damage or loss of functionality resulting from an encounter with a threat	由外部威胁造成的功能破坏或丧失
Drift correction 偏航修正、漂移修正	The capability of anticipating or detecting a disruption in advance and performing a corrective action	提前预测或检测扰动，并实施修正的能力
Electromagnetic interference 电磁干扰	The disruptive interference caused by a magnetic field emitted by an electrical component, such as a generator	由电气部件（如发电机）发射的磁场引起的破坏性干扰
Electronic development fixture 电子研制样机	An electronic, that is, computer-generated, version of a development fixture	一种计算机生成的，电子版的研制样机
Element 元件	A generic term to describe any subdivision of the aircraft hierarchy. An element may be a segment, system, subsystem, or component. The aircraft itself is an element	描述飞机层级细分的通用术语。一个元件可以是一个部段、系统、子系统，或者部件。飞机本身也是一个元件
Empennage 尾翼	The entire tail assembly of an aircraft. May or may not include the tail cone depending on the practice of the manufacturer	飞机整个尾部组件。取决于制造商的工艺，可能包括或者不包括尾锥
Enterprise Systems Engineering (ESE) 企业系统工程	The systems engineering of an entire enterprise, such as a commercial aircraft enterprise	整个企业的系统工程，例如商用飞机公司
Environment 环境	The natural and induced conditions experienced by a system including its people, product, and processes (ANSI/EIA 632, 1999)	系统（包括系统自身的人、产品和过程）经历的自然的和引发的环境 (ANSI/EIA 632, 1999)

Examination 检查	A type of verification. A visual confirmation that a requirement has been met. Also called inspection	一种验证类型。通过目视确认需求被满足。也称为做检验
Failure condition 失效状态	A condition having an effect on the aircraft and/or its occupants, either direct or consequential, which is caused or contributed to by one or more failures or errors, considering flight phase and relevant adverse operational or environmental conditions or external events (ARP 4754A, 2010)	在考虑飞行阶段、相关的不利运行或环境条件，或外部事件情况下，有一个或多个失效/错误引起/造成对飞机或乘员的直接或相继影响的状态 (ARP 4754A, 2010)
Firmware 固件	Any electronic device which contains imbedded programming logic	植入编程逻辑的电子器件
Flow-down 向下分解	The process of passing or allocating any parameter, requirement, or function from a higher level of the aircraft hierarchy to a lower level	飞机顶层向下传递或分配的过程，包括：参数、需求或功能
Function 功能	A task, action, or activity performed to achieve a desired outcome	为了达到期望的结果而执行的任务、行动或者活动
Functional allocation 功能分配	Assignment of requirements to lower-level functions	将需求分配给下一层级的功能
Functional analysis 功能分析	Examination of a defined function to identify all the subfunctions necessary to the accomplishment of that function	对已定义功能的检查，识别实现该功能所必需的所有子功能
Functional hazard assessment (FHA) 功能危害性评估	A systematic, comprehensive examination of functions to identify and classify Failure Conditions of those functions according to their severity (ARP 4754A, 2010)	对功能进行系统的、综合的检查，根据其严重程度对功能进行失效状态识别和归类 (ARP 4754A, 2010)
Fuselage 机身	The body of an aircraft. May or may not include the nose or the tail cone depending on the practice of the manufacturer	飞机机身。取决于制造商的工艺，可能包括或者不包括机头和尾椎
Hierarchy 层次结构	The layered arrangement of a system, especially the abstract hierarchy, or mental model of a system	系统的分层布置，特别是抽象的层级或者系统的心智模型
Holistic 整体的	Treating the system as a whole taking into account the interactions among the elements as opposed to treating the elements separately	把系统当作一个整体，考虑元素间的交互，而不是把元素区别开来对待
Human in the loop 人在环	The capability of having humans in the system where needed	必要时将人纳入到系统中的能力

IDEF0 功能建模	Integrated Definition for Function Model, a method of functional analysis (see Reductionism)	功能建模一种功能分析的方法（参见简化论）
INCOSE	International Council on Systems Engineering	国际系统工程协会
Inspection 检验	Alternative term for examination	检查的另一说法
Integrated product development (IPD) 集成产品研制	A systematic management approach to the development of products such as aircraft and aircraft subsystems. Key elements of IPD include cross-functional integrated product teams (IPTs) and integrated and concurrent activities to develop products and processes	产品研制中的系统化的管理方法，例如飞机研制和飞机子系统研制。IPD 元素包括跨职能的产品协同设计组，以及产品研制和过程中的集成和并行活动
Interface 接口	A boundary between two system elements.	两个系统元素之间的边界
Interface control drawing (ICD) 接口控制文件	A document which captures all basic information about interfaces between two elements, including the type of interface (electrical, pneumatic, hydraulic, and so on) and the interface characteristics (functional or physical)	捕获两个元素间接口的所有基本信息的一份文件，包括接口的类型（电气，气源、液压等）和接口的特征（功能性或者物理的）
Inter-node interaction 节点内交互	The capability of two or more components of a system to interact with each other, such as communicate	系统内两个或多个部件之间的交互能力，如通信
Issue 问题	A risk that has already been realized or whose consequence is inevitable regardless of the mitigation step	已发现的风险，或不论有无缓解措施结果都不可避免的风险
Item 项目	One or more hardware and/or software elements treated as a unit (see Product)	被当作一个单元对待的一个或多个硬件和 / 或软件元素集（见产品）
Large-scale system integration 大规模系统集成	The integration of many systems containing complex interfaces	很多系统的集成，包含复杂的接口
Latent fault or error 潜在故障或错误	Design flaw that lies undetected until a catastrophic event occurs	设计缺陷，直到灾难性的事件发生才被发现
Layered defense 多重防护	The capability of having multiple means of withstanding a disruption; also called defense in depth	多重方式承受破坏的能力，也叫纵深防护
Limit degradation 限制退化	The capability of arresting the degradation of the absorption capability due to aging or lack of maintenance	阻止由于老化或缺乏维修导致吸收能力退化的能力

Margin 余量	An capability of withstanding increased levels of a disruption due to uncertainty in the threat level	承受因危险程度的不确定性导致的更高程度的损坏的能力
Master minimum equipment list (MMEL) 主最低设备清单（MMEL）	A document established by the manufacturer which lists what aircraft equipment can be inoperative (and under what conditions) and still fly the aircraft safely (see MEL)	由制造商编制，列举出在哪些飞机设备不工作的情况下（在何种条件下）飞机仍然可以安全飞行的文件（见 MEL）
Mean time between failures (MTBF) 平均无间隔时间（MTBF）	Mathematical expectation of the time interval between two consecutive failures of a hardware item. NOTE: The definition of this statistic has meaning only for repairable items. For non-repairable items, the term Mean Time To Failure (MTTF) is used (ARP 4754A, 2010)	硬件项目两个连续故障间隔时间的数学期望值。注：此统计定义仅对可维修项目有意义。对于不可维修的项目，使用平均故障时间（MTTF）（ARP 4754A，2010）
Mean time between unscheduled removals (MTBUR) 平均非计划拆换间隔时间	Time interval between two consecutive unscheduled removals of an item. An unscheduled removal is a removal of an item brought about as a result of a known or suspected malfunction and/or defect	两个连续非计划拆换一个项目的时间间隔。非计划拆换是指某一项目因已知或疑似的功能异常或 / 和缺陷而被拆换
Metric 指标	A measure, usually quantitative, of the value of a process, such as SE	对过程价值的度量，通常是定量的，如系统工程
Minimum equipment list (MEL) 最低设备清单	A document established by the airline which lists what aircraft equipment can be inoperative (and under what conditions) and still fly the aircraft safely; a subset of the MMEL (see MMEL)	由航空公司编制，列举出在哪些飞机设备不工作的情况下（在何种条件下）飞机仍然可以安全飞行的文件，是 MMEL 的子集（见 MMEL）
Needs 需要	Those desires usually from the commercial aircraft customer that will result in an economically viable aircraft. These needs will be translated into verifiable product requirements	通常来说，期望来自于商用飞机的客户，有助于飞机经济性的实现。需要将被转化成可验证的产品需求
Off-the-shelf 货架产品	Pertaining to a commercially available product which meets the specified requirements	满足指定要求的市售产品
Performance requirement 性能需求	The extent to which a mission, operation, or function must be executed, generally measured in terms of quantity, quality, coverage, timeliness, or readiness(ANSI/EIA 632, 1999)	任务、运行或功能必须达到的执行程度，一般用数量、质量、范围、及时性或状态进行衡量（ANSI/EIA 632，1999）

Peripheralization 边缘化心理	A complex psychological state which results from a shift in the pilot role from direct contact and control of the aircraft to one of system monitor (Satchell, 1993)	一种复杂的心理状态，由驾驶员角色从直接接触和控制飞机转变到系统监控者而导致（Satchell, 1993）
Preliminary system safety assessment (PSSA) 初步系统安全性评估（PSSA）	A systematic evaluation of a proposed system architecture and its implementation, based on the Functional Hazard Assessment and Failure Condition classification, to determine safety requirements for systems and items (ARP 4754A, 2010)	基于功能危害性评估和失效状态分类，对提议的系统架构及其实现进行系统性的评估，用来确定系统及其项目的安全性要求（ARP 4754A, 2010）
Process assurance 过程保证	The set of activities that ensure that the development of the aircraft, its subsystems, and the supporting processes are appropriate, maintained, and followed	一系列的活动，用于确保飞机、飞机子系统及支持过程的研制是适当的、可维持的和被遵循的
Product 产品	A generic term used for any item, either hardware or software, which will be the end result of the SE process. A product can be a system, subsystem, or component. The term "item" is also used	通用术语，用于任何项目，无论是硬件或软件，系统工程过程的最终结果。产品可以是一个系统、子系统或组件。也可使用术语"项目"
Product Systems Engineering (PSE) 产品系统工程（PSE）	The systems engineering of products containing hardware or software, such as an aircraft	产品的系统工程，包括硬件和软件，如一架飞机
Reductionism 简化论	The concept that the parts of a system should be treated separately so that a system is built up of the sum of its parts (see Holism)	系统部件分开处理的概念，因此系统即是部件的总和（见 Holism）
Redundancy, functional 功能冗余	Having two independent and physically different means to accomplish a function, also called design diversity	用两种独立的、物理上不同的方法来实现一个功能，也被称为设计的多样性
Redundancy, physical 物理冗余	Having two independent and physically identical means to accomplish a function, also called design redundancy	用两种独立的，物理上完全相同的方法来实现一个功能，也被称为设计冗余
Reliability 可靠性	The probability that an item will perform a required function under specified conditions, without failure, for a specified period of time (ARP 4754A, 2010)	项目在规定条件下和规定的时间内，执行所要求的功能而不发生失效的概率（ARP 4754A, 2010）

Reorganization 重构	The capability of restructuring a system when needed to recover from a disruption	需要从扰动中恢复时重构系统的能力
Requirement 需求	A statement of required performance or design constraint to which a product must conform. A requirement must be verifiable	对产品必须满足的性能要求或设计约束的陈述。需求必须是可验证的
Resilience 恢复力	The capability of anticipating or detecting a disruption, surviving that disruption, and recovering all or part of the initial functionality	预测或检测扰动，在扰动后留存，并恢复所有或部分初始功能的能力
Risk 风险	An undesirable situation or circumstance that has a realistic probability of occurring and an unfavorable consequence	一种不期望的状况或环境，它有现实的发生概率和不利后果
Safety 安全性	The state in which risk is acceptable (ARP 4754A, 2010)	风险处于可接受范围的状态（ARP 4754A, 2010）
Segment 部段	A major collection of aircraft equipment. A segment may be the wing, a subsystem, or simply a collection of elements having similar functions	飞机设备的大集合。可以是机翼，子系统或仅仅是有相同功能的元件的集合
Service Systems Engineering (SSE) 服务系统工程	The systems engineering of a service, such as the maintenance of an aircraft	服务系统工程，例如飞机的维修
Similarity 相似性	In systems engineering a type of verification by analysis. In ARP 4754A (2010) applicable to systems similar in characteristics and usage to systems used on previously certificated aircraft. In principle, there are no parts of the subject system more at risk (due to environment or installation) and that operational stresses are no more severe than on the previously certificated aircraft	系统工程中一种通过分析进行验证的类型。在 ARP 4754A(2010) 中适用于在特征与用途方面与已经合格审定的飞机的系统相类似的系统。原则上，与已经合格审定飞机系统相比，该系统上的所有零部件不会处于更大风险（由于环境或安装的关系），其运营压力也不会更为苛刻
Software 软件	Computer programs, procedures, rules, and any associated documentation pertaining to the operation of a computer system (ARP 4754A, 2010)	计算机程序、算法、规则及任何与计算机系统运行相关的文档（ARP 4754A, 2010）

Specialty requirement 专业需求	A requirement set by one of the various specialty engineering disciplines, such as human factors, reliability, maintainability, safety, environments, mass properties, and software. Specialty requirements can either be performance requirements or constraints	特定工程领域设定的需求，例如人为因素、可靠性、可维修性、安全性、环境、质量特性及软件。专业需求可以是性能需求或约束
Specification 产品规范	The collection of requirements which, when taken together, constitute the set of criteria which define the functions and attributes of an item (ARP 4754A, 2010)	需求的集合，它们一起构成规定某个系统、部件或产品的功能和属性的准则 (ARP 4754A, 2010)
Stress 压力	The emotional state, either detrimental or beneficial, which results from various stressors	由各种压力源导致的好的或不好的情绪状态
Stressor 压力源	Any stimulus which may result in stress	可能导致压力的任何刺激
Subsystem 子系统	A subdivision of the aircraft hierarchy of Figure 2.1, one level below the aircraft. The elements of a subsystem, when viewed together, satisfy the definition of a system. Traditionally called a system in the aircraft industry	图 2.1 飞机层级的细分部分，飞机的下一个层级。子系统的所有元素合起来满足系统的定义。在飞机工业界通常称其为系统
Subsystem-level 子系统级	Pertaining to functions, requirements, or trade-offs within a given subsystem	指定子系统内的功能、需求或权衡
Supply chain 供应链	The collection of suppliers of commercial aircraft suppliers and their relationships	商用飞机供应商及供应商关系的集合
Synthesis 综合	The translation of input requirements into possible solutions satisfying those inputs (ANSI/EIA 632, 1999)	需求（输入）到满足需求的可能的解决方案的转化 (ANSI/EIA 632, 1999)
System 系统	An interacting combination of elements, viewed in relation to function (official INCOSE definition). In ARP 4754A (2010) and DO-178B (1992), for example, refers to a subsystem	相互作用的元素的组合，从功能结合来看（INCOSE 官方定义）。参考 ARP 4754A（2010）和 DO-178B（1992）中子系统定义
System analysis 系统分析	Trade-offs and other activities leading to a system synthesis	实现系统综合的权衡分析及其他活动

System safety assessment (SSA) 系统安全性评估	A systematic, comprehensive evaluation of the implemented system functions to show that relevant safety requirements are met (ARP 4754A, 2010)	为表明所实现的系统满足相关安全性要求而进行的系统化、综合化的评估（ARP 4754A，2010）
System synthesis 系统综合	The process of creating a design. System synthesis begins with the development of the system architecture (Section 2.3) and the system functions (Section 3.2) and ends with the assignment of hardware and software to the requirements. System synthesis is discussed in Chapter 7	创造设计的过程。系统综合开始于系统架构（2.3 节）和系统功能（3.2 节）的开发，结束于将硬件和软件分配给相应的需求。第 7 章中讨论系统综合
Systems architecting 系统架构设计	A process for creating unprecedented, complex systems. Focuses on six core concepts or ideas: the systems approach, purpose orientation, ultraquality, modeling, experienced-based heuristics, and certification (Rechtin, 1991)	一种创造前所未有的复杂系统的过程。专注在六大核心概念或想法：系统解决方案、目标导向、超品质、建模、基于经验的探索和合格审定（Rechtin，1991）
Systems engineering 系统工程	The interdisciplinary approach and means to enable the successful realization of successful systems (official INCOSE definition) (see System)	使成功系统成功实现的跨学科的方法和手段（INCOSE 官方定义）（见**系统**）
Systems Engineering Body of Knowledge (SEBoK) 系统工程知识体系	An electronic compendium of facts about systems engineering, in Wiki format	关于系统工程事实的电子纲要，基于维基百科格式
System validation 系统确认	The assurance that the entire system meets its mission objectives	确保整个系统满足其任务目标
Technical performance measurement (TPM) 技术性能指标	The continuing verification of the degree of anticipated and actual achievement of technical parameters (ANSI/ EIA 632, 1999)	对技术参数预期的和实际达到的程度的持续验证（ANSI/EIA 632，1999）
Test 试验	A type of verification which requires instrumentation. Includes both laboratory and flight tests	一种需要仪器设备的验证类型。包括实验室试验和试飞
Top-level 顶层	Pertaining to the highest level of either the aircraft or aircraft system as defined by the aircraft system architecture of Figure 2.1. Also pertains to analyses or relations among two or more subsystems	飞机或飞机系统的最高级别，如图 2.1 所示的飞机系统架构。可以是关于两个或多个子系统的分析或关系

Traceability 可追溯性	The recorded relationship established between two or more elements of the development process, for example between a requirement and its source or between a verification method and its requirement (ARP 4654A, 2010). In systems engineering traceability more commonly refers to the characteristic by which requirements at one level of a design may be related to requirements at another level. Traceability also encompasses the relationship between a performance requirement and the function from which the performance requirement was derived	研制过程中，两个或两个以上元素之间建立的记录关系。例如，需求及其来源之间的关系，或验证方法和需求之间的关系（ARP 4654A，2010）。在系统工程中，可追溯性通常指的是一个层级的设计需求可能与另一个层级的需求相关的特性。可追溯性还包括性能需求和功能的关系，性能需求从功能中衍生出来
Trade-off 权衡	An analysis conducted to determine the preferred option among two or more options, such as the number of engines, based on a figure of merit, such as cost or weight. Trade-offs can be either top-level or subsystem-level	为了从两个或更多的选项中选出最优选项而进行的分析。例如：发动机数量的权衡基于品质因素，如成本和重量。权衡可以是顶层的或子系统级别的
Turnaround time 过站时间	The time between arrival at a gate to departure from a gate of an aircraft	飞机从到达某一登机口到离开该登机口的时间
Validation 确认	The determination that the requirements for a product are sufficiently correct and complete (ARP 4754A, 2010) (see Product) (see also System validation)	确定产品需求充分正确和完整（ARP 4754A，2010）（参见产品）（参见系统验证）
Variability 变异性	The lack of stability in the relationship between the components of a system; a contributor to the complexity of the system; also known as information entropy	系统组件之间的关系缺乏稳定性；系统复杂性的贡献者，也被称为信息熵
Verification 验证	The evaluation of an implementation of requirements to determine that they have been met (ARP 4754A, 2010)	对需求实现的评估，确定需求已经被满足（ARP 4754A，2010）
Vigilance 警觉	Sustained attention, or the ability of observers to maintain their attention to remain alert to stimuli over prolonged periods of time (Satchell, 1993)	持续关注，或观察者在较长时间对刺激保持警觉和注意力（Satchell，1993）
Winglet 翼梢小翼	A vertical surface located on the wingtip of an aircraft to reduce the wingtip vortex effect and thus improve lift. Winglets have become common on modern aircraft	位于飞机翼尖的垂直面，用来降低翼尖涡效应，提高升力。翼梢小翼常见于现代飞机

| Workload 工作负荷 | The number of things an operator has to do within any particular time period modified by their level of difficulty (Shafer, 1987) | 操作人员在任意特定时间内（根据难度等级）需完成的事情的数量（Shafe，1987） |

参 考 文 献

9/11 Commission. 9/11 Commission Report. Edited by T. H. Kean, 2004. Adams, Charlotte. Shop Data Loaders for the Boeing 777, *Avionics Magazine*, September 1996, pp. 42, 43.

Airbus. Fly by Wire. Airbus [cited 26 December 2013]. Available from http://www. airbus.com/tools/airbusfor/pilots/fly-by-wire/?contentId=%5B_TABLE%3Att_ content%3B_FIELD%3Auid%5D%2C&cHash=22935adfac92fcbbd4ba4e144 1d13383, 2013.

Alexander, Christopher. *Notes on the Synthesis of Form*. Cambridge, MA, Harvard University Press, 1964.

ANSI/EIA 632, *Processes for the Engineering of a System*, Electronic Industries Alliance (EIA), Arlington, VA, 1999.

Army Field Manual. FM-770-78 Field Manual: *System Engineering*, Headquarters, Department of the Army, 1979.

Army Technical Manual. TM 38-760-1, *A Guide to System Engineering*, Department of the Army, TM 38-760-1, 1973.

Applegate, John. Systems Engineering in Developing Nations. INCOSE Symposium, 1998.

ATA Specification 100 - Specification for Manufacturers' Technical Data, Revision No. 37, Air Transport Association of America, 1999.

BBC. New Rules to Aid Ash Flight Chaos, 18 May 2010 [cited 31 May 2011]. Available from http://news.bbc.co.uk/2/hi/uk_news/8688517.stm, 2011.

BBC News. Easyjet to Trial Volcanic Ash Detection System, 10 June 2010 [cited 4 April 2011]. Available from http://www.bbc.co.uk/news/10234553, 2011.

Berry, Dennis L. Civil Aircraft Propulsion Integration—Present and Future, SAE Technical Paper Series, Number 932624, 27–30 September 1993.

Billings, Charles. *Aviation Automation: A Concept and Guidelines*. Moffett Field, CA: National Aeronautics and Space Administration (NASA), 1991.

Billings, Charles E. *Aviation Automation: The Search for a Human-Centered Approach*. Mahwah, NJ: Lawrence Erlbaum Associates, 1997, pp. 232–262. Birch, Stuart. Technology Update, *Aerospace Engineering*, December 1995, pp. 9–10.

Bowers, Al. Blended-Wing-Body: Challenges for the 21st Century. NASA Dryden Flight Research Center 2000 [cited 9 January 2014]. Available from http:// www. twitt.org/BWBBowers.html. 2014.

Buede, Dennis M. *The Engineering Design of Systems*. Edited by A. Sage, *Wiley Series in Systems Engineering*. Hoboken, NJ: John Wiley & Sons, Inc., 2000.

Campbell, CADRAT Tool. Personal communication. UK, 2013.

Carson, Ronald S. A Set Theory Model for Anomaly Handling in System Requirements Analysis, INCOSE Proceedings, 1995, pp. 515–522.

Carson, Ronald S. Global System Architecture Optimization: Quantifying System Complexity. In *International Council on Systems Engineering*: INCOSE, 2000. Chapanis, Alphonse. *Human Factors in Systems Engineering*, New York: Wiley, 1996, pp. 14, 206, 207, 277.

Checkland, Peter. *Systems Thinking, Systems Practice*. New York: John Wiley & Sons. 1999.

Commercial Aviation Safety Team (CAST) 2011 [cited 16 February 2014]. Available from http://www.cast-safety.org/about_vmg.cfm, 2014.

Commercial Aviation Safety Team (CAST) Safer Skies Safety Enhancements for Manufacturers. 4 September 2012 [cited 15 February 2014].Available from http:// www.cast-safety.org/pdf/2012-09-04_Safer_Skies_Safety_Enhancements_for_ Manufacturer.pdf, 2012.

Commercial Aviation Safety Team (CAST) Safer Skies Safety Enhancements Commercial Aviation Safety Team, 2014.

Conrow, Edmund H. *Effective Risk Management: Some Keys to Success*. Second edition. Reston, VA: American Institute of Aeronautics and Astronautics, 2003.

Corning, Gerald, *Supersonic and Subsonic CTOL and VTOL Airplane Design*. College Park, Maryland, published by author, 1977.

Defense Specifications, Department of Defense, MIL-STD-961D, 22 August 1995 (supersedes MIL-STD-490A).

Dekker, Sidney. Resilience Engineering: Chronicling the Emergence of Confused Consensus. In *Resilience Engineering*, edited by E. Hollnagel, D. D. Woods and N. Leveson. Aldershot, UK: Ashgate Publishing Limited, 2006.

Dekker, Sidney, Erik Hollnagel, David D. Woods, and Richard Cook. *Resilience Engineering: New Directions for Measuring and Maintaining Safety in Complex Systems*. Lund, Sweden: Lund University, 2008.

Department of Defense (DoD). Standard Practice: System Safety. Washington, DC:

Department of Defense, 2012.

EasyJet. AVOID Volcanic Ash Detector. EasyJet [cited 12 June 2014]. Available from http://corporate.easyjet.com/corporate-responsibility/avoid-volcanic-ash- detection. aspx?sc_lang=en, 2014.

Electronic Industries Association (EIA), *System Engineering*, Engineering Bulletin SYSB-1, 1989.

Electronic Industries Alliance (EIA), Processes for the Engineering of a System, ANSI/ EIA 632, January 1999.

Federal Air Regulation, Part 25, Airworthiness Standards, Transport Category: Airplanes; Federal Aviation Administration, Department of Transportation, March 1993.

Federal Aviation Administration (FAA). Fuel Tank Flammability Reduction Means. In *Advisory Circulars*, edited by Ali Bahrami. Washington DC: Federal Aviation Administration, 2008.

Federal Aviation Administration (FAA). Applicability/Compatibility of STPA with FAA Regulations and Guidance. Edited by Institute of Engineering and Technology. Seattle, WA: Federal Aviation Administration 2012.

Federal Aviation Administration (FAA). Fact Sheet—Commercial Aviation Safety Team. Federal Aviation Administration, 11 July 2013 [cited 15 February 2014]. Available from http://www.faa.gov/news/fact_sheets/news_story. cfm?newsId=15214, 2014.

Federal Aviation Administration (FAA). *FAA Systems Engineering Manual*. Edited by Kimberly Gill. Washington, DC: Federal Aviation Administration, 2014. Generic Open Architecture (GOA) Framework, AS4893, Society of Automotive Engineers, January 1996.

Giachetti, Ronald E. *Design of Enterprise Systems: Theory, Architecture, and Methods*. Boca Raton, FL: CRC Press, 2010.

Grady, Jeffrey O. *System Requirements Analysis*, New York, McGraw-Hill, Inc. This book presents a comprehensive view of the development of requirements for a system,1993.

A Guide to System Engineering, TM 38-760-1, US Army, 1973.

Guidelines for the Certification of Highly-Integrated and Complex Aircraft Systems, Society of Automotive Engineers (SAE) in cooperation with the Federal Aviation Administration (FAA), ARP 4754, November 1996.

Haddon-Cave, Charles. *An Independent Review into the Broader Issues Surrounding*

the Loss of the RAF Nimrod MR2 Aircraft XV230 in Afganistan in 2006. London: The House of Commons, 2009.

Haimes, Yacov Y. On the Definition of Resilience in Systems. *Risk Analysis* Vol. 29, No. 43, pp. 498–501, 2009.

Hall, Arthur D. *A Methodology for Systems Engineering*. Princeton, NJ: D. Van Nostrand Co., Inc., 1962.

Hamzeh, Osama N., W. Woytek Tworzydlo, and Hsien J. Chang. Analysis of Friction-Induced Instabilities in a Simplified Aircraft Brake. In *SAE 1999 Brake Colloquium*. San Diego, CA: Society of Automotive Engineers, 1999.

Hitchins, Derek. *Putting Systems to Work*. Hoboken, NJ: Wiley, 1993.

Hitchins, Derek. *Advanced Systems Thinking, Engineering, and Management*. Norwood, MA: Archtech House, 2003.

Hollnagel, Erik, Jean Pariès, David D. Woods, and John Wreathhall, eds. *Resilience Engineering in Practice: A Guidebook*. Edited by E. Hollnagel, S. Dekker, C. P. Nemeth and Y. Fujita, *Studies in Resilience Engineering*. Farnham, Surrey, UK: Ashgate Publishing Limited, 2011.

Hollnagel, Erik and David D. Woods. Epilogue: Resilience Engineering Precepts. In *Resilience Engineering: Concepts and Precepts*, edited by E. Hollnagel, D. D. Woods and N. Leveson. Farnham, UK: Ashgate Publishing Limited, 2006. Hollnagel, Erik, David D. Woods, and Nancy Leveson, eds. *Resilience Engineering: Concepts and Precepts*. Aldershot, UK: Ashgate Publishing Limited, 2006. *Holy Bible*, revised standard version, New York: Thomas Nelson and Sons, 1952, The New Testament, pp. 195–196.

Honour, Eric C. Requirements Management Cost/Benefit Selection Criteria, *Proceedings of NCOSE*, 1994.

Hooks, Ivy, and Kristan A. Farry. *Customer-Centered Products: Creating Successful Products Through Smart Requirements Management*. New York: American Management Association, 2001.

Hopkin, Paul. *Holistic Risk Management in Practice*. Livingston, UK: Witherby & Co, Ltd, 2002.

Hypersonic Aircraft Propulsion. *Aerospace Engineering*, June 1996.

IEEE Guide for Developing System Requirements Specifications, IEEE Std 1233-1996, 6 June 1996.

IEEE Standard for Application and Management of the Systems Engineering Process, IEEE Std 1220-2005.

INCOSE. *Systems Engineering Handbook.* Edited by SE Handbook WG. Seattle, WA: International Council on Systems Engineering, 2010.

INCOSE Fellows. A Consensus of the INCOSE Fellows. International Council on Systems Engineering [cited 2 June 2006]. Available from http://www.incose. org/practice/fellowsconsensus.aspx, 2006.

Jackson, Scott. Systems Engineering and the Bottom Line, INCOSE Proceedings, 1995.

Jackson, Scott. Introducing Systems Engineering into a Traditionally Commercial Organization, INCOSE Proceedings, 1996.

Jackson, Scott. *Systems Engineering for Commercial Aircraft.* Aldershot, UK: Ashgate Publishing Limited (in English and Chinese), 1997.

Jackson, Scott. *Architecting Resilient Systems: Accident Avoidance and Survival and Recovery from Disruptions.* Edited by A. P. Sage, *Wiley Series in Systems Engineering and Management.* Hoboken, NJ, USA: John Wiley & Sons, 2010.

Jackson, S. and Brtis, J. Overview of Resilience and Theme Issue on the Resilience of Systems. *Insight*, 18 (2015, April).

Jackson, Scott, and Timothy Ferris. Resilience Principles for Engineered Systems. *Systems Engineering* Vol. 16, No. 2, pp. 152–164, 2013.

Jamshidi, M. Systems of Systems Engineering: Innovations for the 21st Century. In *System of Systems Engineering: Innovation for the 21st Century*, edited by M. Jamshidi. Hoboken, NJ: John Wiley & Sons, 2009.

Kehlet, Alan. Major Advances in Aircraft Technologies Expected in the Future, *Innovate* bulletin, McDonnell Douglas, Vol. 26, No. 4, 4th Quarter, 1995.

Kossiakoff, Alexander, and William N. Sweet. *Systems Engineering: Principles and Practice.* Edited by Andrew Sage, *Wiley Series in Systems Engineering and Management.* Hoboken, NJ: John Wiley & Sons. 2003.

Lano, Robert J. *Techniques for Software and System Design.* Vol. 3, *TRW Series on Software Technology.* Amsterdam: North-Holland Publishing Co., 1979.

Leveson, Nancy. *Safeware: System Safety and Computers.* Reading, MA: Addison Wesley, 1995.

Leveson, Nancy. *A New Approach to System Safety Engineering.* Cambridge, MA: Massachusetts Institute of Technology, 2002.

Leveson, Nancy, Nicolas Dulac, David Zipkin, Cutcher-Gershenfeld, John Carroll, and Berry Barrett. Engineering Resilience into a Safety-Critical System. In *Resilience Engineering: Concepts and Precepts*, edited by

E. Hollnagel, D. D. Woods and N. Leveson. Aldershot, UK: Ashgate Publishing

Limited, 2006.

Lin, Kuen, Eric Cheung, Wendy Liu, and Luke Richard. Disbond/Delamination Arrest Features in Aircraft Composite Structures. In *2013 Technical Review*: Joint Advanced Materials & Structures Center of Excellence, 2013.

Littlewood, Bev, and Stringini, Lorenzo, The Risks of Software, *Scientific American*, November 1992, pp. 62–75.

Mackey, Dr William F. Conducting a Technology Management Assessment, INCOSE Proceedings, 1996.

Madni, Azad and Scott Jackson. Towards a Conceptual Framework for Resilience Engineering. *Institute of Electrical and Electronics Engineers (IEEE) Systems Journal* Vol. 3, No. 2, pp. 181–191, 2009.

Martínez-Val, Rodrigo, E. Pérez, T. Muñoz, and Cristina Cuerno. Design Constraints in the Payload-Range Diagram of Ultrahigh Capability Transport Airplanes, *Journal of Aircraft* Vol. 31, No. 6, November–December 1994.

Marczyk, Jacek. *Practical Complexity Management.* Trento, Italy: Editrice/UNI Service, 2009.

Marczyk, Jacek. Complexity Reduction. Como, Italy, 14 June 2012.

MIL-STD-499B. *Systems Engineering*, Washington DC: Department of Defense, 1994 (cancelled draft).

MIL-STD-1808B. *Interface Standard System Subsystem Sub-subsystem Numbering*, Department of Defense, 1 August 2007, Washington, DC. NASA. *Concept of Operations for Commercial and Business Aircraft Synthetic Vision Systems.* Edited by Daniel M. Williams. Hampton, VA: National Aeronautics and Space Administration, 2001.

NASA. *Concept of Operations for Commercial and Business Aircraft Synthetic Vision Systems*, Version 1.0, NASA/TM-2001-211058, Langley Research Center, Hampton, VA, 2001.

NASA. *Columbia Accident Investigation Report.* Washington, DC: National Aeronautics and Space Administration (NASA), 2003.

NASA. Technology Readiness Levels. NASA 2012 [cited 2 March 2014]. Available from http://www.nasa.gov/content/technology-readiness-level/, 2014.

NASA Systems Engineering Handbook, SP-6105. This handbook is a good manual for conducting the systems engineering process. It is also cited as a good source for risk analysis, pp. 7, 37–44, June 1995.

National Transportation Safety Board (NTSB). Safety recommendation. National

Transportation Safety Board 1990 [cited 14 December 2009]. Available from http://www.ntsb.gov/recs/letters/1990/A90_167_175.pdf, 2009.

Next-Generation SST: Technology Requirements, *Aerospace Engineering*, April 1994, pp. 29–31.

Oehmen, Josef, ed. *The Guide for Lean Enablers for Managing Engineering Programs*: Joint MIT-PMI-INCOSE Community of Practice on Lean in Program Management, 2012.

Oxford English Dictionary (OED). In *The Shorter Oxford English Dictionary on Historical Principles*, edited by C. T. Onions. Oxford: Oxford Univeristy Press. Original edition, 1933, 1973.

Page, Scott E. *Diversity and Complexity*. Princeton, NJ: Princeton University Press, 2011.

Pariès, Jean. Lessons from the Hudson. In *Resilience Engineering in Practice: A Guidebook*, edited by E. Hollnagel, J. Pariès, D. D. Woods and J. Wreathhall. Farnham, UK: Ashgate Publishing Limited, 2011.

Paté-Cornell, M. Elisabeth. Organizational Aspects of Engineering System Safety: The Case of Offshore Platforms, *Science*, Vol. 250, November 30, 1990, pp. 1210–1216.

Pirsig, Robert. *Zen and the Art of Motorcycle Maintenance*. New York: Bantam Books, 1974.

Perrow, Charles. *Normal Accidents: Living With High Risk Technologies*. Princeton, NJ: Princeton University Press, 1999.

Petersen, Thomas J. and Sutcliffe, Peter L. Systems Engineering as Applied to the Boeing 777, AIAA 1992 Aerospace Design Conference, Irvine, California, 1992.

Pyster, Arthur, ed. *Systems Engineering Body of Knowledge*. First edition. Stephens Institute, Hoboken, NJ and the Naval Postgraduate School, Monterery, CA, 2012.

RAF. Proceedings of a Board of Inquiry into an Aircraft Accident. Royal Air Force, 2007.

Ramo, Simon, 1973, quoted in Rechtin, 1991, p. 28.

Reason, James. *Human Error*. Cambridge, UK: Cambridge University Press, 1990.
Reason, James. *Managing the Risks of Organisational Accidents*. Aldershot, UK: Ashgate Publishing Limited, 1997.

Rechtin, Eberhardt. *Systems Architecting: Creating and Building Complex Systems*. Englewood Cliffs, NJ; Prentice-Hall. This book, written by Professor Rechtin of the University of Southern California, describes the process of synthesizing any system at the highest level of the system architecture, 1991.

Rijpma, Jos A. Complexity, Tight Coupling and Reliability: Connecting Normal Accidents Theory and High Reliability Theory, *Journal of Contingencies and Crisis Management* Vol. 5, No. 1, pp. 15–23, 1997.

Satchell, Paul. *Cockpit Monitoring and Alerting Systems*. Aldershot, UK: Ashgate Publishing Limited, 1993, pp. 10, 17, 52, 53, 59, 120.

Shafer, John B. Practical Workload Assessment in the Development Process, *Proceedings of the Human Factors Society 31st Annual Meeting*, Santa Monica, California, pp. 1408–1410.

Skybrary. Non Avian Wildlife Hazards to Aircraft. Eurocontrol, 16 October 2013 [cited 4 January 2014].Available from http://www.skybrary.aero/index.php/ Non_Avian_Wildlife_Hazards_to_Aircraft?utm_source=SKYbrary&utm_ campaign=ee75eb4949-SKYbrary_Highlight_01_01_2014&utm_ medium=email&utm_term=0_e405169b04-ee75eb4949-276526209, 2013.

Skybrary. 4D Trajectory Concept. Eurocontrol [cited 9 January 2014]. Available from http://www.skybrary.aero/index.php/4D_Trajectory_Concept? utm_ source=SKYbrary&utm_campaign=b60834be85-SKYbrary_Highlight_ 06_01_2014&utm_medium=email&utm_term=0_e405169b04-b60834be85- 276526209#Benefits_of_4D_Trajectory_Operations, 2014.

Skybrary. SE 120: Terrain Awareness and Warning System. Eurocontrol, 21 August [cited 4 January 2014]. Available from http://www.skybrary.aero/index.php/ SE120:_Terrain_Awareness_and_Warning_System_(TAWS)_Improved_ Functionality, 2014.

Sillitto, Hillary G. Design Principles for Ultra-Large-Scale Systems. In *International Council on Systems Engineering International Symposium*. Chicago, IL, 2010.

Society of Automotive Engineers (SAE). ARP 4754, Certification Considerations for Highly-Integrated or Complex Aircraft Systems Society of Automotive Engineers, 1996.

Society of Automotive Engineers (SAE). ARP 4754A, Guidelines for the Development of Civil Aircraft and Systems, edited by John Dalton: Society of Automotive Engineers, 2010.

Stevens, Richard, Peter Brook, Ken Jackson and EliotArnold. *Systems Engineering: Coping With Complexity*. London: Prentice Hall, 1998.

Software Considerations in Airborne Systems and Equipment Certification. RTCA/DO-178B, RTCA, Inc., 1 December 1992.

Specification for Manufacturers' Technical Data. Air Transport Association of America

(ATA) Specification 100, Revision 28, 15 March 1989.

Specification Practices. Department of Defense, MIL-STD-490A, 4 June 1985. (superseded by MIL-STD-961D).

Specification Practices. MIL-STD-490A, October 30, 1968.

System Engineering, Department of the Army, US Army Field Manual, FM 770-78, April 27, 1979.

System Engineering. Electronic Industries Association (EIA), SYSB-1, December 1989.

Systems Engineering. Draft military standard, MIL-STD-499B, 6 May 1992 (cancelled).

Vaughn, Diane. *The Challenger Launch Decision: Risky Technology, Culture, and Deviance at NASA.* Chicago, IL: University of Chicago Press. Original edition, 1996, 1997.

Woods, David D. Essential Characteristics of Resilience. In *Resilience Engineering: Concepts and Precepts*, edited by E. Hollnagel, Woods, David D., and Leveson, Nancy. Aldershot, UK: Ashgate Publishing Limited, 2006.

Zarboutis, Nikos, and Peter Wright. Using complexity theories to reveal emerged patterns that erode the resilience of complex systems. Paper read at Second Symposium on Resilience Engineering, 8–10 November, at Juan-les-Pins, France, 2006.

Zimmermann, Kyla, Jean Pariès, René Amalberti, and Daniel H. Hummerdal. Is the Aviation Industry Ready for Resilience? Mapping Human Factors Assumptions across the Aviation Sector. In *Resilience Engineering in Practice: A Guidebook*, edited by E. Hollnagel, J. Pariès, D. D. Woods and J. Wreathhall. Farnham, UK: Ashgate Publishing Limited, 2011.

索　引

E